Daniel Ab

RapidIO®
The Embedded System Interconnect

RapidIO®
The Embedded System Interconnect

Sam Fuller
RapidIO Trade Association, USA

with contributions from

Alan Gatherer
CTO, Wireless Infrastructure, Texas, USA
Charles Hill
Motorola, Inc., Arizona, USA
Victor Menasce
Applied Micro Circuits Corporation, Ontario, Canada
Brett Niver
EMC Corporation, Hopkinton, USA
Richard O'Connor
Tundra Semiconductor Corporation, Ontario, Canada
Peter Olanders
Ericsson AB, Stockholm, Sweden
Nupur Shah
Xilinx Inc., California, USA
David Wickliff
Lucent Technologies, Illinois, USA

John Wiley & Sons, Ltd

Other Wiley Editorial Offices

John Wiley & Sons Inc., 111 River Street, Hoboken, NJ 07030, USA

Jossey-Bass, 989 Market Street, San Francisco, CA 94103-1741, USA

Wiley-VCH Verlag GmbH, Boschstr. 12, D-69469 Weinheim, Germany

John Wiley & Sons Australia Ltd, 33 Park Road, Milton, Queensland 4064, Australia

John Wiley & Sons (Asia) Pte Ltd, 2 Clementi Loop #02-01, Jin Xing Distripark, Singapore 129809

John Wiley & Sons Canada Ltd, 22 Worcester Road, Etobicoke, Ontario, Canada M9W 1L1

Wiley also publishes its books in a variety of electronic formats. Some content that appears
in print may not be available in electronic books.

British Library Cataloguing in Publication Data

A catalogue record for this book is available from the British Library

ISBN 0-470-09291-2

Typeset in 10/12pt Times by Graphicraft Limited, Hong Kong
Printed and bound in Great Britain by TJ International, Padstow, Cornwall
This book is printed on acid-free paper responsibly manufactured from sustainable forestry
in which at least two trees are planted for each one used for paper production.

Contents

PREFACE

This book is a collaborative enterprise and is based on the RapidIO specifications, white papers, articles and other contributions from members of the RapidIO Trade Association. As writer/editor, I have tried to offer a good single reference source for the RapidIO technology, explaining not only the technology itself but also its various applications and the reasoning behind its development. I have also tried to structure the presentation of the material to more naturally show how the technology works. This book does not replace the RapidIO specifications themselves and if there are conflicts between what is written here and what the specifications say, the specifications are right. This book is a good introduction for someone who is going to use the RapidIO technology, or contemplate its use at either the chip or system level.

Special thanks must go the Dan Bouvier and Bryan Marietta from Motorola, now Freescale, who led the initial technical development along with Bob Frisch of Mercury Computer Systems. Dan provided the vision and leadership, Bryan provided engineering insight and the words of the specification and Bob provided experience and unique customer insights into what works and what doesn't work in the world of high-speed interconnects. While the work started with the three initial developers, the contributors and contributions to the technology have continued to grow over the years. As befits a true, open standard, contributions have come from dozens of individuals and hundreds of participants. Through them the scope and power of the technology has grown tremendously. Applications we hadn't envisioned have emerged to capture the attention of many of the leading vendors of processors, DSPs, FPGAs, systems, software and test equipment. They have also managed to keep the RapidIO Trade Association busily developing new standards long after most of us were ready to put our pencils down believing we were finished.

When we started this work, we were focused on solving problems unique to embedded systems connectivity. While many other technologies have come and gone (most purporting to be 'the standard' solution to all connectivity requirements), none has been able to offer the full value proposition for embedded systems that RapidIO offers. When you are focused on your customer's requirements you will be able to offer a better solution to their problem. That is what we have tried to accomplish with RapidIO and the embedded systems market. The challenge has been in seeing all of the various embedded systems vendors as a single market or as a group of closely related markets. Then working to develop a great solution for the problems of this market.

When you step back and look at these embedded systems you begin to realize that most all electronics systems perform three basic functions. They process data, they 'move data around' and while they are doing this processing and moving they also save the data in various places along the way. You process data with processors like CPUs, DSPs and Network Processors, you store data in RAM, ROM, FLASH and hard disks, and you move data around on buses and interconnects like PCI, Ethernet, Fibre Channel, and RapidIO. In the area of 'moving data around' standards can help tremendously in simplifying system design. RapidIO was developed to provide standard ways to move data around within embedded systems based on a switched interconnect rather than a bus. It has the capability of replacing lots of existing bus or switch based technologies such as PCI, VME, Utopia, Infiniband, Ethernet (for in-system use) or SPI4. As well as the many proprietary interconnects that systems vendors develop because there just isn't the right solution available on the market.

Thanks should also go to: Craig Lund for his insight and perseverance, Alan Gatherer for suggesting that someone should write a book on this technology and for putting me in touch with the good people at John Wiley and Sons, Bill Quackenbush for his technical contributions, Gary Robinson for leading us through the standards setting process, David Wickliff, Louis-Francois Pau, Steve MacArthur, and Peter Olanders for the customer's insight that they provided, Rick O'Connor, Tom Cox, Victor Menasce, David Somppi and the other good folks at Tundra for all of their many contributions to the effort. Thanks to Greg Shippen for taking the baton and running the next leg. And to Kalpesh, Nupur, Jason, Sarah, John, Andy, Travis, Jim, Kathy, Greg and Stephane for making sure that the journey was always fun.

Sam Fuller
Austin, Texas

1

The Interconnect Problem

This chapter discusses some of the motivations leading to the development of the RapidIO interconnect technology. It examines the factors that have led to the establishment of new standards for interconnects in embedded systems. It discusses the different technical approach to in-system communications compared with existing LAN and WAN technologies and legacy bus technologies.

The RapidIO interconnect technology was developed to ease two transitions that are occurring in the embedded electronics equipment industry. The first transition is technology based. The second transition is market based. The technology transition is a move towards high-speed serial buses, operating at signaling speeds above 1 GHz, as a replacement for the traditional shared bus technologies that have been used for nearly 40 years to connect devices together within systems. This transition is driven by the increasing performance capabilities and commensurate increasing bandwidth requirements of semiconductor-based processing devices and by device level electrical issues that are raised by semiconductor process technologies with feature sizes at 130 nm and below. The second transition is a move towards the use of standards based technologies.

1.1 PROCESSOR PERFORMANCE AND BANDWIDTH GROWTH

Figure 1.1 shows the exponential growth of processor performance over the last 30 years. It also depicts the slower growth of processor bus frequency over that same period of time. The MHz scale of the chart is logarithmic and the difference between the core CPU performance, represented by the clock frequency, and the available bandwidth to the CPU, represented by the bus frequency, continues to grow. The use of cache memory and more advanced processor microarchitectures has helped to reduce this growing gap between CPU performance and available bus bandwidth. Increasingly processors are being developed with large integrated cache memories and directly integrated memory controllers. However multiple levels of

RapidIO® The Embedded System Interconnect.   S. Fuller
© 2005 John Wiley & Sons, Ltd   ISBN: 0-470-09291-2

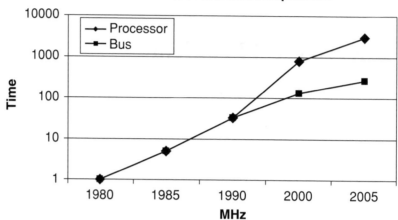

Figure 1.1   Frequency and bandwidth growth

on- and off-chip cache memory and directly integrated memory controllers, while useful for reducing the gap between a processor's data demands and the ability of its buses to provide the data, does little to support the connection of the processor to external peripheral devices or the connection of multiple processors together in multiprocessing (MP) systems.

In addition to the increasing performance of processors, the need for higher levels of bus performance is also driven by two other key factors. First, the need for higher raw data bandwidth to support higher peripheral device performance requirements, second the need for more system concurrency. The overall system bandwidth requirements have also increased because of the increasing use of DMA, smart processor-based peripherals and multiprocessing in systems.

1.2 MULTIPROCESSING

Multiprocessing is increasingly seen as a viable approach to adding more processing capability to a system. Historically multiprocessing was used only in the very highest end computing systems and typically at great cost. However, the continuing advance of semiconductor process technology has made multi-processing a more mainstream technology and its use can offer advantages beyond higher processing performance.

Figure 1.2 is a photograph of a multiprocessing computer system. This computer system uses 76 microprocessors connected together with RapidIO to solve very complex signal processing problems.

Multiprocessing can also be used to reduce cost while achieving higher performance levels. Pricing of processors is often significantly lower for lower-speed parts. The use of multiprocessing may also reduce overall system power dissipation at a given performance point. This occurs because it is often possible to operate a processor at a reduced frequency and achieve a greatly reduced power dissipation. For example, the Motorola 7447 processor has a rated maximum power dissipation of 11.9 W at an operating frequency of 600 MHz.

Figure 1.2 A multiprocessing computer system

The same processor has a maximum power dissipation of 50 W at an operating frequency of 1000 MHz[1]. If the processing work to be done can be shared by multiple processors, overall power dissipation can be reduced. In this case reducing the frequency by 40% reduces maximum power dissipation by 76%. When performance per watt is an important metric, multiprocessing of lower performance processors should always be considered as part of a possible solution.

Operating system technology has also progressed to the point where multiprocessing is easily supported as well. Leading embedded operating systems such as QNX, OSE, and Linux are all designed to easily support multiprocessing in embedded environments.

While these previous obstacles to multiprocessing have been reduced or removed, the processor interconnect has increasingly become the main limiting factor in the development of multiprocessing systems. Existing multiprocessor bus technologies restrict the shared bandwidth for a group of processors. For multiprocessing to be effective, the processors in a system must be able communicate with each other at high bandwidth and low latency.

1.3 SYSTEM OF SYSTEMS

Traditional system design has busied itself with the task of connecting processors together with peripheral devices. In this approach there is typically a single processor and a group of

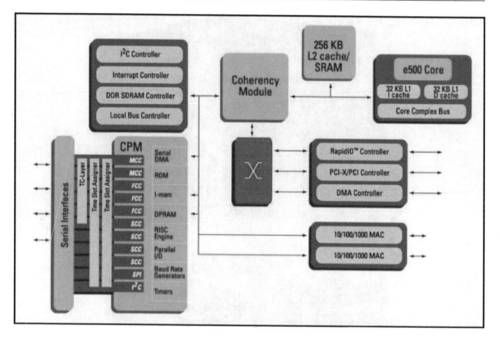

Figure 1.3   Motorola PowerQUICC III block diagram

peripherals attached to it. The peripherals act as slave devices with the main processor being the central point of control. This master/slave architecture has been used for many years and has served the industry well.

With the increasing use of system on a chip (SoC) integration, the task of connecting processors to peripherals has become the task of the SoC developer. Peripheral functionality is increasingly being offered as part of an integrated device rather than as a standalone component.

Most microprocessors or DSPs designed for the embedded market now contain a significant amount of integration. L2 cache, Memory controllers, Ethernet, PCI, all formally discrete functions, have found their way onto the integrated processor die. This integration then moves the system developer's task up a level to one of integrating several SoC devices together in the system. Here a peer-to-peer connection model is often more appropriate than a master/slave model.

Figure 1.3 shows a block diagram of the Motorola PowerQUICC III communications processor. This processor offers a wealth of integrated peripheral functionality, including multiple Ethernet ports, communications oriented ATM and TDM interfaces, a PCI bus interface, DDR SDRAM memory controller and RapidIO controller for system interconnect functionality. It is a good example of a leading edge system on a chip device.

1.4 PROBLEMS WITH TRADITIONAL BUSES

The connections between processors and peripherals have traditionally been shared buses and often a hierarchy of buses (Figure 1.4). Devices are placed at the appropriate level in

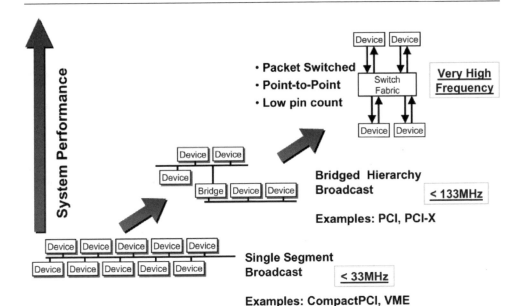

Figure 1.4 Higher system performance levels require adoption of point-to-point switched interconnects

the hierarchy according to the performance level they require. Low-performance devices are placed on lower-performance buses, which are bridged to the higher-performance buses so they do not burden the higher-performance devices. Bridging may also be used to address legacy interfaces.

Traditional external buses used on more complex semiconductor processing devices such as microprocessors or DSPs are made up of three sets of pins, which are soldered to wire traces on printed circuit boards. These three categories of pins or traces are address, data and control. The address pins provide unique context information that identifies the data. The data is the information that is being transferred and the control pins are used to manage the transfer of data across the bus.

For a very typical bus on a mainstream processor there will be 64 pins dedicated to data, with an additional eight pins for parity protection on the data pins. There will be 32–40 pins dedicated to address with 4 or 5 pins of parity protection on the address pins and there will be approximately another 30 pins for control signaling between the various devices sharing the bus. This will bring the pin count for a typical bus interface to approximately 150. Because of the way that semiconductor devices are built there will also be a large complement of additional power and ground pins associated with the bus. These additional pins might add another 50 pins to the bus interface pin requirement, raising the total pin count attributed to the bus alone to 200. This 200 pin interface might add several dollars to the packaging and testing cost of a semiconductor device. The 200 wire traces that would be required on the circuit board would add cost and complexity there as well. If the bus needed to cross a backplane to another board, connectors would need to be found that would bridge the signals between two boards without introducing unwanted noise, signal degradation and cost. Then, if you assume that your system will require the connection of 20 devices to achieve the desired functionality, you begin to understand the role that the bus can play in limiting the functionality and feasibility of complex embedded systems.

For the sake of simplicity, we will discuss peak data bandwidth as opposed to sustained data bandwidth which is often quite a bit lower than peak. The peak data bandwidth of this bus would be the product of the bus frequency and the data bus width. As an example the PCI-X bus is the highest performance general purpose peripheral bus available. If we assume that we are operating the PCI-X bus at 133 MHz and using the 64-bit data path, then the peak data bandwidth is $133 \times 64 = 8.5$ Gbit/s or approximately 1 Gbyte/s.

To increase the performance of the interface beyond 1 Gbyte/s we can either increase the frequency or we can widen the data paths. There are versions of PCI-X defined to operate at 266 and 533 MHz. Running at these speeds the PCI-X bus can support only one attached device.

When compared with the original bus interface on the Intel 8088 processor used by IBM in the first IBM PC, we find that available bus performance has increased significantly. The original 8088 processor had an 8 bit wide data bus operating at 4.77 MHz. The peak data bandwidth was therefore 419 Mbit/s or 4.77 Mbyte/s. When compared with this bus the current PCI-X peripheral bus has widened by a factor of 8 and its signaling speed has increased by a factor of 28 for an overall improvement in peak bandwidth of approximately 2000%. Owing to improvements in bus utilization the improvement of actual bandwidth over the last 20 years has been even more dramatic than this, as has the improvement in actual processor performance.

While the growth in bus performance over the last several years has been impressive there are many indications that a new approach must be taken for it to continue. Here are four important reasons why this is the case.

1.4.1 Bus Loading

Beyond 133 MHz it becomes extremely difficult to support more than two devices on a bus. Additional devices place capacitive loading on the bus. This capacitive loading represents electrical potential that must be filled or drained to reach the desired signal level, the additional capacitance slows the rise and fall time of the signals.

1.4.2 Signal Skew

Because a traditional bus is a collection of parallel wires with the signal valid times referenced to a clock signal, there are limits to how much skew can exist between the transition of the clock and the transition of a signal. At higher speeds the length of the trace as well as the signal transition times out of and into the devices themselves can affect the speed at which the bus is clocked. For a 133 MHz bus the cycle time is 7.5 ns, the propagation delay in FR4 printed circuit board material is approximately 180 ps/inch. For a quarter cycle of the bus (1875 ps) this would be approximately 10 inches of trace.

1.4.3 Expense of Wider Buses

The traditional solution of widening the buses has reached the point of diminishing returns. 128 bit wide data buses have not been well accepted in the industry, despite their use on several processors. The wider buses further reduce the frequency at which the buses may run. Wider buses also increase product cost by increasing the package size and pin count requirements of devices. They may also increase system costs by forcing more layers in the printed circuit boards to carry all of the signal trace lines.

1.4.4 Problems with PCI

PCI is a very common peripheral bus used in computing systems. PCI plays a limited role in embedded systems for attachment of peripheral devices. PCI introduces several additional performance constraints to a system.

1. PCI doesn't support split transactions. This means that the bus is occupied and blocked for other uses for the entire time a transaction is being performed. When communicating with slower peripheral devices this could be for a relatively long time.
2. The length of a PCI transaction isn't known *a priori*. This makes it difficult to size buffers and often leads to bus disconnects. Wait states can also be added at any time.
3. Transactions targeting main memory typically require a snoop cycle to assure data coherency with processor caches.
4. Bus performance is reduced to the least common denominator of the peripherals that are attached. Typically this is 33 MHz, providing peak transfer rates of only 266 Mbyte/s and actual sustained transfer rates less than 100 Mbyte/s.

1.5 THE MARKET PROBLEM

Among the products of the various companies that together supply electronic components to the embedded marketplace, there is no single ubiquitous bus solution that may be used to connect all devices together. Many vendors offer proprietary bus or interconnect technology on their devices. This creates a market for glue chips that are used to bridge the various devices together to build systems. Common glue chip technologies are ASIC and FPGA devices, with the choice of solution typically guided by economic considerations.

The number of unique buses or interconnects in the system increases the complexity of the system design as well as the verification effort. In addition to the device buses, system developers often also develop their own buses and interconnect technologies because the device buses do not offer the features or capabilities required in their systems.

The embedded market is different from the personal computer market. In the personal computer market there is a single platform architecture that has been stretched to meet the needs of notebook, desktop and server applications. The embedded equipment market is quite different in character from the personal computer market. It is also generally older with legacy telecommunications systems architectures stretching back forty years. The technical approaches taken for embedded equipment reflect the availability of components, historical system architectures and competitive pressures. The resulting system designs are quite different from that of a personal computer.

The embedded market also does not have as much dependency on ISA compatibility, favoring architectures such as ARM, PowerPC and MIPS as opposed to the X86 architecture that is predominant on the desktop.

Despite the disparate technical problems being solved by embedded equipment manu-facturers and the variety of components and architectures available to produce solutions, there is still a desire for the use of standard interconnects to simplify the development task, reduce cost and speed time to market. This desire for standard embedded system interconnects is the primary impetus behind the development of the RapidIO interconnect technology.

1.6 RAPIDIO: A NEW APPROACH

The RapidIO interconnect architecture is an open standard which addresses the needs of a wide variety of embedded infrastructure applications. Applications include interconnecting microprocessors, memory, and memory mapped I/O devices in networking equipment, storage subsystems, and general purpose computing platforms.

This interconnect is intended primarily as an intra-system interface, allowing chip-to-chip and board-to-board communications with performance levels ranging from 1 to 60 Gbit/s performance levels.

Two families of RapidIO interconnects are defined: a parallel interface for high-performance microprocessor and system connectivity and a serial interface for serial back-plane, DSP and associated serial control plane applications. The serial and parallel forms of RapidIO share the same programming models, transactions, and addressing mechanisms.

Supported programming models include basic memory mapped IO transactions; port-based message passing and globally shared distributed memory with hardware-based coherency. RapidIO also offers very robust error detection and provides a well-defined hardware and software-based architecture for recovering from and reporting transmission errors.

The RapidIO interconnect is defined as a layered architecture which allows scalability and future enhancements while maintaining backward compatibility.

1.6.1 Why RapidIO?

RapidIO is categorized as an intra-system interconnect as shown in Figure 1.5. Specifically, RapidIO is targeted at intra-system interconnect applications in the high-performance embed-ded equipment market. This market has distinct requirements when compared with the desk-top and server computer spaces. The embedded market has historically been served by a number of different vendors. The openness of the RapidIO Trade Association is well suited to this environment. The RapidIO Trade Association counts among its current members nearly two dozen leading vendors of microprocessors, DSPs, FPGAs, ASICs and embedded memories.

InfiniBand is targeted as a System Area Network (SAN) interconnect. A SAN is used to cluster systems together to form larger highly available systems. SANs usually connect whole computers together within distances of up to 30 m. Operations through a SAN are typically handled through software drivers using message channels or remote direct memory access (RDMA). InfiniBand competes more directly with Fibre Channel, a storage area network technology and Ethernet-based system area networking technologies such as iSCSI.

HyperTransport and PCI Express share some common characteristics with RapidIO, but are more appropriately described as point-to-point versions of PCI. While they maintain compatibility with the PCI interconnect architecture from a software viewpoint, which is very

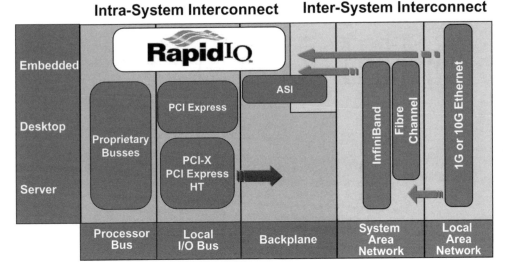

Figure 1.5 RapidIO connects processors, memory, and peripherals within a subsystem or across a backplane

important for desktop computer markets, they do not offer the scalability, robustness and efficiency required by embedded systems developers.

Ethernet has found opportunities as an intra-system interconnect in some applications. Ethernet at 100 Mbps and more recently at 1 Gbps is often integrated directly onto embedded microprocessors. The abundance of low-cost switching devices and relatively low pin count requirements make it an interesting choice for systems requiring a low-cost interconnect with modest (< 50 Mbyte/s) bandwidth requirements. At higher speeds, Ethernet becomes relatively less attractive in intra-system applications due to its higher software overhead, especially when used in conjunction with a TCP/IP protocol stack.

1.7 WHERE WILL IT BE USED?

The RapidIO interconnect is targeted for use in environments where multiple devices must work in a tightly coupled environment. Figure 1.6 illustrates a generic system containing memory controllers, processors, and I/O bridges connected using RapidIO switches.

In computing applications the PCI bus[2] is frequently used. Enterprise storage applications, for example, use PCI to connect multiple disk channels to a system. As disk throughput has increased so has the need for higher system throughput. To meet the electrical requirements of higher bus frequencies the number of supportable devices per bus segment must be decreased. In order to connect the same number of devices, more bus segments are required. These applications require higher bus performance, more device fan-out, and greater device separation. PCI-to-PCI bridge devices could be used to solve this problem, but only within a tree-shaped hierarchy that increases system latency and cost as more PCI devices are added to the system.

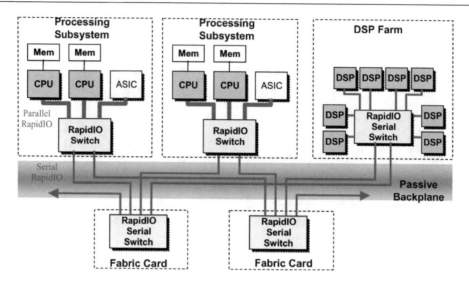

Figure 1.6   RapidIO allows a variety of devices to be interconnected

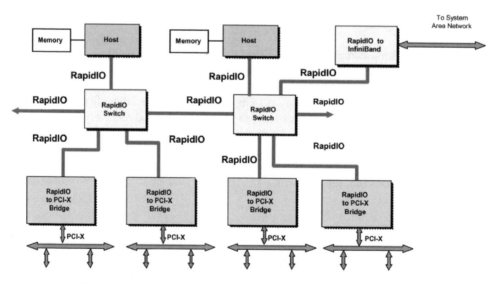

Figure 1.7   RapidIO as a PCI bridging fabric reduces device pin count

RapidIO can be used for transparent PCI-to-PCI bridging, allowing for a flattened architecture utilizing fewer pins with greater transmission distances. Figure 1.7 shows one such
bridged system. In this example, several PCI-X bridges are connected together using RapidIO
switches.

Many systems require the partitioning of functions into field replaceable units. These
printed circuit boards have traditionally been interconnected using multi-drop interfaces
such as VME or PCI. Higher system level performance can be achieved by using RapidIO.
RapidIO is well suited for hot swap applications, because RapidIO's point-to-point topology

enables the removal of devices with little or no electrical impact to neighboring devices or subsystems.

RapidIO was developed by a group of leading embedded system and semiconductor developers. It was conceived as an open standard approach to device connectivity that solved the technical challenges posed by buses. The RapidIO Trade Association rather than any single company oversees the development of RapidIO technology. This is done to ensure that the technology will support the needs of the embedded market as a whole rather than the needs of a single company.

1.8 AN ANALOGY

The basic problem that RapidIO is designed to solve is the movement of data and control information between semiconductor devices within a relatively complex system. The RapidIO architecture is targeted at applications where there will be tens to hundreds, perhaps thousands of devices, all communicating with one another within a single system. It is optimized for this size of connectivity problem. It also assumes that efficiency is important. By this we mean that the overhead introduced by using RapidIO should be minimal. There are two dimensions of overhead. There are the extra bits of control information that accompany the data that is being sent across the interconnect. This includes the destination address of the data, the type of transaction that is intended, the error checking code that is used to validate that the data was received correctly and other important pieces of information. Overhead also includes the amount of work that needs to be done to send the data across the interconnect. The RapidIO controller might do this work, but the software that interacts with the RapidIO controller might also do it. In a system that includes hundreds or thousands of communication links, efficiency becomes an important consideration.

We have discussed RapidIO as a technology that replaces buses. We have also mentioned that Ethernet, a local area network (WAN) technology, may also be sometimes used as a bus replacement technology. The following analogy helps to understand the differences between RapidIO and other LAN and wide area network (WAN) communications technologies.

This analogy compares interconnect technologies to freight shipping technologies. SONET and ATM are used to move data (mostly encoded voice data for telephone calls), but these technologies may also be used to send Internet data as well, around the world. These technologies are similar to freight trains and container ships. They can efficiently move data long distances at relatively low cost, but require a fairly significant and rigid infrastructure to operate in. Ports and rail stations are expensive propositions, but they have the supporting information and technology infrastructure needed to ensure that the rail cars get to their proper destinations and that the container ships are properly loaded with their cargo.

Ethernet represents a LAN technology that relies on a more intelligent driver to get the payload to its proper destination. With Ethernet the intelligence resides more in the packet (at least the packet contains more information) and the infrastructure can be simpler to build and maintain. In this analogy, Ethernet is more akin to our modern highway system. While the connectivity of Ethernet, especially in conjunction with IP, might provide the ability to move anywhere in the country (at least) on common roads, economics and local regulations might require that things get broken up and bridged across other technologies such as ATM or SONET to move data longer distances.

You might ask, at this point where does RapidIO play? Which technology does it replace? The answer is: none of them. RapidIO in this analogy is much more like a forklift that would be used in a factory or distribution warehouse. In an effort to make warehouses more efficient, a group of warehouse and factory equipment vendors have gotten together to establish standards for moving material within warehouses. Hopefully you would agree that this problem is quite different from that of moving material between warehouses.

What has changed is that, in times past the factory and warehouse builders could deliver the best return on invested capital by building their own warehouses and material movement equipment. Now they believe that warehouse material movement technology is not a differentiating technology and they are better off buying it from suppliers whose business it is to build the best in-warehouse material movement technology possible. By establishing standards for this technology they get three additional benefits. First, they expand the market for the technology. This should reduce prices as economies of scale come in to play. Second, they create market competition, which should also reduce prices as well as increase quality. Third, they can ensure that products from different vendors will interoperate well, thus reducing their own engineering and development expenses.

In times past, monopolistic telecommunications vendors operating as vertically integrated companies provided all of the services and equipment needed for communications. They could easily afford to develop their own internal interconnect technologies, because they were the largest and perhaps only market for the technology. There were no economies of scale to be had by opening up the internal standards.

In this analogy we see that RapidIO is a complementary technology to technologies such as Ethernet and ATM. We also see that RapidIO is designed to solve a different class of interconnect problem in an area where no standards have previously existed.

On occasion, technologies designed to solve one problem are repurposed to solve a different problem. In the absence of RapidIO some vendors have resorted to using Ethernet to solve some of the in-system connectivity requirements of embedded equipment. For some applications this is a perfectly acceptable solution, for others it is horrendously inefficient. Depending on the problem being solved Ethernet or RapidIO may or may not be an acceptable solution.

REFERENCES

1. Motorola MPC7xxx microprocessor documentation.
2. PCI Local Bus Specification, Rev.2.2, PCI Special Interest Group 1999.

2

RapidIO Technology

This chapter provides a more detailed explanation of the RapidIO interconnect technology. It describes the overall philosophy that drove the development of RapidIO. It also describes the partitioning of the specifications, as well as the structure of the packets and control symbols that are the building blocks of the RapidIO protocol. The features of the parallel and serial physical layers are discussed. The chapter concludes with discussions on maintenance, error management and system performance.

2.1 PHILOSOPHY

The architecture of RapidIO was driven by certain principal objectives. These objectives were related to a set of requirements unique to high-performance embedded systems. These embedded systems are the unseen technology supporting the telephone and data networks that most of us use on a daily basis. The principal objectives were:

- Focus on applications that connect devices that are within the box or chassis: this includes connectivity between devices on a printed circuit board, mezzanine board connections and backplane connections of subsystems.
- Limit the impact on software: many applications are built on a huge legacy of memory-mapped device I/O. In such applications the interconnect must not be visible to software. Abstracted interfaces such as InfiniBand require a large software re-engineering effort.
- Confine protocol overhead: because system performance is highly dependent on latency and bandwidth, it is important that the protocol impose no more overhead than is absolutely necessary.
- Partition the specifications: this limits design complexity while increasing reuse opportunities. It also enables the deployment of future enhancements without impacting the entire ecosystem.

RapidIO® The Embedded System Interconnect. S. Fuller
© 2005 John Wiley & Sons, Ltd ISBN: 0-470-09291-2

- Manage errors in hardware: in meeting the high standards of availability it is important that the interconnect be able to detect and recover from errors. The interconnect should have the capability of detecting any errors that may occur. Further, the interconnect must utilize hardware mechanisms to recover automatically from errors.
- Limit silicon footprint: many applications require the use of a high-performance interconnect, but have a limited transistor budget. It is important that the interface be able to fit within a small transistor count. As examples, it is important that it fit within commonly available FPGA technology or require only a fraction of the available die area of an ASIC.
- Leverage established I/O driver/receiver technology: since this interface is targeted for intra-system communications, it should not require separate discrete physical layer hardware. It is intended that all of the logic and circuit technology required for communication using RapidIO be implementable on commonly available CMOS technology. Further, the interface should leverage existing industry standard I/O technology and the support infrastructure associated with such standards.

2.2 THE SPECIFICATION HIERARCHY

RapidIO uses a three-layer architectural hierarchy. This hierarchy is shown in Figure 2.1. The logical specifications, at the top of the hierarchy, define the overall protocol and packet formats. They provide the information necessary for end points to initiate and complete transactions. The transport specification, on the middle layer of the hierarchy, defines the necessary route information for a packet to move from end point to end point. The physical layer specifications, at the bottom of the hierarchy, contain the device level interface details, such as packet transport mechanisms, flow control, electrical characteristics, and low-level error management.

This partitioning provides the flexibility to add new transaction types to any of the specifications without requiring modification to the other specification layers. For example,

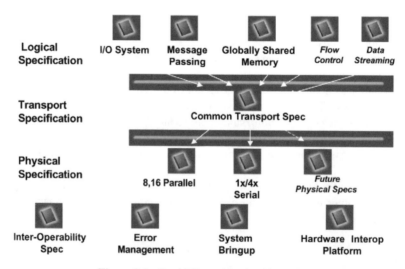

Figure 2.1 RapidIO specification hierarchy

new logical layer specifications can be added and the transactions described in those specifications will operate in a backward compatible fashion across existing RapidIO networks.

2.3 RAPIDIO PROTOCOL OVERVIEW

2.3.1 Packets and Control Symbols

RapidIO operations are based on request and response transactions. Packets are the communication element between end point devices in the system. A master or initiator generates a request transaction, which is transmitted to a target. The target then generates a response transaction back to the initiator to complete the operation. The RapidIO transactions are encapsulated into packets, which include all of the necessary bit fields to ensure reliable delivery to the targeted end point. RapidIO end points are typically not connected directly to each other, but instead will have an intervening fabric. We use the word fabric to mean a collection of one or more switch devices which provides the system connectivity. Control symbols are used to manage the flow of transactions in the RapidIO physical interconnect. Control symbols are used for packet acknowledgement, flow control information, and maintenance functions. Figure 2.2 shows how transactions progress through a RapidIO system.

In the example, shown in Figure 2.2, the initiator begins an operation in the system by generating a request transaction. The request packet is sent to a fabric device, typically a

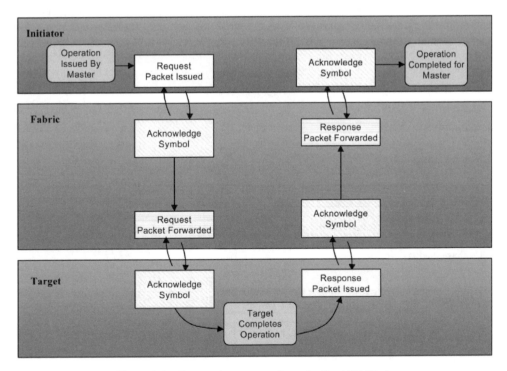

Figure 2.2   Transaction progress through a RapidIO fabric

switch, which is acknowledged with a control symbol. The packet is forwarded to the target through the fabric device. This completes the request phase of the operation. The target completes the request transaction and generates a response transaction. The response packet associated with this transaction returns through the fabric, using control symbols to acknowledge each hop. Once the response packet reaches the initiator and is acknowledged, the operation is considered complete.

2.4 PACKET FORMAT

The RapidIO packet is comprised of fields from the three-level specification hierarchy. Figure 2.3 shows typical request and response packets. These examples assume the use of the parallel physical layer, packets sent over the serial physical layer are slightly different. Certain fields are context dependent and may not appear in all packets.

The request packet begins with physical layer fields. The **S** bit indicates whether this is a packet or control symbol. The **AckID** indicates which packet the fabric device should acknowledge with a control symbol. The **PRIO** field indicates the packet priority used for flow control. The **TT**, **Target Address**, and **Source Address** fields indicate the type of transport address mechanism used, the device address where the packet should be delivered, and where the packet originated. The **Ftype** and **Transaction** indicate the transaction that is being requested. The **Size** is an encoded transaction size. RapidIO transaction data payloads range from 1 to 256 bytes in size. The **srcTID** indicates the transaction ID. RapidIO devices may have up to 256 outstanding transactions between two end points. For memory mapped transactions the **Device Offset Address** follows. For write transactions a Data Payload completes the transaction. All packets end with a 16-bit CRC.

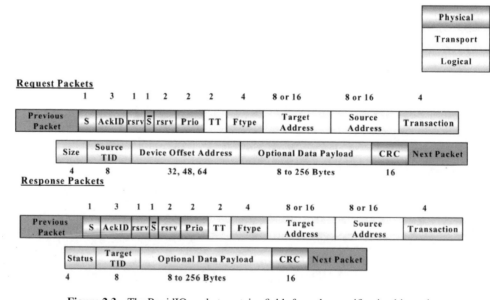

Figure 2.3   The RapidIO packet contains fields from the specification hierarchy

Response packets are very similar to request packets. The **Status** field indicates whether the transaction was successfully completed. The **TargetTID** field corresponds to the request packet source transaction ID field.

2.5 TRANSACTION FORMATS AND TYPES

A fully software transparent interconnect requires a rich set of transaction types. A RapidIO transaction is described through two fields: the packet format type **Ftype**, and the **Transaction**. To ease the burden of transaction deciphering, transactions are grouped by format as shown in Table 2.1.

Table 2.1   RapidIO has a rich set of transaction formats

Functions	Transaction types
I/O non-coherent functions	NREAD (read non-sharable memory)
	NWRITE, NWRITE_R, SWRITE (write non-sharable memory)
	ATOMIC (read-modify-write to non-sharable memory)
Port-based functions	DOORBELL (generate an interrupt)
	MESSAGE (write to port)
System support functions	MAINTENANCE (read or write configuration, control, and status registers)
User-defined functions	Open for application specific transactions
Cache coherence functions	READ (read globally shared cache line)
	READ_TO_OWN (write globally shared cache line)
	CASTOUT (surrender ownership of globally shared cache line)
	IKILL (instruction cache invalidate)
	DKILL (data cache invalidate)
	FLUSH (return globally shared cache line to memory)
	IO_READ (read non-cacheable copy of globally shared cache line
OS support functions	TLBIE (TLB invalidate)
	TLBSYNC (force completion of TLB invalidates)

2.6 MESSAGE PASSING

When data must be shared among multiple processing elements in a system, a protocol must be put in place to maintain ownership. In many embedded systems this protocol is managed through software mechanisms. If the memory space is accessible to multiple parties, then locks or semaphores are used to grant access to devices in proper order. In other cases, processing elements may only have access to locally owned memory space. In these 'shared nothing' systems, a mechanism is needed to pass data from the memory of one processing element to another. This can be done by using message passing mailboxes.

RapidIO provides a useful message passing facility. The RapidIO message passing protocol describes transactions that enable mailbox and doorbell communications. Details on RapidIO message passing are contained in Chapter 5. A RapidIO mailbox is a port through which one device may send a message to another device. The receiving device determines what to do with the message after it arrives. A RapidIO message can range in length from zero to 4096 bytes. A receiver can have 1 to 4 addressable message queues to capture inbound messages.

The RapidIO doorbell is a lightweight port-based transaction, which can be used for in-band interrupts. A doorbell message has a 16-bit software definable field, which can be used for a variety of messaging purposes between two devices.

2.7 GLOBALLY SHARED MEMORY

One of the protocol extensions offered in RapidIO is support for a globally shared distributed memory system. This means that memory may be physically located in different places in a system, yet will be properly cached amongst different processing elements.

For RapidIO, a directory-based coherency scheme is specified to support this. For this method each memory controller is responsible for tracking where the most current copy of each data element resides in the system. A directory entry is maintained for each device participating in the coherency domain. Simple coherency states of modified, shared, or local (MSL) are tracked for each element. For more information on the RapidIO globally shared memory (GSM) extensions refer to Chapter 13.

2.8 FUTURE EXTENSIONS

RapidIO is partitioned to support future protocol extensions. This can be done at the user level through the application of user definable transaction format types, or through future or reserved fields. RapidIO is architected so that switch fabric devices do not need to interpret packets as they flow through, making future forward compatibility much easier to achieve.

2.9 FLOW CONTROL

Flow control is an important aspect of any interconnect. Flow control refers to the rules and mechanisms that the interconnect uses to decide which of possibly several available trans-actions to send at any given time. The objective is for devices to be able to complete transactions in the system without being blocked by other transactions. Bus-based interconnects use arbitration algorithms to be sure that devices make forward progress and that higher priority transactions take precedence over lower priority ones. With switch-based interconnects, transactions enter at different points in the system and there is no opportunity to employ a centralized arbitration mechanism. This creates the need for alternate methods to manage transaction flow in the system.

RapidIO uses several complementary mechanisms to provide for the smooth flow of data in the system and to avoid system deadlocks.

2.9.1 Link Level Flow Control

One of the objectives of RapidIO is to limit overhead and complexity as much as possible, especially in the area of flow control. For RapidIO, flow control is specified as part of the physical specification. This is because transaction flow is largely dependent on the physical interconnect and system partitioning. Each RapidIO packet carries with it a transaction priority. Each transaction priority is associated with a transaction request flow. There are currently three transaction request flows defined. Transaction request flows allow higher priority requests to move ahead of lower priority requests. Transaction ordering rules are managed within a transaction request flow, but not between flows.

Transaction request flows make it possible for system implementers to structure traffic through the machine to guarantee deterministic behavior. For example, an application may require that certain time critical operations be delivered with a guaranteed latency. This data can be identified as a high-priority request flow. To handle critical data, the switch fabric may contain queuing structures that allow the movement of higher-priority traffic ahead of lower-priority traffic. These queuing structures may include interval timers to guarantee that flows from different source ports targeted to a common destination port are given opportunity to move critical data through. This sort of approach can provide isochronous delivery behavior for certain data classes.

At the link level, RapidIO describes three types of flow control mechanisms; retry, throttle, and credits. The retry mechanism is the simplest mechanism and is required not only for flow control, but also as a component of hardware error recovery. A receiver, unable to accept a packet because of a lack of resources or because the received packet was corrupt, may respond with a retry control symbol. The sender will retransmit the packet.

The throttle mechanism makes use of the idle control symbol. Idle control symbols may be inserted into a packet during packet transmission. This allows a device to insert wait states in the middle of packets. A receiving device can also send a throttle control symbol to the sending device requesting that it slow down by inserting idle control symbols.

The credit-based mechanism is useful for devices that implement transaction buffer pools, especially switch fabric devices. In this scheme certain control symbols contain a buffer status field that represents the current status of the receiver's buffer pool for each transaction flow. A sender sends packets only when it knows the receiver has available buffer space.

2.9.2 End-to-end Flow Control

In addition to the link level flow control mechanisms; RapidIO has defined a set of end-to-end flow control mechanisms. The link level mechanisms manage the flow of information between adjacent devices. In more complex systems, situations may arise where, for periods of time, the combined traffic from multiple sources towards one or more destinations might cause serious degradations in overall system performance. The end-to-end flow control mechanism uses a special congestion control packet that can be generated by a switch or an end point. The congestion control packet is sent back through a fabric to the sources of congesting traffic and will shut off the congesting traffic output for a period of time. This will have the effect of reducing traffic congestion in the system by reducing traffic at its source.

Unlike the link-level flow control mechanisms, which are specified as part of the physical layer and are implemented with control symbols, the end-to-end flow control mechanisms are specified as logical layer transactions and are implemented with packets. Legacy switch fabric devices will pass the new flow control packets, although they will not generate or respond to the new packets.

2.10 THE PARALLEL PHYSICAL LAYER

The RapidIO logical packet is defined as a serial chain of bits and is independent of physical layer implementation. This means that the RapidIO protocol can correctly operate over anything from serial to parallel interfaces, from copper to optical fiber media. The first physical interface considered and defined is known as the 8- or 16-bit link protocol end point specification with low-voltage differential signaling (8/16 LP-LVDS). This specification defines the use of 8 or 16 data bits in each direction along with clock and frame signals in each direction (Figure 2.4) using the IEEE standard LVDS signaling technology. As all signals are transmitted using a differential pair of wires, the pin count required to implement this interface is higher than might be assumed for an 8 or 16 bit data path. For the 8-bit interface there are 40 signal pins required. For the 16-bit interface there are 76 signal pins required. These pin counts are still considerably lower than the approximately 200 pins required for a traditional bus of comparable or lower performance.

The 8/16 LP-LVDS interface is a source synchronous interface. This means that a clock is transmitted along with the associated data. This also means that there is no defined phase relationship between the sender's and the receiver's clock, and they may also be at different frequencies. Source synchronous clocking allows longer transmission distances at higher frequencies. Two clock pairs are provided for the 16-bit interface to help control skew. The receiving logic is able to use the receive clock for resynchronization of the data into its local clock domain.

The FRAME signal is used to delineate the start of a packet or control symbol. It operates as a no-return-to-zero (NRZ) signal where any transition marks an event.

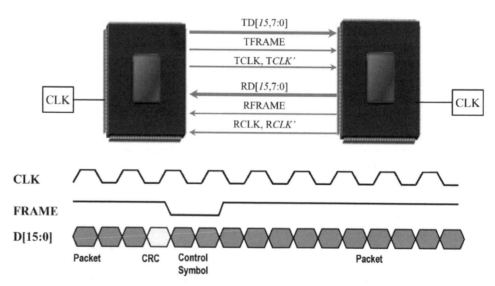

Figure 2.4   The RapidIO 8/16 LP-LVDS physical layer provides for full duplex communications between devices

2.10.1 Parallel Electrical Interface

As mentioned, RapidIO adopts the IEEE 1596.3 low-voltage differential signals (LVDS) standard as basis for the electrical interface[1]. LVDS is a low-swing (250–400 mV) constant-current differential signaling technology targeted toward short-distance printed circuit board applications. LVDS is technology independent and can be implemented in CMOS. Differential signals provide improved noise margin, immunity to externally generated noise, lower EMI, and reduced numbers of power and ground signal pins. LVDS has a simple receiver-based termination of 100 Ω. At higher frequencies the RapidIO specification recommends additional source termination to reduce signal reflection effects.

The target frequencies of operation are from 250 MHz to more than 1 GHz. Data is sampled on both edges of the clock. The resulting bandwidth scales to 4 Gbyte/s for the 8-bit and 8 Gbyte/s for the 16-bit interfaces. These bandwidths are considerably higher than those available for most existing bus technologies.

2.11 THE SERIAL PHYSICAL LAYER

Since the Serial RapidIO specification is defined only in the physical layer (the RapidIO technology defines the physical layer as the electrical interface and the device to device link protocol), most of the RapidIO controller logic remains the same. As a result, much of the design knowledge and verification infrastructure are preserved. This also eases system level switching between parallel and serial links.

During the initial development stages of the Serial RapidIO specifications the committee decided to try and preserve as many of the concepts found in the RapidIO parallel specification as possible. The parallel specification includes the concept of packets and in-band control symbols. These were delineated and differentiated by both a separate frame signal and an S bit in the header. In the serial link specification this delineation is accomplished using spare characters (K codes) found in the 8B/10B encoding. In this way, the sending device indicates to the receiving link partner the start of a packet, end of packet or start of control symbol using these codes as delimiting markers. More details on this are provided in Chapter 7.

2.11.1 PCS and PMA Layers

The Serial RapidIO specification uses a physical coding sublayer (PCS) and physical media attachment (PMA) sublayer to organize packets into a serial bit stream at the sending side and to extract the bit stream at the receiving side (this terminology is adopted from IEEE 802.3 Ethernet standards working group).

Besides encoding and decoding the packets before transmission and after reception, the PCS function is also responsible for idle sequence generation, lane striping, lane alignment, and lane de-striping on reception. The PCS uses an 8B/10B encoding for transmission over the link. This notation means that 8 bits of data are transformed in to 10 bits of coded data. The coded data includes the original data, as well as recoverable clock information. More details on this function are included in Chapter 7.

The PCS layer also provides the mechanisms for automatically determining the operational mode of the port as either 1-lane or 4-lane. The PCS layer also provides compensation for clocking differences between the sender and receiver.

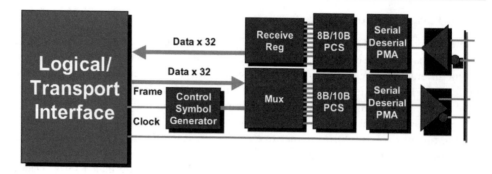

Figure 2.5   The RapidIO serializer/deserializer function integrates directly with the logic and transport layers

The PMA function is responsible for serializing 10-bit parallel code-groups to/from the serial bit stream on a lane-by-lane basis. Upon receiving the data, the PMA function provides alignment of the received bit stream on to 10-bit code-group boundaries, independently on a lane-by-lane basis. It then provides a continuous stream of 10-bit code-groups to the PCS, one stream for each lane. The 10-bit code groups are not observable by layers higher than the PCS.

2.11.2 Electrical Interface

Serial RapidIO uses differential current steering drivers based on those defined in the 802.3 XAUI specifications. This signaling technology was developed to drive long distances over backplanes.

For the Serial RapidIO technology, two transmitter specifications were designated: a short-run transmitter and a long-run transmitter. The short-run transmitter is used mainly for chip-to-chip connections either on the same printed circuit board or across a single connector such as that for a mezzanine card. The minimum swings of the short-run specification reduce the overall power used by the transceivers.

The long-run transmitter uses larger voltage swings that are capable of driving across backplanes. This allows a user to drive signals across two connectors and common printed circuit board material. The signaling is intended to support common backplane environments such as the PICMG 3.0 AdvancedTCA mechanical chassis standard for fabric backplanes.

To ensure interoperability between drivers and receivers of different vendors and technologies, AC coupling must be used at the receiver input.

2.12 LINK PROTOCOL

An important feature of RapidIO technology is that packet transmission is managed on a link-by-link basis. In the past, with synchronous buses, a mastering device had explicit handshake signals with the target device. These signals indicated whether a transaction was acknowledged and accepted by the target device.

With a source synchronous interface like the RapidIO specification defines, it is not practical to rely on a synchronous handshake since the target port of a transaction is decoupled from the sending port. Sometimes this decoupling, meaning the distance measured in device count between the sending and target port, can be quite large. Many interconnects have ignored this issue and continue to rely on an end-to-end handshake to guarantee delivery. This has the disadvantage of not allowing precise detection and recovery of errors and also forces far longer feedback loops for flow control.

To address this issue, RapidIO technology uses control symbols for link level communication between devices. Packets are explicitly tagged between each link with a sequence number otherwise known as the AckID. The AckID is independent of the end-to-end transaction ID. Using control symbols, the receiving device indicates for each packet whether it has been received or not. The receiver will also provide buffer status information. Receiving devices can immediately detect a lost packet, and through control symbols, can re-synchronize with the sender and recover the lost packet without software intervention. The receiving device then forwards the packet to the next switch in the fabric, and so on, until the final target of the packet is reached.

2.13 MAINTENANCE AND ERROR MANAGEMENT

RapidIO steps beyond traditional bus-based interfaces by providing a rich set of maintenance and error management functions. These enable the initial system discovery, configuration, error detection and recovery mechanisms. A discovery mechanism with similarities to PCI is used to find devices in the system and configure them.

2.13.1 Maintenance

A maintenance operation is supported by all RapidIO devices and is used to access predefined maintenance registers within each device. The registers contain information about the device and its capabilities. Also included are error detection and status registers such as watchdog timer settings. For more complex error situations, such as a failing component, software may make use of these registers to recover or shut down a RapidIO device.

2.13.2 System Discovery

Through the use of the maintenance operation, a RapidIO host may traverse a system and configure each device in the network. Since RapidIO allows complex system topologies such as meshes, a mechanism is provided so that the host can determine if it has been to a node before. For high-reliability applications, RapidIO allows more than one host to configure the machine.

2.13.3 Error Coverage

The mean time between failure (MTBF) and mean time to repair (MTTR) of a system are often critical considerations in embedded infrastructure applications. In these systems reliable packet delivery is required. It is intolerable for an error to go undetected or for a packet to be dropped. Further, once detected, it is desirable to be able to recover from the errors with minimal system interruption. Because of these factors and because RapidIO operates at very high frequencies, where errors are more likely, it is necessary to provide strong error coverage. Much of the RapidIO error coverage is handled through the physical layer specification. This allows different coverage strategies, depending on the physical environment without affecting other specification layers.

The 8/16 LP-LVDS physical layer can detect moderate burst errors. Because this physical layer is intended as a board level interconnect, it is assumed that the interface will not suffer from as high a bit error rate as it might if it were traversing a cable. Several error detection schemes are deployed to provide coverage. Packets are covered using a CCITT16 cyclic redundancy check (CRC), and control symbols are covered through transmission of the control symbol and its complement. Packets are transmitted and acknowledged following a strict sequence using the AckID field. The FRAME signal is covered against inadvertent transitions by treating it as a no-return-to-zero NRZ signal. It must stay in the new state for more than 1 clock to be considered valid.

The serial physical layer includes all of these capabilities and adds additional error coverage through the use of 8B/10B coding. The 8B/10B coding, described in more detail in Chapter 7, imposes additional constraints on the bit streams sent across the link. Violations of these constraints are detected as error conditions in the receiver and may be recovered by the RapidIO error recover techniques.

2.13.4 Error Recovery

The control symbol is at the heart of the hardware error recovery mechanism. Should a packet be detected with a bad checksum, control symbols are sent to verify that both sender and receiver are still synchronized and a packet retry is issued. Should a transaction be lost, a watchdog time-out occurs and the auto recovery state machines attempt to re-synchronize and retransmit.

If an interface fails severely, it may not be able to recover gracefully in hardware. For this extreme case, RapidIO hardware can generate interrupts so that system software can invoke a higher-level error recovery protocol. Software may query the maintenance registers to reconstruct the current status of the interface and potentially restart it. An in-band device reset control symbol is provided.

2.14 PERFORMANCE

One of RapidIO's intended uses is as a processor and memory interface where both latency and bandwidth are important considerations. RapidIO may also be used to carry data streams where deterministic delivery behavior is required. RapidIO offers several features aimed at improving system performance. The high-speed nature of the interface allows it to offer bandwidth equivalent to or higher than that available from existing buses. The parallel physical layer uses separate clock and frame signals to reduce clock and packet delimitation overhead. Source routing and transaction priority tagging reduce the blocking of packets, especially those of a critical nature. Large data payloads of up to 256 bytes and responseless stream write operations move larger volumes of data with less transaction overhead while not blocking links for extreme periods of time.

2.14.1 Packet Structures

The RapidIO packet is structured to promote simplified construction and parsing of packets within a wider on-chip parallel interface. This then limits the amount of logic operating on the narrower high-frequency interface. Packets are organized in byte granularities with 32-bit word alignments. In this way RapidIO header fields will land consistently in specific byte lanes on the receiving device. This also limits the need for complex reassembly, lane steering and parsing logic.

2.14.2 Source Routing and Concurrency

Traditional bus-based systems, such as those using PCI, have relied on address broadcasting to alert targets of an operation. In these systems, this is effective since all devices can monitor a common address bus and respond when they recognize a transaction to their address domain.

Switch-based systems can employ two methods to route transactions: broadcast or source routing. In a broadcast scheme a packet is sent to all connected devices. It is expected that one and only one device will actually respond to the packet. The advantage of the broadcast scheme is that the master does not need to know where the receiver resides. The disadvantage is that there is a large amount of system bandwidth wasted on the paths for which the transaction was not targeted.

To take full advantage of available system bandwidth, RapidIO employs source routing. This means that each packet has a source-specified destination address that instructs the fabric specifically where the transaction is to be routed. With this technique only the path between the sender and receiver is burdened with the transaction. The switch devices make decisions on a much smaller number of bits, resulting in smaller route tables and lower latency. This method leaves open bandwidth on other paths for other devices in the system to communicate with each other concurrently.

2.14.3 Packet Overhead

A performance concern in any tightly coupled intra-system interconnect is the transaction overhead required for the interface. Such overhead includes all bits needed to complete a transaction. Some examples of overhead include arbitration, addresses, transaction type, acknowledgements, error coverage, etc. Figure 2.6 shows some typical RapidIO operations and the number of bytes required to complete each operation. It is important to remember that RapidIO is a full duplex interface and therefore the interface can be fully pipelined with several outstanding operations at various stages of completion. Reply overhead does not contend with sending overhead. This is different from traditional buses, which require turnaround cycles and arbitration phases that add to the overhead. For most typical RapidIO transactions the total overhead will be about 28 bytes. This includes the overheard of the request packet, the response packet, the control symbol acknowledgements and the error checking codes. Streaming write (SWRITE) transactions have lower overhead, as do most messages.

2.15 OPERATION LATENCY

In memory transaction environments, where processors or other devices are dependent on the results of an operation before proceeding, latency becomes a key factor to system performance. For latency-sensitive environments the parallel RapidIO interface is preferred. Serial RapidIO will offer higher, and in some cases significantly higher latencies than that offered by

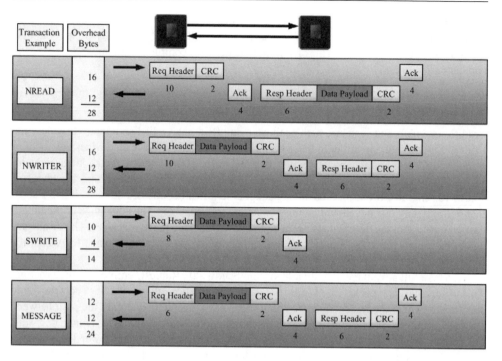

Figure 2.6 Operation overhead includes the total number of bytes sent in each direction to complete an operation (assumes 8-bit route and 32-bit offset address)

parallel RapidIO or traditional buses. It is assumed that the operations of interest here are small byte or cache-line-oriented operations. Since a RapidIO interface is narrower than traditional parallel buses, an operation will require more clock cycles for data transmission; RapidIO also has extra overhead associated with routing and error management. However relative to a bus RapidIO has virtually no contention or arbitration overhead, a significantly higher operating frequency, and a separate and concurrent reply path.

While there is complexity and additional latency created by converting a memory or message transaction into a RapidIO packet, the simplicity of the RapidIO protocol ensures that this conversion process can be accomplished in times of the order of 100 nanoseconds in commonly available technologies. Latency through RapidIO switch fabric devices is also typically measured at around 100 nanoseconds.

The result is that a parallel RapidIO interface can offer typical transaction latencies equivalent to or better than existing bus technologies while offering significantly higher link and system bandwidths.

REFERENCE

1. IEEE Std. 1596.3-1996, IEEE Standard for Low-Voltage Differential Signals (LVDS) for Scalable Coherent Interface (SCI), IEEE Computer Society, 31 July 1996.

3

Devices, Switches, Transactions and Operations

This chapter introduces the types of devices that would use RapidIO as a system interconnect. It then presents the concepts of transactions and operations and the transaction ordering rules that RapidIO employs to ensure proper system operation.

3.1 PROCESSING ELEMENT MODELS

Figure 3.1 shows a simple example of a RapidIO-based computing system. This is a multi-processing system where RapidIO is used to connect the four processing elements together

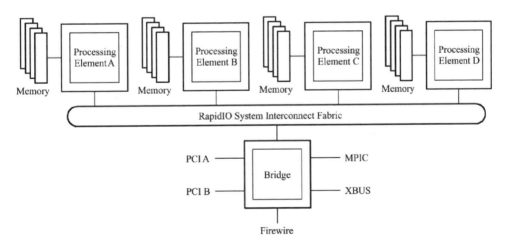

Figure 3.1   A possible RapidIO-based computing system

RapidIO® The Embedded System Interconnect.   S. Fuller
© 2005 John Wiley & Sons, Ltd   ISBN: 0-470-09291-2

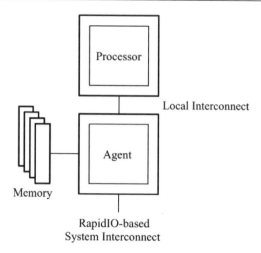

Figure 3.2   Processing element example

with a single peripheral bridge element. The processing elements, in this system provide the following functions: processing, memory control and connection to a RapidIO interconnect controller. The bridge part of the system provides I/O subsystem services such as high-speed PCI interfaces, gigabit Ethernet ports, interrupt control, and other system support functions. In this example there are several memory banks, each associated with an individual processor. In a situation like this, the latency of the RapidIO interconnect is less critical to system performance as the processors are able to satisfy most of their data requirements from the directly associated memories.

Figure 3.2 shows an example of how a processing element might consist of a processor connected to an agent device. The agent carries out several services on behalf of the processor. Most importantly, it provides access to a local memory that has much lower latency than memory that is local to another processing element (remote memory accesses). It also provides an interface to the RapidIO interconnect to service remote memory or peripheral accesses. In this example the local interconnect is typically a proprietary processor bus interface. This approach to a processing element may be used for incorporating legacy processor devices into a system.

3.1.1 Integrated Processor-memory Processing Element Model

Another form of a processor-memory processing element is a fully integrated component that is designed specifically to connect to a RapidIO interconnect system as shown in Figure 3.3. This type of device integrates the memory system controller and other support logic with a processor on the same piece of silicon or within the same package. This approach to processing is becoming more and more prevalent in the industry. The Motorola PowerQUICC III device is an example of this type of device. Future DSP devices from a number of vendors are also taking this approach to processor integration and development.

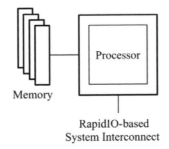

Figure 3.3 Integrated processor-Memory processing element example

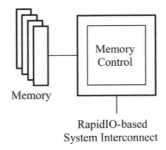

Figure 3.4 Memory-only processing element example

3.1.2 Memory-only Processing Element Model

A different processing element may not contain a processor at all, but may be a memory-only device as shown in Figure 3.4. This type of device is much simpler than a processor; it responds only to requests from the external system, not to local requests as in the processor-based model. As such, its memory is remote for all processors in the system.

3.2 I/O PROCESSING ELEMENT

This type of processing element is shown as the bridge in Figure 3.1. This device has distinctly different behavior than a processor or a memory device. An I/O processing element converts RapidIO transactions into other I/O transactions such as PCI or Ethernet. This conversion is typically a very straightforward operation as the I/O peripheral or I/O bus element typically will present itself to the system as a set of memory mapped registers and a range of read/write memory addresses. One of RapidIO's core objectives is to appear as a software transparent interconnect to I/O peripheral devices. The RapidIO Trade Association has created an inter-operability specification that details how RapidIO devices should interoperate not only with each other but with other technologies like PCI as well. A discussion of this topic appears in Chapter 9.

3.3 SWITCH PROCESSING ELEMENT

A switch processing element is a device that provides connectivity between multiple devices in a system. A switch may be used to connect a variety of RapidIO-compliant processing elements. A possible switch is shown in Figure 3.5. Behavior of the switches, and the interconnect fabric in general, is addressed in Chapter 6. Switches are not required in RapidIO-based systems, it is possible to use RapidIO as a point-to-point link between two devices, however the use of switches may reduce overall system cost and improve system performance. The RapidIO interconnect architecture assumes that switches will be an important component of system design. Consequently the RapidIO specifications provide significant descriptions of proper switch operations.

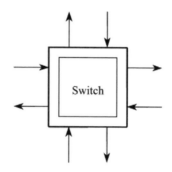

Figure 3.5 Switch element example

3.4 OPERATIONS AND TRANSACTIONS

RapidIO devices use operations to communicate with each other. Operations are composed of transactions. As part of the discussion of switch, memory and processing elements, it is useful to understand the concept of operation and transaction ordering. For a simple system of two devices, ordering is a simple concept. For a more complex system where hundreds of devices might be issuing thousands of transactions, providing a system that ensures that all operations will be completed in a consistent and proper manner is more difficult. The following sections describe operation and transaction ordering and system deadlock considerations in a system using RapidIO.

3.4.1 Operation Ordering

Most operations in a system don't have a requirement placed on them with respect to the order in which they are completed; changes to the temporal relationships between the initiation of two or more operations and their completion will not adversely affect proper system operation. Figure 3.6 shows a system where the NREAD RESPONSE transactions return in a different order than the NREAD transactions that generated them. The Read operations are also completed in a different order. For most applications this behavior is acceptable and often

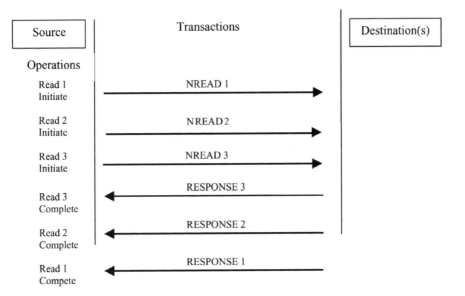

Figure 3.6 Unordered operations

encouraged as it may increase system performance. The Read transactions, in this example, may be targeted at single or multiple destinations.

There are, however, several tasks that require events to occur in a specific order. As an example, a processing element may wish to write a set of registers in another processing element. The sequence in which those writes are carried out may be critical to the correct operation of the target processing element. Without the application of specific system rules the transactions might become reordered in transit between devices and the completion ordering might be different from the initiation ordering. The RapidIO interconnect constrains ordering only between a specific source and destination pair operating at a given priority level, there is no other valid ordering concept within the RapidIO architecture.

In certain cases a processing element may communicate with another processing element or set of processing elements in different contexts. A set or sequence of operations issued by a processing element may have requirements for completing in order at the target processing element. That same processing element may have another sequence of operations that also requires a specific completion order at the same target processing element. However, the issuing processing element may have no requirements for completion order between the two sequences of operations. Further, it may be desirable for one of the sequences of operations to complete at a higher priority than the other sequence.

A transaction request flow is defined as an ordered sequence of request transactions from a given source (as indicated by the source identifier) to a given destination (as indicated by the transaction destination identifier). Each packet in a transaction request flow has the same source identifier and the same destination identifier and the same priority.

The flows between each source and destination pair are prioritized. There are three available priority flows. The limitation to three flows is due to constraints imposed by the physical layer specifications. When multiple transaction request flows exist between a given source and destination pair, transactions of a higher priority flow may pass transactions of a lower priority flow, but transactions of a lower priority flow may not pass transactions of a higher priority flow. Transactions will not overtake each other on the RapidIO links themselves. Typically, a RapidIO end point will maintain buffering for several packets at each priority level. Within a priority level, packets will typically be sent in the order in which they were received. Between priority levels, packets of higher priority will generally be sent before packets of lower priority. The actual techniques that are used to determine which packet to send next on a RapidIO link are implementation dependent.

Response transactions are not part of any transaction request flow. There is no ordering between any pair of response transactions and there is no ordering between any response transaction and any request transaction that did not cause the generation of the response.

To support transaction request flows, all devices must comply with the following fabric delivering ordering and end point completion ordering rules.

Fabric Delivery Ordering Rules

1. Non-maintenance request transactions within a transaction request flow (same source identifier, same destination identifier, and same priority) shall be delivered to the logical layer of the destination in the same order that they were issued by the logical layer of the source.
2. Non-maintenance request transactions that have the same source (same source identifier) and the same destination (same destination identifier) but different priorities shall be delivered to the logical layer of the destination as follows.
   - A transaction of a higher priority transaction request flow that was issued by the logical layer of the source before a transaction of a lower priority transaction request flow shall be delivered to the logical layer of the destination before the lower priority transaction.
   - A transaction of a higher priority transaction request flow that was issued by the logical layer of the source after a transaction of a lower priority transaction request flow may be delivered to the logical layer of the destination before the lower priority transaction.
3. Request transactions that have different sources (different source identifiers) or different destinations (different destination identifiers) are unordered with respect to each other.

End point Completion Ordering Rules

1. Write request transactions in a transaction request flow shall be completed at the logical layer of the destination in the same order that the transactions were delivered to the logical layer of the destination.
2. A read request transaction with source A and destination B shall force the completion at the logical layer of B of all write requests in the same transaction request flow that were received by the logical layer of B before the read request transaction.

Read request transactions need not be completed in the same order that they were received by the logical layer of the destination. As a consequence, read response transactions do not need to be issued by the logical layer of the destination in the same order that the associated read request transactions were received.

Write response transactions will likely be issued at the logical layer in the order that the associated write request was received. However, since response transactions are not part of any flow, they are not ordered relative to one another and may not arrive at the logical layer of their destination in the same order as the associated write transactions were issued. Therefore, write response transactions do not need to be issued by the logical layer in the same order as the associated write request was received.

3.4.2 Transaction Delivery

There are two basic types of delivery schemes that can be built using RapidIO processing elements: unordered and ordered. The RapidIO logical protocols assume that all outstanding transactions to another processing element are delivered in an arbitrary order. In other words, the logical protocols do not rely on transaction interdependencies for operation. RapidIO also allows completely ordered delivery systems to be constructed. Each type of system puts different constraints on the implementation of the source and destination processing elements and any intervening hardware. The specific mechanisms and definitions of how RapidIO enforces transaction ordering are discussed in the appropriate physical layer specification.

3.4.3 Ordered Delivery System Issues

Ordered delivery systems place additional implementation constraints on both the source and destination processing elements as well as any intervening hardware. Typically an ordered system requires that all transactions between a source/destination pair be completed in the order generated, not necessarily the order in which they can be accepted by the destination or an intermediate device. For example, if several requests are sent and the first of the several requests is found to have an error and must be retried, then the system must 'unroll' all the transmitted requests and retransmit all of previously sent requests starting from the first one to maintain the proper system ordering.

3.4.4 Deadlock Considerations

A deadlock can occur if a dependency loop exists. A dependency loop is a situation where a loop of buffering devices is formed, in which forward progress at each device is dependent upon progress at the next device. If no device in the loop can make progress then the system is deadlocked.

The simplest solution to the deadlock problem is to discard a packet. This releases resources in the network and allows forward progress to be made. RapidIO is designed to be a reliable fabric for use in real time tightly coupled systems, therefore discarding packets is not an acceptable solution.

In order to produce a system with no chance of deadlock it is required that a deadlock-free topology be provided for responseless operations. Dependency loops to single-direction packets can exist in unconstrained switch topologies. Often the dependency loop can be avoided with simple routing rules. Topologies such as hypercubes or three-dimensional meshes physically contain loops. In both cases, routing is done in several dimensions (x, y, z). If routing is constrained to the x dimension, then y, then z (dimension ordered routing), topology related dependency loops can be avoided in these structures.

In addition, a RapidIO end point design must not form dependency links between its input and output ports. A dependency link between input and output ports would occur if a processing element is unable to accept an input packet until a waiting packet can be issued from the output port.

RapidIO supports operations, such as read operations, that require response transactions to be received before the operation is complete. These operations can lead to a dependency link between a processing element's input port and output port.

As an example of an input to output port dependency, consider a memory element where the output port queue is full. The memory element cannot accept a new request at its input port since there is no place to put the response in the output port queue. No more transactions can be accepted at the input port until the output port is able to free entries in the output queue by issuing packets to the system.

Descriptions of how a RapidIO system can maintain a deadlock free environment are described in Chapters 7 and 8 on the serial and parallel physical layers.

4

I/O Logical Operations

4.1 INTRODUCTION

This chapter describes the set of input and output (I/O) logical operations and their associated transactions as defined in Part I of the RapidIO specification. I/O logical operations provide support for basic reads and writes to and from memory spaces located within a RapidIO system. Details of the transaction types, packet formats, and other necessary transaction information are presented.

The I/O operations are achieved through the use of request/response transaction pairs. The transaction pairs operate through a RapidIO fabric. The RapidIO fabric does not track the transactions as they travel through the fabric and, from the perspective of the fabric, there is no explicit or implicit relationship between request transactions and the responses to those requests. Although there may be intermediate switch devices within a system and hence multiple packet transmissions, from the perspective of the RapidIO logical layer there will be a single request transaction and a single response transaction if required. Intermediate switch devices do not differentiate between request and response transactions. Their role is simply to forward a transaction towards its final destination.

There are six basic I/O operations defined in the RapidIO architecture. Table 4.1 presents the defined operations, the transactions that are used to perform the operation and a description of the operation.

4.2 REQUEST CLASS TRANSACTIONS

We will start by examining the Request class operation. We will look at the NREAD transaction first. An NREAD is a request by one RapidIO device to another to deliver the contents of a region of memory to the requester. The memory request can be between 1 and 256 bytes in length. There are some alignment restrictions on the data that is returned. For data over

RapidIO® The Embedded System Interconnect.   S. Fuller
© 2005 John Wiley & Sons, Ltd   ISBN: 0-470-09291-2

Table 4.1   I/O logical operations

Operation	Transactions used	Description
Read	NREAD, RESPONSE	Non-coherent read from system memory
Write	NWRITE	Non-coherent write to system memory
Write-with-response	NWRITE_R, RESPONSE	Non-coherent write to system memory that waits for a response before signaling operation completion
Streaming-write	SWRITE	Non-coherent write optimized for large DMA transfers
Atomic (read–modify–write)	ATOMIC, RESPONSE	Read–modify–write operation useful for multiprocessor semaphoring
Maintenance	MAINTENANCE	Transactions targeting RapidIO specific registers

8 bytes, it must be double-word aligned and a multiple of 8 bytes in length. For data below 8 bytes, the data must still be properly aligned in memory and the rdsize and wdptr fields are used to generate byte masks to indicate which bytes of data are valid. The minimum payload size is 8 bytes. This helps to ensure that RapidIO packets remain a multiple of 8 bytes in length. More detail on data alignment restrictions in RapidIO is found in Section 4.8.

4.2.1 Field Definitions for Request Class Transactions

Figure 4.1 shows the Type 2 packet with all of its fields. More detail on these fields is provided in the example below. The FTYPE field refers to the format type of the transaction. The FTYPE for Request class transactions is 2. Type 2 transactions include NREAD and several ATOMIC operations. The actual transaction type is specified in the TTYPE (transaction type) field. The combination of FTYPE and TTYPE uniquely identify the transaction format. Table 4.2 lists the defined Type 2 transactions.

RDSIZE is used in conjunction with the Address field and the W and XADD fields to specify the location, size and alignment of the data to be read. We will look at the generation of the RDSIZE, and W fields in more detail in the following example. The Address and XADD fields provide a double-word aligned address for the requested data. The address field is 29 bits long, however, because these 29 bits specify double-words rather than bytes, they

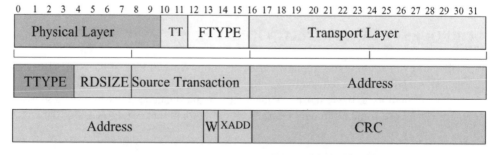

Figure 4.1   RapidIO FTYPE 2 (Request class) field descriptions

Table 4.2 Type 2 transactions

Encoding	Transaction Field
0b0000–0011	Reserved
0b0100	NREAD transaction
0b0101–1011	Reserved
0b1100	ATOMIC inc: post-increment the data
0b1101	ATOMIC dec: post-decrement the data
0b1110	ATOMIC set: set the data (write 0b11111 . . .')
0b1111	ATOMIC clr: clear the data (write 0b00000 . . .')

support addressability to a 4 Gbyte address space. XADD (also referred to as extended address most significant bits or XADSMS) provides two more bits of address space. These two bits would be used as the most significant address bits and would increase the address space support size to 16 Gbyte per device.

This discussion has assumed the small address model. RapidIO also supports a medium and large address model. The medium address model adds 16 bits to the address field. The large model adds 32 bits to the address field. There is no information in the packet to indicate which model is employed for any given packet. The choice of address model is negotiated at system bringup between the sender and the receiver. If a target device advertises that it has a 64 bit address space then the large address model must be used for all packets destined for that device. The extra address space information is added to the packet in a contiguous big-endian fashion after the Source Transaction ID field and before the end of the packet.

In RapidIO it is assumed that senders and receivers know what they are doing and that senders are prepared to send properly configured transactions that match the characteristics of the receiver. Appendix A of this book describes RapidIO capability attribute registers which provide information to the system describing the specific capabilities of a RapidIO device.

4.3 RESPONSE CLASS TRANSACTIONS

A response transaction is issued by a RapidIO end point when it has completed a request made to it by an initiating RapidIO end point. Response transaction packets are always directed and are transmitted in the same way as request transaction packets. Broadly, format Types 12, 13, 14, and 15 are considered Response class transactions. Currently, Type 12 and 14 are reserved and Type 15 is implementation defined. Type 13 is the main Response class transaction.

The Type 13 packet format returns status, data (if required), and the requestor's transaction ID. A RESPONSE packet with an 'ERROR' status or a response that is not expected to have a data payload never has a data payload. The Type 13 format is used for response packets to all request packets except maintenance and responseless writes. Maintenance requests are responded to by maintenance responses.

4.3.1 Field Definitions for Response Packet Formats

Figure 4.2 illustrates the format and fields of Type 13 packets. The field value 0b1101 in bit positions 12–15 specifies that the packet format is Type 13.

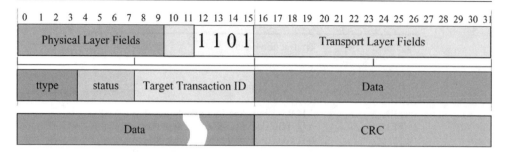

Figure 4.2　Response packet format

Table 4.3　Field definitions and encodings for all response packets

Field	Encoding	Sub-field	Definition
Transaction	0b0000		RESPONSE transaction with no data payload
Type	0b0001–0111		Reserved
	0b1000		RESPONSE transaction with data payload
	0b1001–1111		Reserved
TargetTID			The corresponding request packet's transaction ID
Status	Type of status and encoding		
	0b0000	DONE	Requested transaction has been successfully completed
	0b0001–0110		Reserved
	0b0111	ERROR	Unrecoverable error detected
	0b1000–1011		Reserved
	0b1100–1111	Implementation	Implementation defined—can be used for additional information such as an error code

Table 4.3 presents the field definitions and encodings for all response packets. There are only three unique fields in the Response packet format. The Transaction Type is used to indicate what type of Response transaction is being sent. There are two defined types, a response with a data payload and a response without a data payload. Other encodings are reserved. The Target Transaction ID is identical to the Request Transaction ID that the response packet is responding to. The requesting device would use this ID to match the response with the request.

If the response includes data, the data is included after the Target Transaction ID. Note that there is no size field to indicate the length of the data included in the response. The minimum size of this data field would be 8 bytes and the maximum size would be 256 bytes. It is up to the receiver to determine the end of the packet based on an end of packet symbol that is part of the physical layer information. *The length of the data field in a response packet cannot be determined solely by the fields of the response packet.* We will cover this in more detail in Chapters 7 and 8 describing how the RapidIO physical layer works.

The requesting device should retain information on the size and alignment of the data that it requested as the response transaction will deliver the data in the format requested.

Two possible response transactions can be delivered to a requesting processing element:

- A DONE response indicates to the requestor that the desired transaction has completed and it also returns data for read-type transactions as described above.

- An ERROR response means that the target of the transaction encountered an unrecoverable error and could not complete the transaction.

For most transaction types, a request transaction sent into the system is marked with a transaction ID that is unique for each requestor and responder processing element pair. This transaction ID allows a response to be easily matched to the original request when it is returned to the requestor. An end point cannot reuse a transaction ID value to the same destination until the response from the original transaction has been received by the requestor. The number of outstanding transactions that may be supported is implementation dependent.

Transaction IDs may also be used to indicate sequence information if ordered reception of transactions is required by the destination processing element and the interconnect fabric can reorder packets. The receiving device can either retry subsequent out-of-order requests, or it can accept and not complete the subsequent out-of-order requests until the missing transactions in the sequence have been received and completed.

4.4 A SAMPLE READ OPERATION

Having introduced the Type 2 and Type 13 transaction packet formats, we can now understand how a Read operation is accomplished in the RapidIO architecture. The Read operation (Figure 4.3) is quite straightforward and serves as a good demonstration of how the RapidIO protocol works. A Read operation is composed of an NREAD transaction and a RESPONSE transaction.

Let's assume that a processor intends to read the contents of a memory-mapped register on a peripheral device across a RapidIO interconnect. To do this, the processor will produce an NREAD transaction and send it to the peripheral. The peripheral will retrieve the contents of the register and use it as the payload of a RESPONSE transaction that will be sent back to the processor. We will make the following assumptions for this example.

1. The processor or source has a RapidIO deviceID of 0x10
2. The peripheral or target has a RapidIO deviceID of 0x36
3. The memory address to be read on the peripheral is 0xffff0004 and is 4 bytes in length. The address space on the peripheral is addressable with 32 bits.

The RapidIO interface logic on the processor prepares the packet for the transaction that will be sent to the peripheral. Figure 4.4 shows all 12 bytes of this transaction. Working from left to right and from top to bottom, the fields in this transaction are as follows. The physical layer fields make up the first ten bits of this packet. They will be described in detail in a later

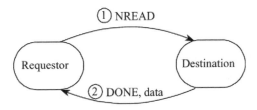

Figure 4.3 Read operation

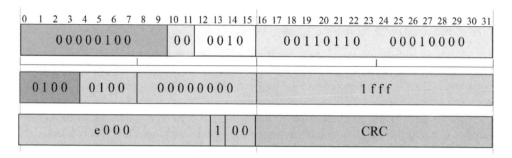

Figure 4.4   Sample Read transaction

section. The next 2 bits are the Transport Type (TT) fields, they are used to decode the Transport Layer Fields. In this case, we assume that we are using 8-bit deviceIDs, this will lead us to placing the value 0b0011011000010000 (0x3610) in the Transport Layer Fields section of the packet. Bits 12–16 contain the Function Type of the transaction. In the case of a Request transaction the FTYPE is 2 (0b0010). The first four bits of the second word contain the specific transaction type. In the case of an NREAD transaction this field will be 4 (0b0100).

The next field is the RDSIZE field. This field will be described later in conjunction with the address, WDPTR and XADD field.

The Source Transaction ID field is an 8-bit field that is used to uniquely identify the transaction to the receiver. This 8-bit field enables there to be up to 256 transactions in process between any source and destination pair (although devices are not required to support this many outstanding transactions and actual implementations will likely support much fewer than 256 outstanding transactions). For this example the Source Transaction ID is 0.

The remaining fields are the Address field, the word pointer field (WDPTR or W), the Extended address most significant bits (XAMSBS or XADD) field and the CRC field. These fields are not as straightforward to generate as their names might imply. RDSIZE does not specify the size of the read in a direct manner, instead it encodes the length of the data as well as the alignment of the data in cases where the data is shorter than 8 bytes. For data longer than 8 bytes the data is assumed to be an integer multiple of 8 bytes and to be aligned on 8 byte boundaries. WDPTR is used for some transactions to identify which word of a double-word aligned data field is targeted. WDPTR may also be used to help in specifying data sizes greater than 8 bytes.

RDSIZE and WDPTR are most easily generated via a table lookup with the transaction size and the three least significant address lines as inputs.

XAMSBS provides for two additional bits of most significant address. This extends the byte addressable address range for a RapidIO device to 34 bits, enough to support up to a 16 Gbyte address space. XAMSBS extends medium and large address fields of 45 and 63 bits in the same manner.

Figure 4.5 shows how the target address and data length of the read request are processed to build the RapidIO packet header.

The RapidIO approach reduces a typical bus transaction that would have used 32 bits of address, 8 bits of byte lane enables, and 8 bits of transaction size (for a total of 48 bits of information) to 36 bits. These 36 bits contain all of the information necessary to unambiguously

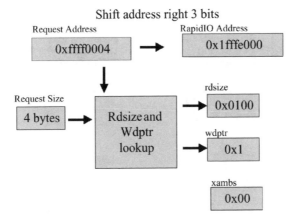

Figure 4.5   Address field generation

describe all legal RapidIO read or write transactions between 1 byte and 256 bytes in length to or from a 34 bit address space. This is a 25% reduction in the overhead of communicating the address and length of a transaction.

The CRC field covers the entire packet except for the first 6 bits. These bits are part of the physical layer information and may change as the packet moves through the system. They are excluded from the CRC calculation to allow the single CRC to provide complete end-to-end coverage of the packet as it moves through the system. The CRC is required not to be recalculated as a packet moves through the system (although it must be checked by every receiver). The actual CRC information is not included in this example. CRC calculation details are covered in a later section. The completely formed request packet is now ready to be sent.

Figure 4.6 shows the flow of the operation from the initiator to the target and then back to the initiator. Between the initiator and the fabric and between the fabric and the target, as the request is forwarded, an acknowledge symbol is sent back to the transmitter to indicated that the packet was correctly received. At this point the transmitter can release the transmit buffer that should be holding a copy of the transaction. If an error occurs in the packet in the link between two devices, the receiver would detect the error and send a NACK symbol back to the transmitter. Typically the transmitter would then resend the transaction from its buffer. Chapter 11 discusses error management in more detail.

The target, upon receiving a transaction, would perform the designated operation. In this case, it would read 4 bytes of data from the memory mapped registers at 0xffff0004. These 4 bytes would be placed in an 8 byte payload (with proper alignment) and a response transaction of Type 13 would be generated and sent back through the fabric to the transmitter.

The actual number of intermediate devices in the fabric are unknowable to the transaction itself. The transactions target device IDs in the system. The intermediate fabric devices are responsible for forwarding the transaction packet to the appropriate target device.

When the operation initiator (master) receives the response packet and determines that it is the target of this response transaction, it will compare the Source Transaction ID with its list of outstanding transactions. If there is a match, the initial operation is completed with the associated data. The response transaction does not contain enough information for the operation initiator to complete the operation. For example, the response packet returns the data, but does

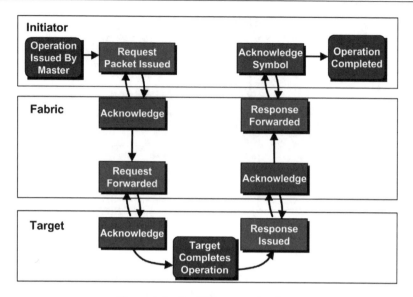

Figure 4.6   RapidIO transaction flow

not return the address or length or alignment of the returned data. The operation initiator must retain the context of the original request transaction to understand the context of the data returned in the response transaction and to complete the entire operation. In this case, returning the 4 bytes of register data to the requesting processor.

If the response source transaction ID does not match any outstanding transaction ID tracked by the operation initiator then an error is signaled.

4.5 WRITE OPERATIONS

The Type 5 packet format supports write (Figure 4.7), write with response (Figure 4.8) and atomic (read–modify–write) operations. Type 5 transaction packets always contain a data payload. A data payload that consists of a single double-word or less has sizing information. The WRSIZE field specifies the maximum size of the data payload for multiple double-word transactions. The fields, WRSIZE and WDPTR, are used to specify the size of the data payload and are encoded the same way that the RDSIZE and WDPTR fields specify the size of the data for a read operation. Aside from the data payload, the fields of the Type 5 Write transaction packet (Figure 4.9) are essentially the same as for a Type 2 Request transaction.

① NWRITE or SWRITE, data

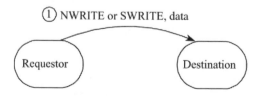

Figure 4.7   Write operation

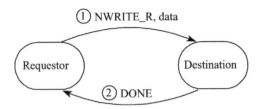

Figure 4.8   Write with response operation

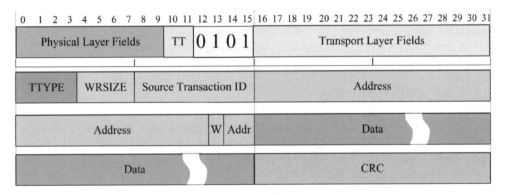

Figure 4.9   Type 5 packet format

Table 4.4   Write class transaction types

Encoding	Transaction field
0b0000–0011	Reserved
0b0100	NWRITE transaction
0b0101	NWRITE_R transaction
0b0110–1101	Reserved
0b1110	ATOMIC test-and-swap: read and return the data, compare with 0, write with supplied data if compare is true
0b1111	Reserved

Write transactions may request the sending of a response or not. Table 4.4 shows the defined write class transaction types. Transaction Type 4 is a basic NWRITE transaction that does not expect a write completion response. Transaction Type 5 is an NWRITE_R transaction that expects a completion response. The normal response would be a DONE response packet with no data payload.

The ATOMIC increment, decrement, set and clear operations fall in the Request or Read transaction packet format. These operations assume that the data to be operated on resides in memory and that status will be reported on the operation, but no new data will be written to the memory.

The ATOMIC test-and-swap transaction is a more complex transaction and adds a data field because the data in memory may be updated as a result of the transaction. The operation of the test-and-swap operation is to read the data from memory, compare it with zero, if the read value is zero then a new value, supplied in the data field is written to the memory. This is performed as a single operation at the destination. Basically, if the memory contents are zero then they are updated with the value of the data field, otherwise they are left unchanged. The status field of a response transaction is used to report the results of the operation as well as to return the data being tested. The data field size is limited to one double-word (8 bytes) of data payload. The addressing scheme defined for the write transactions also controls the size of the atomic operation in memory. For ATOMIC transactions the bytes must be contiguous in memory and must be of byte, half-word (2 bytes), or word (4 bytes) size, and be aligned to the natural boundary and byte lane as with a regular write transaction. Double-word (8-byte) and 3, 5, 6, and 7-byte ATOMIC test-and-swap transactions are not allowed.

4.6 STREAMING WRITES

A slight variant of the NWRITE transaction is the streaming write (SWRITE) transaction. SWRITE transactions are of format Type 6. This transaction makes some simplifying assumptions about alignment and payload size to reduce header overhead. SWRITE is used for moving large quantities of data in DMA type operations. SWRITE transactions can offer extremely efficient movement of data within a RapidIO system. The header and CRC overhead of SWRITE operations is only 10 bytes, for 256 byte payloads this is less than 5% of the payload so the efficiency is over 95%.

The SWRITE transaction (Figure 4.10), like the response transaction, does not contain a data size length field so the actual length of the packet cannot be determined from the contents of the packet alone. The data field for a SWRITE transaction is at least 8 bytes long and no more than 256 bytes long. It is double-word aligned and is an integer multiple of double-words in length. It begins after the XAMSBS field and extends to the CRC field. The receiver finds the length of the SWRITE data field by finding the end of the packet, this is delimited by a control symbol, and then working back two or four bytes over the CRC and optional PAD fields.

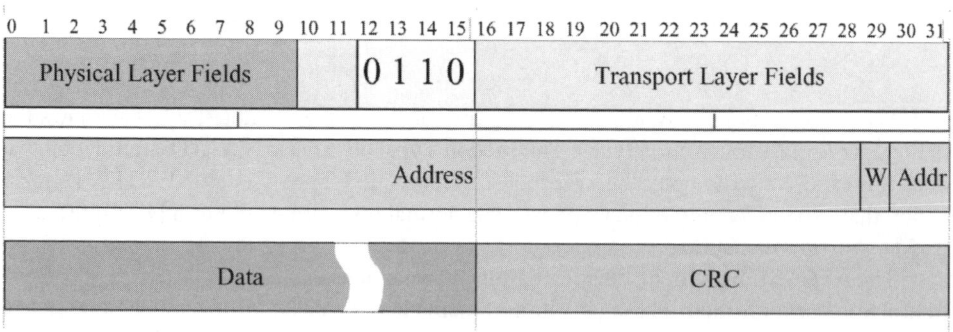

Figure 4.10   SWRITE transaction format

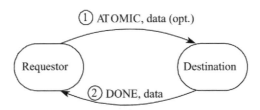

Figure 4.11 Read–modify–write operation

4.7 ATOMIC OPERATIONS

The read–modify–write operation (Figure 4.11), consisting of the ATOMIC and RESPONSE transactions (typically a DONE response), is used by a number of co-operating processing elements to perform synchronization using non-coherent memory. The allowed specified data sizes are one word (4 bytes), one half-word (2 bytes) or one byte, with the size of the transaction specified in the same way as for an NWRITE transaction. Double-word (8-byte) and 3, 5, 6, and 7-byte ATOMIC transactions may not be specified.

The atomic operation is a combination read and write operation. The destination reads the data at the specified address, returns the read data to the requestor, performs the required operation to the data, and then writes the modified data back to the specified address without allowing any intervening activity to that address. Defined operations are increment, decrement, test-and-swap, set, and clear. Of these, only test-and-swap requires the requesting processing element to supply data. The target data of an atomic operation may be initialized using an NWRITE transaction.

If the atomic operation is to memory, data is written to the memory, regardless of the state of any system-wide cache coherence mechanism for the specified cache line or lines, although it may cause a snoop of any caches local to the memory controller.

4.8 MAINTENANCE OPERATIONS

The Type 8 MAINTENANCE packet format (Table 4.5) is used to access the RapidIO capability and status registers (CARs and CSRs) and data structures. Unlike other request formats, the Type 8 packet format serves as both the request and the response format for maintenance operations. Type 8 packets contain no addresses, and contain only data payloads for write requests and read responses. All configuration register read accesses are performed in word (4-byte), and optionally double-word (8-byte) or specifiable multiple double-word quantities up to a limit of 64 bytes. All register write accesses are also performed in word (4-byte), and optionally double-word (8-byte) or multiple double-word quantities up to a limit of 64 bytes.

The WRSIZE field specifies the maximum size of the data payload for multiple double-word transactions. The data payload may not exceed that size, but may be smaller if desired. Both the maintenance read and the maintenance write request generate the appropriate maintenance response.

The maintenance port-write operation is a write operation that does not have guaranteed delivery and does not have an associated response. This maintenance operation is useful for

Table 4.5　Maintenance transaction fields

Type 8 fields	Encoding	Definition
Transaction	0b0000	Specifies a maintenance read request
	0b0001	Specifies a maintenance write request
	0b0010	Specifies a maintenance read response
	0b0011	Specifies a maintenance write response
	0b0100	Specifies a maintenance port-write request
	0b0101–1111	Reserved
Hop count		An 8-bit field that is used to target maintenance transactions at switch devices. RapidIO switch devices are not required to have deviceIDs. Hop count is used as an alternative mechanism to address switch devices
config_offset		Double-word offset into the CAR/CSR register block for reads and writes
srcTID		The type 8 request packet's transaction ID (reserved for port-write requests)
TargetTID		The corresponding type 8 response packet's transaction ID
status	0b0000	DONE. Requested transaction has completed successfully
	0b0001–0110	Reserved
	0b0111	ERROR. Unrecoverable error detected
	0b1000–1011	Reserved
	0b1100–1111	Implementation-defined. Can be used for additional information such as an error code

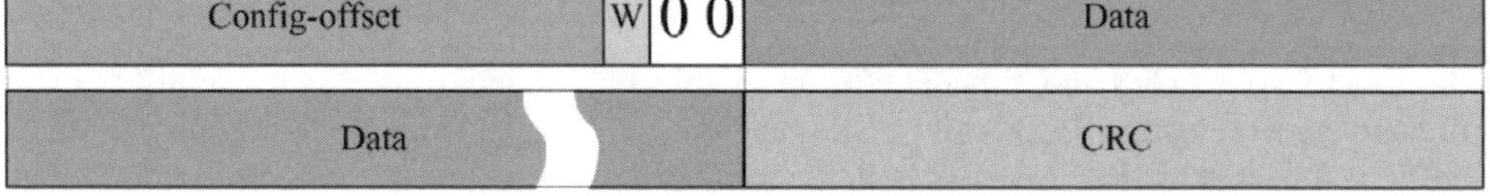

Figure 4.12　Maintenance packet format

sending messages such as error indicators or status information from a device that does not contain an end point, such as a switch. The data payload is typically placed in a queue in the targeted end point and an interrupt is typically generated to a local processor. A port-write request to a queue that is full or busy servicing another request may be discarded.

Figure 4.12 displays a Type 8 maintenance request packet and its associated fields. The field value 0b1000 specifies that the packet format is of Type 8. The srcTID and config_offset fields are reserved for port-write requests.

4.9 DATA ALIGNMENT

To simplify the addressing of data in a system, RapidIO assumes that all transactions are aligned to double-word (8-byte) boundaries (Figure 4.13). If the data being sent does not start or end on a double-word boundary, then a special mask is used to indicate which bytes contain valid data (Figures 4.14, 4.15). Sub-byte data is not supported at all. The minimum data size that can be sent across RapidIO is one byte (8 bits).

For write operations, a processing element will properly align data transfers to a double-word boundary for transmission to the destination. This alignment may require breaking up a data stream into multiple transactions if the data is not naturally aligned. A number of data payload sizes and double-word alignments are defined to minimize this burden. Figure 4.16 shows a 48-byte data stream that a processing element wishes to write to another processing element through the interconnect fabric. The data displayed in the figure to be written is

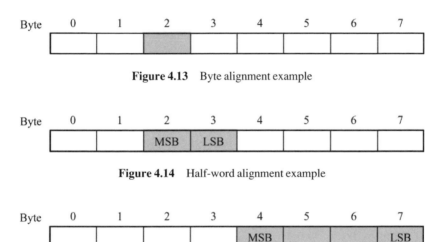

Figure 4.13   Byte alignment example

Figure 4.14   Half-word alignment example

Figure 4.15   Word alignment example

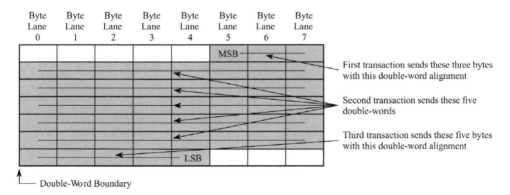

Figure 4.16   Data alignment example

shaded in grey. Because the start of the stream and the end of the stream are not aligned to a double-word boundary, the sending processing element is responsible for breaking the transaction in to at least three separate transactions as shown in the figure.

The first transaction sends the first three bytes (in byte lanes 5, 6, and 7) and indicates a byte lane 5, 6, and 7 three-byte write. The second transaction sends all of the remaining data, except for the final sub-double-word. As a RapidIO packet can contain up to 256 bytes of payload, this 40-byte transaction easily fits. The third transaction sends the final 5 bytes in byte lanes 0, 1, 2, 3, and 4 indicating a five-byte write in byte lanes 0, 1, 2, 3, and 4.

5

Messaging Operations

The RapidIO specification offers several operations designed to be used for sending messages between processing elements in a system. These operations do not describe a specific programming model; rather they represent a standard packet format for direct inter-processor communications.

5.1 INTRODUCTION

A common approach to distributed processing systems is the use of tightly coupled processors connected to distributed memory elements. These processors may be operating under a monolithic operating system. For example a single Linux system can run efficiently on up to several tens of processors. Often a single operating system is tasked with managing the pool of processors and memory. In most of these cases, it is quite efficient for the processors to work out of a common hardware-maintained coherent memory space. This allows processors to communicate initialization and completion of tasks through the use of semaphores, spin locks, and inter-processor interrupts. Memory is managed centrally by the operating system with a paging protection scheme. This approach to multiprocessing is quite mature and has been used for decades.

In other distributed systems, the processors and memory may be more loosely coupled. Several operating systems or kernels may coexist in the system, each kernel being responsible for a small part of the entire system. The kernels may be from different software vendors and may be running on different processor architectures. In these systems it is quite useful to have a simple communication mechanism, which the kernels can use to communicate with other kernels within the system. For example, PowerPC processors running Linux might be communicating with TigerSHARC digital signal processors running QNX. For a given application, there may be no reason to share memory spaces between these devices. In this type of system it is desirable to have a common hardware and software interface mechanism available

RapidIO® The Embedded System Interconnect.   S. Fuller
© 2005 John Wiley & Sons, Ltd   ISBN: 0-470-09291-2

on all devices to accomplish the communication simply, cost-effectively and with high performance. In these systems the processors communicate with each other through message passing.

In these message passing systems, two mechanisms typically are used to move commands or data from one device to another. The first mechanism is called direct memory access (DMA), the second is messaging. The primary difference between the two models is that DMA transactions are steered by the source whereas messages are steered by the target. This means that a DMA source not only requires access to a target, but must also have visibility into the target's address space. The message source requires access only to the target and does not need visibility into the target's address space. In distributed systems it is common to find a mix of DMA and messaging deployed.

The RapidIO architecture contains a packet transport mechanism that can used for messaging. The RapidIO messaging model meets several goals:

- A message is constructed of one or more transactions that can be sent and received through a possibly unordered interconnect
- A sender can have a number of outstanding messages queued for sending
- A sender can send a higher-priority message before a lower-priority message and can also pre-empt a lower-priority message to send a higher-priority message and have the lower-priority message resume when the higher priority message is complete (prioritized concurrency)
- A sender requires no knowledge of the receiver's internal structure or memory map
- A receiver of a message has complete control over its local address space
- A receiver can have a number of outstanding messages queued for servicing if desired
- A receiver can receive a number of concurrent multiple-transaction messages if desired

5.2 MESSAGE TRANSACTIONS

The *RapidIO Message Passing Logical Specification* defines two different packet formats for use in messaging. The Type 10 packet format is used for sending very short, 16-bit data payloads. The Type 10 packet is known as the DOORBELL transaction format. DOORBELL transactions are well suited for sending interrupts between processors. The Type 11 packet is used for sending data payloads of up to 4096 bytes using multiple transactions. The Type 11 packet transaction is called the MESSAGE transaction. These transactions will now be described in more detail.

5.2.1 Type 10 Packet Format (Doorbell Class)

The Type 10 packet format is the DOORBELL transaction format. Type 10 packets never have data payloads. The field value 0b1010 specifies that the packet format is of Type 10.

The doorbell operation (Figure 5.1), consisting of the DOORBELL and RESPONSE transactions (typically a DONE response) as shown below, is used by a processing element to send a very short message to another processing element through the interconnect fabric. The DOORBELL transaction contains the info field which is used to hold the transaction information. The transaction does not have a data payload. The info field is software defined, and can be used for any desired purpose. Typically the operating system running on the processor would define the meaning of the info field used in DOORBELL transactions.

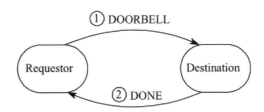

Figure 5.1   Doorbell operation

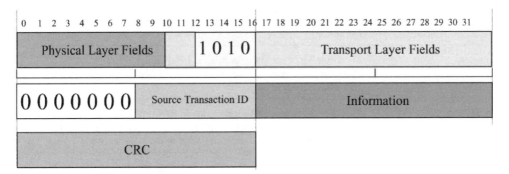

Figure 5.2   Doorbell message packet format (Type 10)

A processing element that receives a doorbell transaction takes the packet and puts it in a doorbell message queue within the processing element. This queue may be implemented in hardware or in local memory. Some examples of queue structures are included in sections 5.3 and 5.4. The local processor is expected to read the queue to determine the sending processing element and the info field and determine what action to take based on that information.

Figure 5.2 displays a Type 10 packet with all its fields. Note that there are 8 reserved bits. These should be set to 0 for all transactions. Next there is a source transaction ID and then there are 16 bits of information. If the information is numerical and longer than 8 bits then the numerical data is considered to be presented in big-endian format, with the most significant byte coming first in the bit stream.

This packet may be used for delivering interrupt information to a processor. In this usage the information field would be used to convey interrupt level and target information to the recipient. It may also be used for sending semaphores between processing devices.

5.2.2 Type 11 Packet Format (Message Class)

The Type 11 packet is the MESSAGE transaction format. Type 11 packets always have a data payload. The data payload is always a multiple of double-words in length. Sub-double-word messages are not specifiable and, if desired, would be managed in software.

The data message operation (Figure 5.3), consisting of the MESSAGE and RESPONSE transactions (typically a DONE response) as shown below, is used by a processing element's message passing support hardware to send a data message to other processing elements. Completing a data message operation can consist of up to 16 individual MESSAGE transactions.

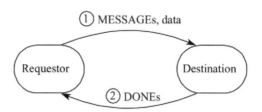

Figure 5.3   Message operation

Table 5.1   Specific field definitions and encodings for Type 11 Packets

Field	Encoding	Definition
msglen		Total number of packets comprising this message operation. A value of 0 indicates a single-packet message. A value of 15 (0b1111) indicates a 16-packet message, etc.
msgseg		Specifies the part of the message supplied by this packet. A value of 0 indicates that this is the first packet in the message. A value of 15 (0b1111) indicates that this is the sixteenth packet in the message, etc.
ssize		Standard message packet data size. This field informs the receiver of a message the size of the data payload to expect for all the packets for a single message operations except for the last packet in the message. This prevents the sender from having to pad the data field excessively for the last packet and allows the receiver to properly put the message in local memory.
	0b0000–1000	Reserved
	0b1001	8 bytes
	0b1010	16 bytes
	0b1011	32 bytes
	0b1100	64 bytes
	0b1101	128 bytes
	0b1110	256 bytes
	0b1111	Reserved
mbox		Specifies the recipient mailbox in the target processing element.
letter		Identifies a slot within a mailbox. This field allows a sending processing element to concurrently send up to four messages to the same mailbox on the same processing element.

MESSAGE transaction data payloads are always multiples of doubleword quantities. The maximum message operation payload is 4096 bytes. This payload would be composed of 16 MESSAGE transactions, each carrying 256 byte data payloads. Definitions and encodings of fields specific to Type 11 packets are provided in Table 5.1.

The combination of the letter, mbox, and msgseg fields uniquely identifies the message packet in the system for each requestor and responder processing element pair in the same way as the transaction ID is used for other request types.

The processing element's message passing hardware that is the recipient of a data message operation will examine a number of packet fields (Figure 5.4) in order to place the MESSAGE packet data in local memory:

- Message length (msglen) field—specifies the number of transactions that comprise the data message operation.
- Message segment (msgseg) field—identifies which part of the data message operation is contained in this transaction. The message length and segment fields allow the individual packets of a data message to be sent or received out of order. Up to 16 message segments may be specified.
- Mailbox (mbox) field—Specifies which mailbox is the target of the data message. Up to four mailboxes may be supported.
- Letter (letter) field—allows receipt of multiple concurrent data message operations from the same source to the same mailbox. Up to four letters may be specified.
- Standard size (ssize) field—specifies the data size of all of the transactions except (possibly) the last transaction in the data message.

Although the RapidIO specification uses the nomenclature of mailbox, letter and message segment, logically, these fields refer to 8 bits of unique message identifier information that can be used to uniquely identify and manage up to 256 different message streams between any two processing elements. From this information, the message passing hardware of the recipient processing element will calculate where in local memory the transaction data should be placed.

For example, assume that the mailbox starting addresses for the recipient processing element are at addresses 0x1000 for mailbox 0, 0x2000 for mailbox 1, 0x3000 for mailbox 2, and 0x4000 for mailbox 3, and that the processing element receives a message transaction with the following fields:

- message length of 6 packets
- message segment is 3rd packet
- mailbox is mailbox 2
- letter is 1
- standard size is 32 bytes
- data payload is 32 bytes (it shall be 32 bytes since this is not the last transaction)

Using this information, the processing element's message passing hardware can determine that the 32 bytes contained in this part of the data message shall be put into local memory at address 0x3040.

Target Address = Mailbox_2_base + (message_segment *standard_size)
0x3040   =   0x3000   +   (0x0002   *   0x0020   )

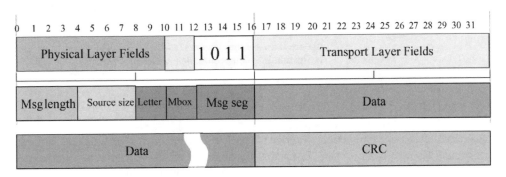

Figure 5.4 Message packet format (Type 11)

This simple addressing structure makes it very easy for the receiving processing element to calculate the address of the target memory structure to store the message data into. While easy to calculate, the location of the receiver target memory structure is not visible to the sender. This allows the development of secure systems that might provide message-based communications only between processing elements.

5.2.3 Response Transactions

All responses to message transactions are generated with Type 13 packets. Type 13 packets have been discussed in Chapter 4. Their use in message operations is similar to their use in I/O operations. Three possible response transactions can be received by a requesting processing element:

- A DONE response indicates to the requestor that the desired transaction has been completed. A DONE response may include the target_info field to return status information to the operation initiator.
- A RETRY response will be generated for a message transaction that attempts to access a mailbox that is busy servicing another message operation, as can a doorbell transaction that encounters busy doorbell hardware. All transactions that are retried for any reason shall be retransmitted by the sender. This prevents a transaction from partially completing and then leaving the system in an unknown state.
- An ERROR response means that the target of the transaction encountered an unrecoverable error and could not complete the transaction.

Table 5.2 presents specific field definitions for Type 13 packets when they are used to respond to messaging transactions.

Table 5.2   Specific field definitions for Type 13 packets

Field	Sub-field	Length	Definition
target_info			When the response contains a target_info field, these three sub-fields are used:
	letter	2 bits	Identifies the slot within the target mailbox. This field allows a sending processing element to concurrently send up to four messages to the same mailbox on the same processing element
	mbox	2 bits	Specifies the recipient mailbox from the corresponding message packet
	msgseg	4 bits	Specifies the part of the message supplied by the corresponding message packet. A value of 0 indicates that this is the response for the first packet in the message. A value of 15 (0xF) indicates that this is the response for the sixteenth (and last) packet in the message, etc.
targetTID		8 bits	Transaction ID of the request that caused this response

5.3 MAILBOX STRUCTURES

The RapidIO message transactions describe the packets that are used to communicate between processing elements in a system. The actual transmit and receive buffer structures are not specified by the RapidIO Trade Association. Nor has the RapidIO Trade Association defined standard software API interfaces to the messaging structures. The simplicity of the transactions makes it a trivial programming effort to develop drivers that would interact with the messaging hardware.

The messages will typically be written to and read from buffers in the local memory of the processing elements. Depending on the expected messaging load, different structures may be used to support the use and reuse of the message buffers associated with the transmit and receive hardware. The following sections describe both simple and more complicated approaches to managing the message buffers associated with RapidIO messaging transactions.

The inbound mailbox refers to the hardware associated with the reception of RapidIO messaging transactions. This mailbox will first check the message for errors then will acknowledge the transaction to the transmitter. The mailbox will then place the data associated with the message into buffer memory and signal to the targeted processor that a message has been received. The processor would then retrieve the message data for further processing.

Because a single RapidIO end point might be receiving messages from several other devices, it is useful to provide an inbox structure that is capable of supporting the reception of several nearly concurrent messages.

5.3.1 A Simple Inbox

Perhaps the simplest inbound mailbox structure is that of a single-register port or direct map into local memory space.

In this structure (Figure 5.5), the inbound single transaction message is posted to either a register, set of registers, or circular queue in local memory. In the case of the circular queue,

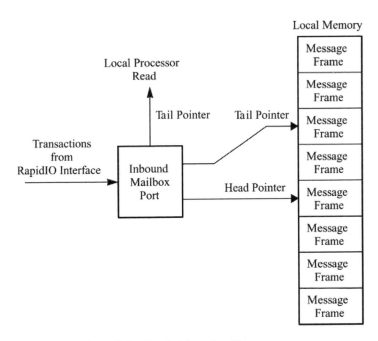

Figure 5.5 Simple inbound mailbox structure

the mailbox hardware maintains head and tail pointers that point at locations within a fixed window of pre-partitioned message frames in memory. Whenever the head pointer equals the tail pointer, the receive buffer is signaled to be full. No more messages can be accepted and they are retried on the RapidIO interface. As messages are received into the message frames, the local processor is interrupted. The interrupt service routine (ISR) of the local processor reads the mailbox port that contains the message located at the tail pointer. The ISR may then increment the tail pointer.

The RapidIO MESSAGE transaction allows up to four such inbound mailbox ports per target address. The DOORBELL transaction is defined as a single mailbox port.

5.3.2 An Extended Inbox

A second more extensible structure for managing the receive message buffers is shown in Figure 5.6. In this system, instead of using a directly addressed circular queue, there are separate post and free lists, which contain pointers to the message frame buffers.

One of these structures would be required for each priority level supported in an implementation. The size of the message frames is equal to the maximum message size that can be accepted by the receiver. As with the simple inbox, the sender specifies only the mailbox, all pointers are managed by the inbound hardware and the local processor. Message priority and letter number are managed by software.

The advantage of the extended inbox structure is that it allows local software to service message frames in any order. It also allows memory regions to be moved in and out of the message structure instead of forcing software to copy the message to a different memory location.

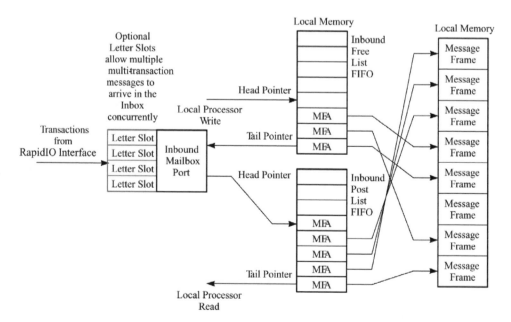

Figure 5.6   Extended inbound mailbox structure

5.3.3 Receiving Messages

When a message transaction is received, the inbound mailbox port takes the message frame address (MFA) pointed at by the inbound free list tail pointer and increments that pointer (this may cause a memory read to prefetch the next MFA), effectively taking the MFA from the free list. Subsequent message transactions from a different sender or with a different letter number are now retried until all of the transactions for this message operation have been received, unless there is additional hardware to handle multiple concurrent message operations for the same mailbox, differentiated by the letter slots.

When the entire message is received and written into memory, the inbound post list pointer is incremented and the MFA is written into that location. If the queue was previously empty, an interrupt is generated to the local processor to indicate that there is a new message pending. This causes a window where the letter hardware is busy and cannot service a new operation between the receipt of the final transaction and the MFA being committed to the local memory.

When the local processor services a received message, it reads the MFA indicated by the inbound post FIFO tail pointer and increments the tail pointer. When the message has been processed (or possibly deferred), it puts a new MFA in the memory address indicated by the inbound free list head pointer and increments that pointer, adding the new MFA to the free list for use by the inbound message hardware.

If the free list head and tail pointer are the same, the FIFO is empty and there are no more MFAs available and all new messages are retried. If the post list head and tail pointers are the same, there are no outstanding messages awaiting service from the local processor. Underflow conditions are fatal since they indicate improper system behavior. This information can be part of an associated status register.

These structures and mechanisms are presented as examples and represent neither requirements nor official recommendations for how RapidIO message mailboxes should be structured.

5.4 OUTBOUND MAILBOX STRUCTURES

The structure of the outbound mailbox, like the inbound mailbox, can vary in complexity in connection with the capabilities and performance of the end point.

5.4.1 A Simple Outbox

Generation of a message can be as simple as writing to a memory-mapped descriptor structure either in local registers or memory. The simple outbound message queue looks similar to the simple inbox message queue (Figure 5.7).

The local processor reads a port in the outbound mailbox to obtain the position of a head pointer in local memory. This read may result in a special data value that indicates that the outbound message queue is full. This might be a technique to efficiently communicate queue status to the processor. Assuming that the queue is not full, then the processor would write a descriptor structure and message to the location pointed to by the head pointer. After the descriptor and message data have been written, the processor would perform a write to the message port hardware to advance the head pointer and mark the message as queued.

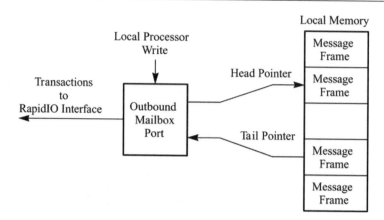

Figure 5.7   Simple outbound queue structure

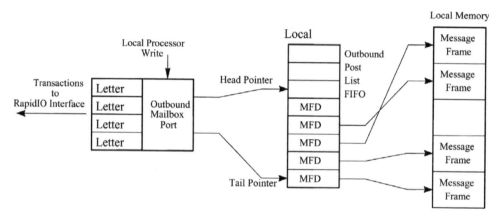

Figure 5.8   Extended outbound queue structure

5.4.2 An Extended Outbox

A more extensible method of queueing messages is again a two-level approach (Figure 5.8). Multiple structures are required if concurrent operation is desired in an implementation. The FIFO is a circular queue of some fixed size. The message frames are of a size that is equal to the maximum message operation size that can be accepted by the receivers in the system. As with the receive side, the outbound slots can be virtual and any letter number can be handled by an arbitrary letter slot.

5.4.3 Transmitting Messages

When the local processor wishes to send a message, it stores the message in local memory, writes the message frame descriptor (MFD) to the outbound mailbox port (which in-turn writes it to the location indicated by the outbound post FIFO head pointer) and increments the head pointer.

The advantage of this method is that software can have one or more pre-set messages stored in local memory. Whenever it needs to communicate an event to a specific end point it writes the address of the appropriate message frame to the outbound mailbox, and the outbound mailbox generates the message transactions and completes the operation.

If the outbound head and tail pointers are not equal, there is a message waiting to be sent. This causes the outbound mailbox port to read the MFD pointed to by the outbound tail pointer and then decrements the pointer (this may cause a memory read to pre-fetch the next MFD). The hardware then uses the information stored in the MFD to read the message frame, packetize it, and transmit it to the receiver. Multiple messages can be transmitted concurrently if there is hardware to support them, differentiated by the letter slots in the extended outbound message queue.

If the free list head and tail pointer are the same, the FIFO is empty and there are no more MFDs to be processed. Underflow conditions are fatal because they indicate improper system behavior. This information can also be part of a status register.

Because the outbound and inbound hardware are independent entities, it is possible for more complex outbound mailboxes to communicate with less complex inboxes by simply reducing the complexity of the message descriptor to match. Likewise simple outboxes can communicate with complex inboxes. Software can determine the capabilities of a device during initial system setup. The capabilities of a device's message hardware are stored in the port configuration registers.

6

System Level Addressing in RapidIO Systems

The RapidIO architecture assumes that there may be hundreds or thousands of devices in a system. It is not feasible to uniquely identify and separate this many devices on the basis of a single global address space. RapidIO uses device IDs to uniquely identify all of the devices that comprise a RapidIO system.

Because of its focus on bit efficiency, RapidIO defines two models of device addressing. There is a small system model that provides 8 bits of device ID. With 8-bit IDs there could be up to 256 individual devices in a system. The large system model provides for 16-bit device IDs. With 16-bit device IDs, there could be up to 65 536 unique devices in a system. Packets using the small system model are 16 bits smaller than packets using the large system model.

The system level addressing is specified in *Part III: Common Transport Specification* of the RapidIO specifications. This transport specification is independent of any RapidIO physical or logical layer specification.

This chapter contains the transport layer packet format definitions. Three fields are added to the packet formats to support system level addressing. The transport formats are intended to be fabric independent so that the system interconnect can be anything required for a particular application; therefore all descriptions of the transport fields and their relationship with the logical packets are shown as bit streams.

6.1 SYSTEM TOPOLOGY

By virtue of the use of device IDs as the system level addressing scheme, RapidIO can support virtually any system topology. The device IDs do not convey any intrinsic information about where they are located. It is the responsibility of the interconnection fabric to discover where the devices are located and to forward packets to them based on the target device ID only.

Similar to a switched Ethernet network, which makes no assumptions about the *a priori* location of any given Ethernet device, a RapidIO network learns, during the system discovery

RapidIO® The Embedded System Interconnect.   S. Fuller
© 2005 John Wiley & Sons, Ltd   ISBN: 0-470-09291-2

phase of system bringup where devices are located in the system. Switches are programmed to understand the direction, although not the exact location of all devices in a system. When device locations change, as might occur during a hot swap or failover situation, only the switches need to be reconfigured to understand the new system topology.

In RapidIO, devices communicate with other devices by sending packets that contain source and destination device IDs. These device IDs, 8- or 16-bit fields in the packet header, are used as keys by the switch fabric devices to forward the packet towards its final destination. In RapidIO, switch devices will typically use lookup tables to hold the associations between device IDs and proper output ports. RapidIO switches will be programmed to understand which of multiple output ports should be used to forward a packet with any given destination device ID. Owing to the relatively small size of the device ID field and the common use of hashing keys to reduce table size, this lookup can be done quickly and need not add to the latency of a RapidIO packet traversing a switch device.

Also, because the source device ID is also included in the packet header, switches and end points are always aware of not only where a packet is going to, but also where it came from. The source device ID can be used by the system to send back a response packet indicating proper operation completion or perhaps an error condition. System level addressing in RapidIO is very similar to, although much simpler, than that of switched Ethernet. Ethernet relies on extended protocol stacks such as TCP/IP to create reliable communications channels between devices. RapidIO achieves these reliable channels solely through hardware constructs, which are more efficient for moving data around within a limited trusted system. RapidIO is not a replacement for Ethernet in a LAN environment. It is far too insecure for such an application. In like manner, Ethernet is not a replacement for RapidIO as an in-system communications technology. At higher speeds it requires far too much software overhead and introduces significant latency into the communications path between adjacent processing devices.

This switch-based approach to packet forwarding is much simpler to manage than alternative path-based routing schemes. In path-based schemes each end point must keep track of the proper path to every other end point in a system. This will be accomplished through the use of a table in each end point, which maps device IDs to ports for every other device in the system. If the system topology changes due to an event such as a path failover, every end point must be made aware of the failover and must have its path mapping tables updated appropriately. All of these tables, of which there might be several thousand, must be properly updated before the system can resume normal operation. A master processor will generally manage the updating of these path tables. The master processor must then keep a master copy of all the tables in all devices, and must be responsible for the topology changeover of the system both before and after the failover event; perhaps also interacting, post-event with all of the devices in the system to test that all of the connectivity paths are properly configured.

6.2 SWITCH-BASED SYSTEMS

As has been discussed, a RapidIO system will typically be organized around switches. Figure 6.1 shows a small system in which five processing elements are interconnected through two switches. Systems such as this can be built today with available RapidIO-based processor and switch products. A packet sent from one processing element to another is routed through

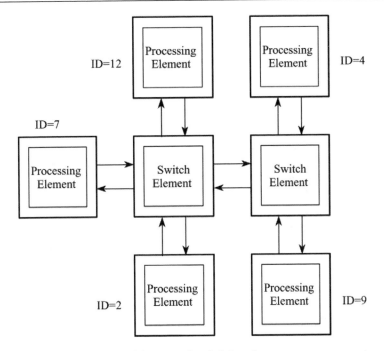

Figure 6.1 A small switch-based system

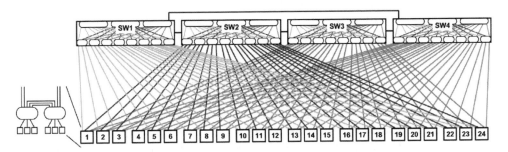

Figure 6.2 A complex switch-based system

the interconnect fabric comprised of the switches by interpreting the transport fields. Because a request usually requires a response, the transport fields include both the destination as well as the source device ID for the packet. The source device ID can be used to determine the return path to the requestor.

Figure 6.2 shows a more complex switch-based system. Here there are multiple line cards, denoted by the numbers 1–24. Each line card contains multiple processing elements, connected together using RapidIO. The line cards are connected to one another through a multi-level switch hierarchy that offers multiple paths between devices. These multiple paths offer additional system bandwidth and could also offer redundant data paths between devices. This particular system offers connectivity for up to 120 devices in a single system.

6.3 SYSTEM PACKET ROUTING

In RapidIO, each directly addressable device in the system will have one or more unique device identifiers. When a packet is generated, the device ID of the destination of the packet is put in the packet header. The device ID of the source of the packet is also put in the packet header for use by the destination when generating response packets. When the destination of a request packet generates a response packet, it swaps the source and destination fields from the request, making the original source the new destination and itself the new source. Packets are routed through the fabric on the basis of the destination device ID.

The most common method used for routing RapidIO packets through a fabric is through the use of routing tables in the switches. Each switch in the interconnect fabric contains a table that instructs the switch how to route every destination ID from an input port to the proper output port. The simplest form of this method allows only a single path from every processing element to every other processing element. More complex forms of this method may allow adaptive routing for redundancy and congestion relief. However, the actual method by which packets are routed between the input of a switch and the output of a switch is implementation dependent.

6.4 FIELD ALIGNMENT AND DEFINITION

The *RapidIO Common Transport Specification* adds a transport type (tt) field to the logical specification packet that allows four different transport packet types to be specified. The tt field indicates which type of additional transport fields are added to the packet.

The three fields (tt, destinationID, and sourceID) added to the logical packets allow for two different sizes of the device ID fields, a large (16-bit), and a small (8-bit), as shown in Table 6.1. The two sizes of device ID fields allow two different system scalability points to optimize packet header overhead, and affix additional transport field overhead only if the additional addressing is required. The small device ID fields allow a maximum of 256 devices to be attached to the fabric. The large device ID fields allow systems with up to 65 536 devices.

Figure 6.3 shows the transport header definition bit stream. The shaded fields are the bits associated with the logical packet definition that are related to the transport bits. Specifically, the field labeled 'Logical ftype' is the format type field defined in the logical specifications. This field comprises the first four bits of the logical packet. The second logical field shown 'Remainder of logical packet' is the remainder of the logical packet of a size determined by the logical specifications, not including the logical ftype field which has already been included in the combined bit stream. The unshaded fields (tt=0b00 or tt=0b01 and destinationID and

Table 6.1   tt Field definitions

tt	Definition
0b00	8-bit deviceID fields
0b01	16-bit deviceID fields
0b10	Reserved
0b11	Reserved

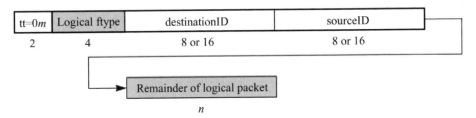

Figure 6.3   Destination–source transport bit stream

sourceID fields) are the transport fields added to the logical packet by the *RapidIO Common Transport Specification*.

6.5 ROUTING MAINTENANCE PACKETS

While RapidIO end points must have device IDs, it is not required that RapidIO switches have device IDs. For most operations this shouldn't be a problem as switches are not typically the targets of an operation. However, there is a case where switches are the targets of operations. This is when the system is being configured or reconfigured or when the system needs to retrieve information from the registers contained within a switch. Because switches may not have device IDs, there is no way for a typical read or write operation to target a switch.

To remedy this situation, an alternative method of addressing for maintenance packets is defined. Maintenance packets include an additional hop_count field in the packet. This field specifies the number of switches (or hops) into the network, the packet should experience from the issuing processing element to the target destination device. Whenever a switch processing element receives a maintenance packet it examines the hop_count field. If the received hop_count is zero, the transaction is targeted at that switch. If the hop_count field is not zero, it is decremented and the packet is sent out of the switch according to the destination ID field. This method allows easy access to any intervening switches in the path between two addressable processing elements. However, since maintenance response packets are always targeted

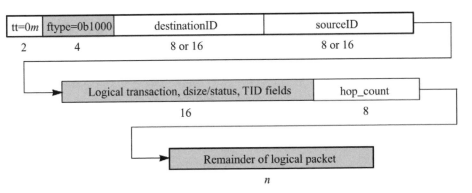

Figure 6.4   Maintenance packet transport bit stream

at an end point, the hop_count field for transactions targeted at non-switch elements, should always be assigned a value of 0xFF by the source of the packets. This will prevent the transaction from being inadvertently accepted by an intervening switch device. Figure 6.4 shows the transport layer fields that are added to a maintenance logical packet to support this. Maintenance logical packets are described in more detail in Chapter 3.

7

The Serial Physical Layer

The RapidIO serial physical layer, typically referred to as Serial RapidIO, addresses the electrical connection of devices on a board or across a backplane. The serial physical layer defines a full duplex serial link between devices using unidirectional differential signals in each direction. The serial physical layer protocol provides support for packet delivery between RapidIO devices. This support includes packet and control symbol transmission, flow control, error management, and other device-to-device functions.

The RapidIO serial physical layer has the following properties:

- Embeds the transmission clock with data using an 8B/10B encoding scheme.
- Supports one serial differential pair, referred to as one lane, or four ganged serial differential pairs, referred to as four lanes, in each direction.
- Uses special 8B/10B codes, called K-codes, to manage the link, including flow control, packet demarcation, and error reporting.
- Allows transferring packets between RapidIO 1x/4x LP-Serial (Serial RapidIO) Ports and RapidIO Physical Layer 8/16 LP-LVDS (Parallel RapidIO) ports without requiring packet manipulation.
- Employs similar retry and error recovery protocols to those of the Parallel RapidIO physical layer.
- Supports transmission rates of 1.25, 2.5, and 3.125 Gbaud (data rates of 1.0, 2.0, and 2.5 Gbps) per lane.

This chapter first defines the individual elements that make up the serial link protocol such as packets, control symbols, and the serial bit encoding scheme. This is followed by a description of the link protocol. Finally, the signal descriptions, and electrical requirements are covered. Registers related to the serial physical layer are discussed in the Appendices.

RapidIO® The Embedded System Interconnect.   S. Fuller
© 2005 John Wiley & Sons, Ltd   ISBN: 0-470-09291-2

Table 7.1   Packet field definitions

Field	Description
ackID[0–4]	Acknowledge ID is the packet identifier for acknowledgments back to the packet sender. The serial physical layer defines five bits here, enough to uniquely identify up to 32 outstanding transactions between two devices
rsvd[0–2]	The reserved bits are set to logic 0 when the packet is generated and ignored when a packet is received
prio[0–1]	Sets packet priority: 0b00 lowest priority 0b01 medium priority 0b10 high priority 0b11 highest priority
crc[0–15]	16-bit code used to detect transmission errors in the packet. This field is added to the end of the packet

7.1 PACKETS

As described in earlier chapters, RapidIO uses packets to send data and control information around within a system. The physical layer specifications describe the portions of the packets that are used to move a packet between adjacent devices. This is often referred to as the data link protocol layer. In both the parallel and serial physical layers, there are 24 bits that are defined as part of the physical layer. These 24 bits are divided into 8 bits of header information and 16 bits of cyclic redundancy check (CRC) code. This section discusses the serial packet format and the fields that are added to a packet by the serial physical layer. Table 7.1 presents the fields added to a packet by the serial physical layer.

7.1.1 Packet Format

Figure 7.1 shows the format of a serial packet and how the physical layer ackID, rsvd, and prio fields are prefixed at the beginning of the packet and the 16-bit CRC field is appended to the end of the packet.

The unshaded fields are the fields added by the physical layer. The shaded field is the combined logical and transport layer bits and fields that are passed to the physical layer. The 3-bit rsvd field is required to make the packet length an integer multiple of 16 bits. The physical layer fields wrap around the transport and logical fields.

Serial RapidIO packets are required to have a length that is an integer multiple of 32 bits. This sizing simplifies the design of the transmit and receive port logic where the internal data

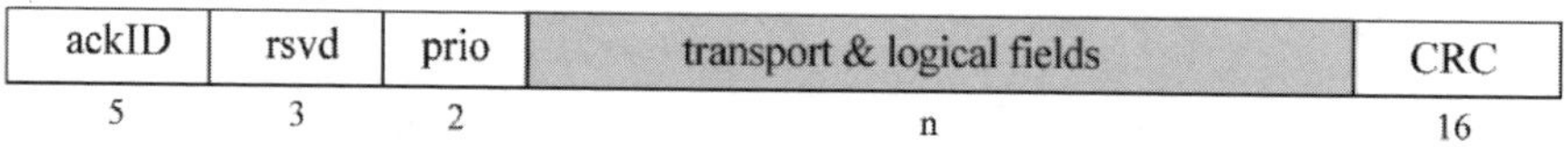

ackID	rsvd	prio	transport & logical fields	CRC
5	3	2	n	16

Figure 7.1   Packet Format

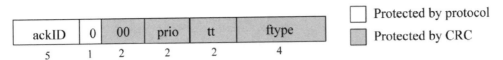

Figure 7.2   Error coverage of first 16 bits of packet header

paths are typically an integer multiple of 32 bits in width. If the length of a packet defined by this combination of specifications is an odd multiple of 16 bits (including the CRC), a 16-bit pad whose value is 0 (0x0000) will be appended at the end of the packet such that the resulting padded packet is an integer multiple of 32 bits in length.

7.1.2 Packet Protection

A 16-bit CRC code is added to each packet by the serial physical layer to provide error detection. The code covers the entire packet except for the ackID field and one bit of the rsvd field, which are considered to be zero for the CRC calculations. Figure 7.2 shows that the first six bits (the ackID and the first reserved bit) of the serial layer header are not covered by the code.

This structure allows the ackID to be changed on a link-by-link basis as the packet is transported across the fabric without requiring that the CRC be recomputed for each link. Since ackIDs on each link are assigned sequentially for each subsequent transmitted packet, an error in the ackID field is easily detected.

7.1.2.1 Packet CRC Operation

The CRC is appended to a packet in one of two ways. For a packet with length, exclusive of the CRC, of 80 bytes or less, a single CRC is appended at the end of the logical fields. For packets with length, exclusive of CRC, of greater than 80 bytes, a CRC is added after the first 80 bytes and a second CRC is appended at the end of the logical layer fields.

The second CRC value is a continuation of the first. The first CRC is included in the running calculation, meaning that the running CRC value is not reinitialized after it is inserted after the first 80 bytes of the packet. This allows intervening devices to regard the embedded CRC value as two bytes of packet payload for CRC checking purposes. If the CRC appended to the end of the logical layer fields does not cause the end of the resulting packet to align to a 32-bit boundary, a 2-byte pad of all logic 0s is post-pended to the packet. The pad of logic 0s allows the CRC check to always be done at the 32-bit boundary.

The early CRC value can be used by the receiving processing element to validate the header of a large packet and start processing the data before the entire packet has been received, freeing up resources earlier and reducing transaction completion latency.

Figure 7.3 is an example of a padded packet of length greater than 80 bytes. This packet includes the two CRC codes and a pad at the end of the packet to bring the total packet length to an integer multiple of 32 bits.

7.1.2.2 16-Bit Packet CRC Code

The ITU (International Telecommunications Union) polynomial $X^{16}+X^{12}+X^{5}+1$ is used to generate the 16-bit CRC for packets. The value of the CRC is initialized to 0xFFFF (all logic

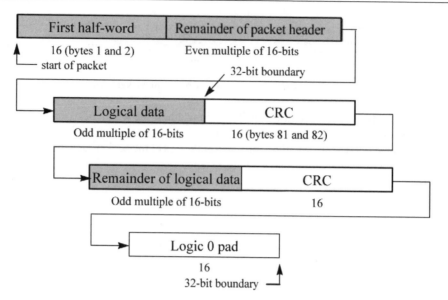

Figure 7.3 Padded packet of length greater than 80 bytes

Table 7.2 Control symbol field definitions

Field	Description
stype0 [0–2]	Encoding for control symbols that make use of parameter0 and parameter1
parameter0 [0–4]	Used in conjunction with stype0 encodings
parameter1 [0–4]	Used in conjunction with stype0 encodings
stype1 [0–2]	Encoding for control symbols which make use of the cmd field
cmd [0–2]	Used in conjunction with the stype1 field to define the link maintenance commands
CRC [0–4]	5-bit code used to detect transmission errors in control symbols

1s) at the beginning of each packet. For the CRC calculation, the uncovered six bits are treated as logic 0s.

7.2 CONTROL SYMBOLS

This section discusses the RapidIO serial physical layer control symbols. Control symbols are the message elements used by ports connected by a serial link to manage all aspects of the serial link operation.

Control symbols (Figure 7.4) can carry two functions, one encoded in the stype0 field and one encoded in the stype1 field. The functions encoded in stype0 are 'status' functions that convey status about the port transmitting the control symbol. The functions encoded in stype1 are either requests to the receiving port or transmission delimiters. The fields parameter0 and parameter1 are used by the functions encoded in the stype0 field. The cmd field is a modifier for the functions encoded in the stype1 field.

0	2	3	7	8	12	13	15	16	18	19	23
stype0 [0–2]		**parameter0** [0–4]		**parameter1** [0–4]		**stype1** [0–2]		**cmd** [0–2]		**CRC** [0–4]	

Figure 7.4 Control symbol format

A control symbol carrying one function is referred to by the name of the function it carries. A control symbol carrying two functions may be referred to by the name of either function that it carries. For example, a control symbol with stype0 set to packet-accepted and stype1 set to NOP is referred to a packet-accepted control symbol. A control symbol with stype0 set to packet-accepted and stype1 set to restart-from-retry is referred to as either a packet-accepted control symbol or a restart-from-retry control symbol, depending on which name is appropriate for the context.

Control symbols are specified with the ability to carry two functions so that a packet acknowledgment and a packet delimiter can be carried in the same control symbol. Packet acknowledgment and packet delimiter control symbols constitute the vast majority of control symbol traffic on a busy link. Carrying an acknowledgment (or status) and a packet delimiter whenever possible in a single control symbol allows a significant reduction in link overhead traffic and an increase in the link bandwidth available for packet transmission.

7.2.1 Stype0 Control Symbol Definitions

The encoding and function of stype0 and the information carried in parameter0 and parameter1 for each stype0 encoding are presented in Table 7.3.

The status control symbol is the default stype0 encoding and is used when the control symbol does not convey another stype0 function. Table 7.4 defines the valid parameters for stype0 control symbols.

The following sections depict various control symbols. Since control symbols can contain one or two functions, shading in the figures is used to indicate which fields are applicable to that specific control symbol function.

Table 7.3 Stype0 control symbol encoding

stype0 [0–2]	Function	Contents of	
		Parameter0	Parameter1
0b000	Packet-accepted	packet_ackID	buf_status
0b001	Packet-retry	packet_ackID	buf_status
0b010	Packet-not-accepted	packet_ackID	cause
0b011	Reserved		
0b100	Status	ackID_status	buf_status
0b101	Reserved		
0b110	Link-response	ackID_status	port_status
0b111	Reserved		

Table 7.4 Stype0 parameter definitions

Parameter	Definition
packet_ackID [0–4]	The ackID of the packet being acknowledged by an acknowledgment control symbol
ackID_status [0–4]	The value of ackID expected in the next packet the port receives. For example, a value of 0b00001 indicates the device is expecting to receive ackID 1
buf_status [0–4]	Specifies the number of maximum length packets that the port can accept without issuing a retry due to a lack of resources. The value of buf_status in a packet-accepted, packet-retry, or status control symbol is the number of maximum packets that can be accepted, inclusive of the effect of the packet being accepted or retried
	Value 0–29: The encoding value specifies the number of new maximum sized packets the receiving device can receive. The value 0, for example, signifies that the downstream device has no available packet buffers (thus is not able to hold any new packets)
	Value 30: Signifies that the downstream device can receive 30 or more new maximum sized packets
	Value 31: The downstream device can receive an undefined number of maximum sized packets, and relies on the retry protocol for flow control

7.2.1.1 Packet-accepted Control Symbol

The packet-accepted control symbol indicates that the receiving device has taken responsibility for sending the packet to its final destination and that resources allocated by the sending device can be released. This control symbol shall be generated only after the entire packet has been received and found to be free of detectable errors.

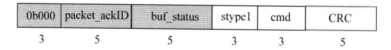

0b000	packet_ackID	buf_status	stype1	cmd	CRC
3	5	5	3	3	5

Figure 7.5 Packet-accepted control symbol format

7.2.1.2 Packet-retry Control Symbol

A packet-retry control symbol indicates that the receiving device was not able to accept the packet because of some temporary resource conflict, such as insufficient buffering, and the sender should retransmit the packet.

0b001	packet_ackID	buf_status	stype1	cmd	CRC
3	5	5	3	3	5

Figure 7.6 Packet-retry control symbol format

7.2.1.3 Packet-not-accepted Control Symbol

The packet-not-accepted control symbol is used to indicate to the sender of a packet why the packet was not accepted by the receiving port. As shown in Figure 7.7, the control symbol contains a cause field that indicates the reason for not accepting the packet and a packet_ackID field. If the receiving device is not able to specify the cause, or the cause is not one of the defined options, the general error encoding shall be used.

The cause field shall be used to display informational fields useful for debug. Table 7.5 displays the reasons a packet may not be accepted, indicated by the cause field.

0b010	packet_ackID	cause	stype1	cmd	CRC
3	5	5	3	3	5

Figure 7.7   Packet-not-accepted control symbol format

Table 7.5   Cause field definitions

Cause [0–4]	Definition
0b00000	Reserved
0b00001	Received unexpected ackID on packet
0b00010	Received a control symbol with bad CRC
0b00011	Non-maintenance packet reception is stopped
0b00100	Received packet with bad CRC
0b00101	Received invalid character, or valid but illegal character
0b00110–0b11110	Reserved
0b11111	General error

7.2.1.4 Status Control Symbol

The status control symbol is the default stype0 encoding and is used when the control symbol does not convey another stype0 function. The status control symbol contains the ackID_status and the buf_status fields. The buf_status field indicates to the receiving port the number of maximum length packet buffers the sending port had available for packet reception at the time the control symbol was generated. The ackID_status field allows the receiving port to determine if it and the sending port are in sync with respect to the next ackID value the sending port expects to receive.

0b100	ackID_status	buf_status	stype1	cmd	CRC
3	5	5	3	3	5

Figure 7.8   Status control symbol format

7.2.1.5 Link-response Control Symbol

The link-response control symbol is used by a device to respond to a link-request control symbol as described in the link maintenance protocol described in Section 7.4. The status reported in the status field is the status of the port at the time the associated input-status link-request control symbol was received.

The possible values of the port_status field of the link-response control symbol are shown in Table 7.6.

0b110	ackID_status	port_status	stype1	cmd	CRC
3	5	5	3	3	5

Figure 7.9 Link-response control symbol format

Table 7.6 Port_status field definitions

Port_status [0–4]	Status	Description
0b00000		Reserved
0b00001		Reserved
0b00010	Error	The port has encountered an unrecoverable error and is unable to accept packets
0b00011		Reserved
0b00100	Retry-stopped	The port has retried a packet and is waiting in the input retry-stopped state to be restarted
0b00101	Error-stopped	The port has encountered a transmission error and is waiting in the input error-stopped state to be restarted
0b00110–0b01111		Reserved
0b10000	OK	The port is accepting packets
0b10001–0b11111		Reserved

7.2.2 Stype1 Control Symbol Definitions

The encoding of stype1 and the function of the cmd field are shown in Table 7.7.

Table 7.7 Stype1 control symbol encoding

stype1 [0–2]	stype1 function	cmd [0–2]	cmd function	Packet delimiter
0b000	Start-of-packet	0b000		Yes
0b001	Stomp	0b000		Yes
0b010	End-of-packet	0b000		Yes
0b011	Restart-from-retry	0b000		*
0b100	Link-request	0b000–0b010	Reserved	*
		0b011	Reset-device	
		0b100	Input-status	
		0b101–0b111	Reserved	
0b101	Multicast-event	0b000		No
0b110	Reserved	0b000		No
0b111	NOP (Ignore)**	0b000		No

* Denotes that restart-from-retry and link-request control symbols may only be packet delimiters if a packet is in progress.
** NOP (Ignore) is not defined as a control symbol, but is the default value when the control symbol does not convey another stype1 function.

7.2.2.1 Start-of-packet Control Symbol

The start-of-packet control symbol format is shown in Figure 7.10.

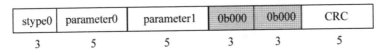

stype0	parameter0	parameter1	0b000	0b000	CRC
3	5	5	3	3	5

Figure 7.10   Start-of-packet control symbol format

7.2.2.2 Stomp Control Symbol

The stomp control symbol is used to cancel a partially transmitted packet. The protocol for packet cancellation is described later. The stomp control symbol format is shown in Figure 7.11.

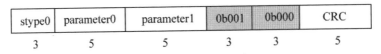

stype0	parameter0	parameter1	0b001	0b000	CRC
3	5	5	3	3	5

Figure 7.11   Stomp control symbol format

7.2.2.3 End-of-packet Control Symbol

The end-of-packet control symbol format is shown in Figure 7.12.

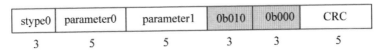

stype0	parameter0	parameter1	0b010	0b000	CRC
3	5	5	3	3	5

Figure 7.12   End-of-packet control symbol format

7.2.2.4 Restart-from-retry Control Symbol

The restart-from-retry control symbol cancels a current packet and may also be transmitted on an idle link. This control symbol is used to mark the beginning of packet retransmission, so that the receiver knows when to start accepting packets after the receiver has requested a packet to be retried. The control symbol format is shown in Figure 7.13.

stype0	parameter0	parameter1	0b011	0b000	CRC
3	5	5	3	3	5

Figure 7.13   Restart-from-retry control symbol format

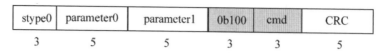

stype0	parameter0	parameter1	0b100	cmd	CRC
3	5	5	3	3	5

Figure 7.14   Link-request control symbol format

Table 7.8   Cmd field definitions

cmd[0–2] encoding	Command name	Description
0b000–0b010		Reserved
0b011	Reset-device	Reset the receiving device
0b100	Input-status	Return input port status; functions as a link-request (restart-from-error) control symbol under error conditions
0b101–0b111		Reserved

7.2.2.5 Link-request Control Symbol

A link-request control symbol is used by a device either to issue a command to the connected device or to request its input port status. A link-request control symbol cancels the current packet and can be sent between packets. Under error conditions, a link-request/input-status control symbol acts as a link-request/restart-from-error control symbol as described in Section 7.6.4. This control symbol format is displayed in Figure 7.14.

The second field of the link-request control symbol is a 3-bit command field. This field contains the command that is sent to the link. Two commands are defined: reset-device and input-status (Table 7.8).

The reset-device command causes the receiving device to go through its reset or power-up sequence. All state machines and the configuration registers reset to the original power on states. The reset-device command does not generate a link-response control symbol.

Owing to the undefined reliability of system designs it is necessary to put a safety lockout on the reset function of the link-request control symbol. A device receiving a reset-device command in a link-request control symbol shall not perform the reset function unless it has received four reset-device commands in a row without any other intervening packets or control symbols, except status control symbols. This will prevent spurious reset commands from inadvertently resetting a device.

The input-status command requests the receiving device to return the ackID value it expects to next receive from the sender on its input port and the current input port operational status for informational purposes. This command causes the receiver to flush its output port of all control symbols generated by packets received before the input-status command. Flushing the output port is implementation dependent and may result in either discarding the contents of the receive buffers or sending the control symbols on the link. The receiver then responds with a link-response control symbol.

7.2.2.6 Multicast-event Control Symbol

The multicast-event control symbol differs from other control symbols in that it carries information not related to the link carrying the control symbol. The multicast-event control

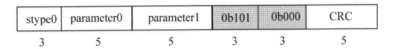

stype0	parameter0	parameter1	0b101	0b000	CRC
3	5	5	3	3	5

Figure 7.15 Multicast-event control symbol format

symbol allows the occurrence of a user-defined system event to be multicast throughout a system.

The format of the multicast-event control symbol format is shown in Figure 7.15.

7.2.3 Control Symbol Protection

A 5-bit CRC is used to provide error protection for the control symbol. The CRC covers control symbol bits 0–18 and provides up to 5-bit burst error detection for the entire 24-bit control symbol.

7.2.3.1 CRC-5 Code

The ITU polynomial $X^5+X^4+X^2+1$ is used to generate the 5-bit CRC for control symbols. The CRC check bits c0, c1, c2, c3, and c4 occupy the last 5 bits of a control symbol. It should be noted that the 5-bit CRC must be generated by each transmitter and verified by each receiver. Before the 5-bit CRC is computed, the CRC should be set to all 1s (0b11111). In order to simplify the CRC implementation for various technologies, a virtual 20th bit has been added to the control symbol fields for purposes of the CRC calculation. For all computations, the 20th bit is the last bit applied and is always set to a logic level of 0 (0b0).

7.3 PCS AND PMA LAYERS

This section describes the functions provided by the Physical coding sublayer (PCS) and Physical media attachment (PMA) sublayer. (The PCS and PMA terminology is adopted from IEEE 802.3). The topics include 8B/10B encoding, character representation, serialization of the data stream, code groups, columns, link transmission rules, idle sequences, and link initialization.

The concept of lanes is used to describe the width of a Serial RapidIO link. A lane is defined as one unidirectional differential pair in each direction. Serial RapidIO currently specifies two link widths. The 1x link is a one-lane link and the 4x link is a 4-lane link. Wider links are possible, but are not currently specified.

Figure 7.16 shows the structure of a typical Serial RapidIO end point. At the top of the diagram are the logical and transport layers that are responsible for the creation of the bulk of the RapidIO packet itself. Immediately below these layers is the serial protocol layer which is responsible for generation and consumption of control symbols and for managing the link through the link management protocol. Below this layer is the PCS layer. The boundary between the link protocol management layer and the PCS layer will typically be where the clock boundary changes from that used internal to the device to that used for the RapidIO end point itself. The PCS layer is responsible for lane striping, idle sequence generation and

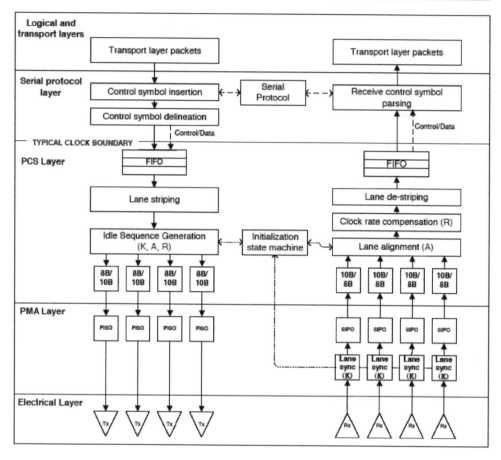

Figure 7.16   Structure of a Serial RapidIO end point

character conversion to the appropriate 8B/10B encoded K-codes and D-codes. Below the PCS layer is the PMA layer which is responsible for the attachment to the electrical layer. This PMA layer is also responsible for ensuring that the various lanes are properly aligned with respect to each other. The electrical layer represents the differential drivers and receivers and the electrical connections between them.

The remainder of this chapter will examine the functionality of the the PCS, PMA and electrical layers in more detail.

7.3.1 PCS Layer Functions

The physical coding sublayer (PCS) function is responsible for idle sequence generation, lane striping, encoding for transmission, decoding, lane alignment, and de-striping on reception. The PCS uses an 8B/10B encoding for transmission over the link. The 8B/10B encoding scheme was originally developed by IBM and is widely used in the industry for combining data and clocking information into a single signal.

The PCS layer also provides mechanisms for determining the operational mode of the port as 4-lane or 1-lane operation, and means to detect link states. It provides for clock difference tolerance between the sender and receiver without requiring flow control.

The PCS layer performs the following transmit functions:

- De-queues packets and delimited control symbols awaiting transmission as a character stream.
- Stripes the transmit character stream across the available lanes.
- Generates the idle sequence and inserts it into the transmit character stream for each lane when no packets or delimited control symbols are available for transmission.
- Encodes the character stream of each lane independently into 10-bit parallel code groups.
- Passes the resulting 10-bit parallel code groups to the PMA.

The PCS layer performs the following receive functions:

- Decodes the received stream of 10-bit parallel code groups for each lane independently into characters.
- Marks characters decoded from invalid code groups as invalid.
- If the link is using more than one lane, aligns the character streams to eliminate the skew between the lanes and reassembles (de-stripes) the character stream from each lane into a single character stream.
- Delivers the decoded character stream of packets and delimited control symbols to the higher layers.

7.3.2 PMA Layer Functions

The PMA (physical medium attachment) function is responsible for serializing 10-bit parallel code groups to/from a serial bitstream on a lane-by-lane basis. Upon receiving data, the PMA function provides alignment of the received bitstream to 10-bit code group boundaries, independently on a lane-by-lane basis. It then provides a continuous stream of 10-bit code groups to the PCS, one stream for each lane. The 10-bit code groups are not observable by layers higher than the PCS.

7.3.3 Definitions

Definitions of terms used in the discussion of the PCS and PMA layers are provided below.

- Byte: an 8-bit unit of information. Each bit of a byte has the value 0 or 1.
- Character: a 9-bit entity comprised of an information byte and a control bit that indicates whether the information byte contains data or control information. The control bit has the value D or K indicating that the information byte contains respectively data or control information.
- D-character: a character whose control bit has the value D.
- K-character: a character whose control bit has the value K. Also referred to as a special character.
- Code group: a 10-bit entity that is the result of 8B/10B encoding a character.
- Column: a group of four characters that are transmitted simultaneously on a 4x (4-lane) link.
- Comma: a 7-bit pattern, unique to certain 8B/10B special code groups, that is used by a receiver to determine code group boundaries.

- Idle sequence: the sequence of characters (code groups after encoding) that is transmitted when a packet or control symbol is not being transmitted. The idle sequence allows the receiver to maintain bit synchronization and code group alignment in-between packets and control symbols.
- Lane alignment: the process of eliminating the skew between the lanes of a 4-lane Serial link such that the characters transmitted as a column by the sender are output by the alignment process of the receiver as a column. Without lane alignment, the characters transmitted as a column might be scattered across several columns of output by the receiver. The alignment process uses special alignment characters transmitted as part of the idle sequence.
- Striping: the method used on a 4x link to send data across four lanes simultaneously. The character stream is *striped* across the lanes, on a character-by-character basis, starting with lane 0, to lane 1, to lane 2, to lane3, and wrapping back with the 5th character to lane 0.

7.3.4 The 8B/10B Transmission Code

The 8B/10B transmission code used by the PCS encodes 9-bit characters (8 bits of information and a control bit) into 10-bit code groups for transmission and reverses the process on reception. Encodings are defined for 256 data characters and 12 special (control) characters.

The code groups used by the code have either an equal number of ones and zeros (balanced) or the number of ones differs from the number of zeros by two (unbalanced). This selection of code groups guarantees a minimum of two transitions, 0 to 1 or 1 to 0, within each code group and it also eases the task of maintaining balance. Characters are encoded into either a single balanced code group or a pair of unbalanced code groups. The members of each code group pair are the logical complement of each other. This allows the encoder, when selecting an unbalanced code group, to select a code group unbalanced toward ones or unbalanced toward zeros, depending on which is required to maintain the 0/1 balance of the encoder output code group stream.

The 8B/10B code has the following properties.

- Sufficient bit transition density (3–8 transitions per code group) to allow clock recovery by the receiver.
- Special code groups that are used for establishing the receiver synchronization to the 10-bit code group boundaries, delimiting control symbols and maintaining receiver bit and code group boundary synchronization.
- Balanced (can be AC coupled).
- Detection of single and some multiple-bit errors.

7.3.5 Character and Code Group Notation

The description of 8B/10B encoding and decoding uses the following notation for characters, code group and their bits.

The information bits ([0–7]) of an unencoded character are denoted by the letters A through H, where H denotes the most significant information bit (RapidIO bit 0) and A denotes the least significant information bit (RapidIO bit 7). This is shown in Figure 7.17.

Each data character has a representation of the form D$x.y$ where x is the decimal value of the least significant 5 information bits EDCBA, and y is the decimal value of the most

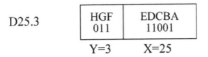

Figure 7.17 Character notation example (D25.3)

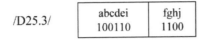

Figure 7.18 Code group notation example (/D25.3/)

significant 3 information bits HGF as shown below. Each special character has a similar representation of the form K$x.y$.

The output of the 8B/10B encoding process is a 10-bit code group. The bits of a code group are denoted by the letters a through j. The bits of a code group are all of equal significance, there is no most significant or least significant bit. The ordering of the code group bits is shown in Figure 7.18.

The code groups corresponding to the data character D$x.y$ are denoted by /D$x.y$/. The code groups corresponding to the special character K$x.y$ are denoted by /K$x.y$/.

7.3.6 Running Disparity

The 8B/10B encoding and decoding functions use a binary variable called running disparity. The variable can have a value of either positive (RD+) or negative (RD−). The encoder and decoder each have a running disparity variable for each lane. For a 4x link, the lane's running disparity values are all independent of each other.

The primary use of running disparity in the encoding process is to keep track of whether the decoder has output more ones or more zeros. The current value of the encoder running disparity is used to select which unbalanced code group will be used when the encoding for a character requires a choice between two unbalanced code groups.

7.3.6.1 Running Disparity Rules

After power-up and before the port is operational, both the transmitter (encoder) and receiver (decoder) must establish current values of running disparity. The transmitter uses a negative value as the initial value for the running disparity for each lane. The receiver may use either a negative or positive initial value of running disparity for each lane.

The following algorithm is used for calculating the running disparity for each lane. In the encoder, the algorithm operates on the code group that has just been generated by the encoder. In the receiver, the algorithm operates on the received code group that has just been decoded by the decoder.

Each code group is divided to two sub-blocks as shown in Figure 7.18, where the first six bits (abcdei) form one sub-block (6-bit sub-block) and the second four bits (fghj) form a second sub-block (4-bit sub-block). Running disparity at the beginning of the 6-bit sub-block is the running disparity at the end of the previous code group. Running disparity at the beginning of the 4-bit sub-block is the running disparity at the end of the 6-bit sub-block. Running disparity at the end of the code group is the running disparity at the end of the 4-bit sub-block.

The sub-block running disparity is calculated as follows:

1. The running disparity is positive at the end of any sub-block if the sub-block contains more 1s than 0s. It is also positive at the end of a 4-bit sub-block if the sub-block has the value 0b0011 and at the end of a 6-bit sub-block if the sub-block has the value 0b000111.
2. The running disparity is negative at the end of any sub-block if the sub-block contains more 0s than 1s. It is also negative at the end of a 4-bit sub-block if the sub-block has the value 0b1100 and at the end of a 6-bit sub-block if the sub-block has the value 0b111000.
3. In all other cases, the value of the running disparity at the end of the sub-block is the same as the running disparity at the beginning of the sub-block (the running disparity is unchanged).

7.3.7 8B/10B Encoding

The 8B/10B encoding function encodes the 9-bit characters into 10-bit code groups. The encodings for the 256 data characters (D$x.y$) and for the 12 special characters (K$x.y$) are contained in the RapidIO specification. These encodings are based almost completely on the encodings defined by the IEEE 802.3 working group for use in the 10 gigabit Ethernet (XAUI) adapter interface standard. For any given 9-bit character there are two columns of encodings, one marked RD– and one marked RD+. When encoding a character, the code group in the RD– column is selected if the current value of encoder running disparity is negative and the code group in the RD+ column is selected if the current value of encoder running disparity is positive. After each character is encoded, the resulting code group shall be used by the encoder to update the running disparity according to the running disparity rules.

7.3.8 Transmission Order

The parallel 10-bit code group output by the encoder is serialized and transmitted with bit 'a' transmitted first and with a bit ordering of abcdeifghj. This is shown in Figure 7.19, which gives an overview of the process of a character passing through the encoding, serializing, transmission, deserializing, and decoding processes. The left side of the figure shows the transmit process of encoding a character stream using 8B/10B encoding and the 10-bit serialization. The right side shows the reverse process of the receiver deserializing and using 8B/10B decoding on the received code groups. The dotted line shows the functional separation between the PCS layer that creates the 10-bit code groups and the PMA layer that serializes the code groups.

Figure 7.19 also shows on the receive side, the bits of a special character containing the comma pattern that is used by the receiver to establish 10-bit code boundary synchronization.

7.3.9 8B/10B Decoding

The 8B/10B decoding function decodes received 10-bit code groups into 9-bit characters, detects received code groups that have no defined decoding and marks the resulting characters in the output stream of the decode as invalid character (INVALID).

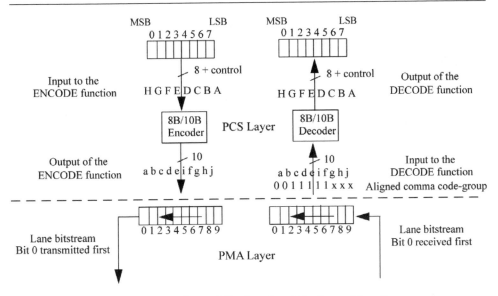

Figure 7.19 Lane encoding, serialization, deserialization and decoding process

The decoding function uses an inverse of the 8B/10B encoding tables and the current value of the decoder running disparity. The decoder then compares the received code group with the code groups in the selected column of both tables. If a match is found, the code group is decoded to the associated character. If no match is found, the code group is decoded to a character that is flagged in some manner as invalid. After each code group is decoded, the decoded code group is then used by the decoder to update the decoder running disparity according to the running disparity rules.

Table 7.9 defines the special characters and columns of special characters used by the Serial RapidIO physical layer. Special characters are used for the following functions:

- Alignment to code group (10-bit) boundaries on lane-by-lane basis.
- Alignment of the receive data stream across four lanes.
- Clock rate compensation between receiver and transmitters.

Table 7.9 Special characters and columns

Code group/column designation	Code group/column use	Number of code groups	Encoding
/PD/	Packet delimiter control symbol	1	/K28.3/
/SC/	Start of control symbol	1	/K28.0/
/I/	Idle symbols		
/K/	1x Sync	1	/K28.5/
/R/	1x Skip	1	/K29.7/
/A/	1x Align	1	/K27.7/
‖I‖	Idle column symbols		
‖K‖	4x Sync column	4	/K28.5/K28.5/K28.5/K28.5/
‖R‖	4x Skip column	4	/K29.7/K29.7/K29.7/K29.7/
‖A‖	4x Align column	4	/K27.7/K27.7/K27.7/K27.7/

7.3.10 Special Code Groups and Characters

This section describes the special characters used to delimit packets, align data columns and match clock domains between devices.

7.3.10.1 Packet Delimiter Control Symbol (/PD/)

PD and /PD/ are aliases for respectively the K28.3 character and the /K28.3/ code group which are used to delimit the beginning of a control symbol that contains a packet delimiter.

7.3.10.2 Start of Control Symbol (/SC/)

SC and /SC/ are aliases for respectively the K28.0 character and the /K28.0/ code group which are used to delimit the beginning of a control symbol that does not contain a packet delimiter.

7.3.10.3 Idle (/I/)

I and /I/ are aliases for respectively any of the idle sequence characters (A, K, or R) and idle sequence code groups (/A/, /K/, or /R/).

7.3.10.4 Sync (/K/)

K and /K/ are aliases for respectively the K28.5 character and the /K28.5/ code group which is used in the idle sequence to provide the receiver with the information it requires to achieve and maintain bit and 10-bit code group boundary synchronization. The /K28.5/ code group was selected as the Sync character for the following reasons:

1. It contains the comma pattern in bits abcdeif which can be easily found in the code group bit stream and marks the code group boundary.
2. The bits ghj provide the maximum number of transitions (i.e. 101 or 010).

A comma is a 7-bit string defined as either b'0011111' (comma+) or b'1100000' (comma−).

7.3.10.5 Skip (/R/)

R and /R/ are aliases for respectively the K29.7 character and the /29.7/ code group which are used in the idle sequence and are also used in the clock compensation sequence.

7.3.10.6 Align (/A/)

A and /A/ are aliases for respectively the K27.7 character and the /27.7/ code group which are used in the idle sequence and are used for lane alignment on 4x links.

7.4 USING THE SERIAL PHYSICAL LAYER

Now that we have covered the bits that make up the serial physical layer header, the control symbols that are used to manage the communications between the ports and the encoding of the bit streams using the 8B/10B technique to include both clocking and DC balance into the signal, we can examine how the link actually works to send data between two end points.

7.4.1 Port Initialization Process

Port initialization is the process that RapidIO uses to initialize and synchronize a pair of ports for communications. This process includes detecting the presence of a partner at the other end of the link (a link partner), establishing bit synchronization and code group boundary alignment and if the port is capable of supporting both 1x and 4x modes (a 1x/4x port), discovering whether the link partner is capable of 4x mode (4-lane) operation, selecting 1x or 4x mode operation and if 1x mode is selected, selecting lane 0 or lane 2 for link reception.

Several state machines control the initialization process. The RapidIO specification presents details on the structure of the state machines. The number and type of state machines depends on whether the port supports only 1x mode (a 1x port) or supports both 1x and 4x modes (a 1x/4x port). In either case, there is a primary state machine and one or more secondary state machines. The use of multiple state machines results in a simpler overall design. As might be expected, the initialization process for a 1x port is simpler than, and a subset of, the initialization process for a 1x/4x port.

7.4.2 Packet Exchange Protocol

Once the ports are initialized, packets may be exchanged. There is a protocol for packet communications between two Serial RapidIO end points. Control symbols are used to manage the flow of packets across the link. Packets are delimited by and acknowledged by control symbols. Error conditions are also communicated via control symbols. Control symbols also are used to support the flow control protocol.

7.4.2.1 Control Symbols

Control symbols are the message elements used by ports connected by a serial link. They are used for link maintenance, packet delimiting, packet acknowledgment, error reporting, and error recovery.

Control symbols are delimited for transmission by a single 8B/10B special (control) character. The control character marks the beginning of the control symbol and immediately precedes the first bit of the control symbol. The control symbol delimiting special character is added to the control symbol before the control symbol is passed to the PCS sublayer for lane striping (if applicable) and 8B/10B encoding. Since control symbol length is fixed at 24-bits, control symbols do not need end delimiters. The combined delimiter and control symbol is referred to as a delimited control symbol.

One of two special characters is used to delimit a control symbol. If the control symbol contains a packet delimiter, the special character PD (K28.3) is used. If the control symbol does not contain a packet delimiter, the special character SC (K28.0) is used. This use of special characters provides the receiver with an 'early warning' of the content of the control symbol.

Any control symbol that does not contain a packet delimiter may be embedded in a packet. An embedded control symbol may contain any defined encoding of stype0 and an stype1 encoding of 'multicast-event' or 'NOP'. Control symbols with stype1 encodings of start-of-packet, end-of-packet, stomp, restart-from-retry, or link-request cannot be embedded as they would terminate the packet.

The manner and degree to which control symbol embedding is used on a link impacts both link and system performance. For example, embedding multicast-event control symbols allows their propagation delay and delay variation through switch processing elements to be minimized and is highly desirable for some multicast-event applications. On the other hand, embedding all packet acknowledgment control symbols rather than combining as many of them as possible with packet delimiter control symbols reduces the link bandwidth available for packet transmission and may be undesirable.

7.4.2.2 Packets

Serial RapidIO packets are delimited for transmission by control symbols. Since packet length is variable, both start-of-packet and end-of-packet delimiters are required. The control symbol marking the end of a packet (packet termination) follows the end of the packet or the end of an embedded control symbol.

The following control symbols are used to delimit packets.

- start-of-packet
- end-of-packet
- stomp
- restart-from-retry
- any link-request

The beginning of a packet (packet start) is marked by a start-of-packet control symbol. A packet is terminated in one of the following three ways:

- The end of a packet is marked with an end-of-packet control symbol.
- The end of a packet is marked with a start-of-packet control symbol that also marks the beginning of a new packet.
- The packet is canceled by a restart-from-retry, stomp or any link-request control symbol.

Each packet requires an identifier to uniquely identify its acknowledgment control symbol. This identifier, the acknowledge ID (ackID), is five bits long, allowing for a range of 1 to 32 outstanding unacknowledged request or response packets between adjacent processing elements, but only up to 31 outstanding unacknowledged packets are allowed at any one time.

The first value of ackID assigned after a reset is 0b00000. Subsequent values of ackID are assigned sequentially (in increasing numerical order, wrapping back to 0 on overflow) to indicate the order of the packet transmission. The acknowledgments themselves are made with control symbols.

The Serial RapidIO link protocol uses retransmission to recover from packet transmission errors. To enable packet retransmission, a copy of each packet transmitted across a Serial link is kept by the sending port until either a packet-accepted packet acknowledgment control symbol is received for the packet from the receiving port, indicating that the port has received the packet without detected error and has accepted responsibility for the packet, or the port determines that the packet has encountered an unrecoverable error condition. Beyond possible errors, a port may also reject a packet if it does not have sufficient input buffer space available at the priority level of the packet.

Each packet has a priority that is assigned by the end point processing element that initiates the packet. The priority is carried in the priority (PRIO) field of the physical layer of the packet and has four possible values: 0, 1, 2, or 3. Packet priority increases with the priority value with 0 being the lowest priority and 3 being the highest. Packet priority is used for several purposes including transaction ordering and deadlock prevention.

7.4.3 Idle Sequence

The idle sequence is a sequence of code groups that will be transmitted continuously over each lane of an LP-Serial link whenever the link doesn't have packets or control symbols to send. An idle sequence may not be inserted in the middle of a packet. The idle sequence is transmitted over each lane as part of the port initialization process as required by the port initialization protocol.

The 1x idle sequence consists of a pseudo-random sequence of the code groups /K/, /A/, and /R/ (the idle code groups) and is used by ports operating in 1x mode. The 4x idle sequence consists of a pseudo-random sequence of the columns ||K||, ||A||, ||R|| (the idle columns) and is used by ports operating in 4x mode. There are no requirements on the length of an idle sequence. An idle sequence may be of any length.

The pseudo-random selection of code groups in the idle sequence results in an idle sequence whose frequency spectrum has no discrete spectral lines which minimizes the EMI generated by long idle sequences.

7.4.4 Data Flow Across a 1x Serial RapidIO Link

Figure 7.20 shows an example of control symbol, packet, and idle sequence transmission on a 1x Serial link. Figure 7.21 shows the same sequence of packets being transmitted across a 4x Serial link. The data is presented in columns that run from left to right. The first code group sent is a /SC/ representing a start control symbol followed by three data code groups containing the 24 bits of control symbol information. This control symbol, whose function is not shown, is followed by four idle characters. The four idle characters are followed by a /PD/ code group, which delimits the start of a packet. Packet information is contained in the next three code groups, which represents the packet delimiter symbol information. The code groups representing the RapidIO packet itself immediately follow this symbol information. This packet is 28 bytes in length. This packet is followed by an end-of-packet control symbol, which is followed by another start-of-packet control symbol and another packet of data. This packet is able to transfer 16 bytes of information before it is terminated with a restart-from-retry control symbol. The packet also endures two other control symbols, which are inserted in the data stream after the eighth data byte and after the twelfth data byte. In the RapidIO protocol, control symbols can be inserted into the communications channel at virtually any time. They do not need to wait until after the packet has completed transmission. After the transmission of the restart-from-retry control symbol another packet is sent across the link, properly completed and then the link becomes idle.

7.4.5 Data Flow Across a 4x Serial RapidIO Link

A serial port operating in 4x mode will stripe the character stream of delimited control symbols and packets across the four lanes before 8B/10B encoding as follows. Packets and delimited control symbols will be striped across the four lanes, beginning with lane 0. The first character

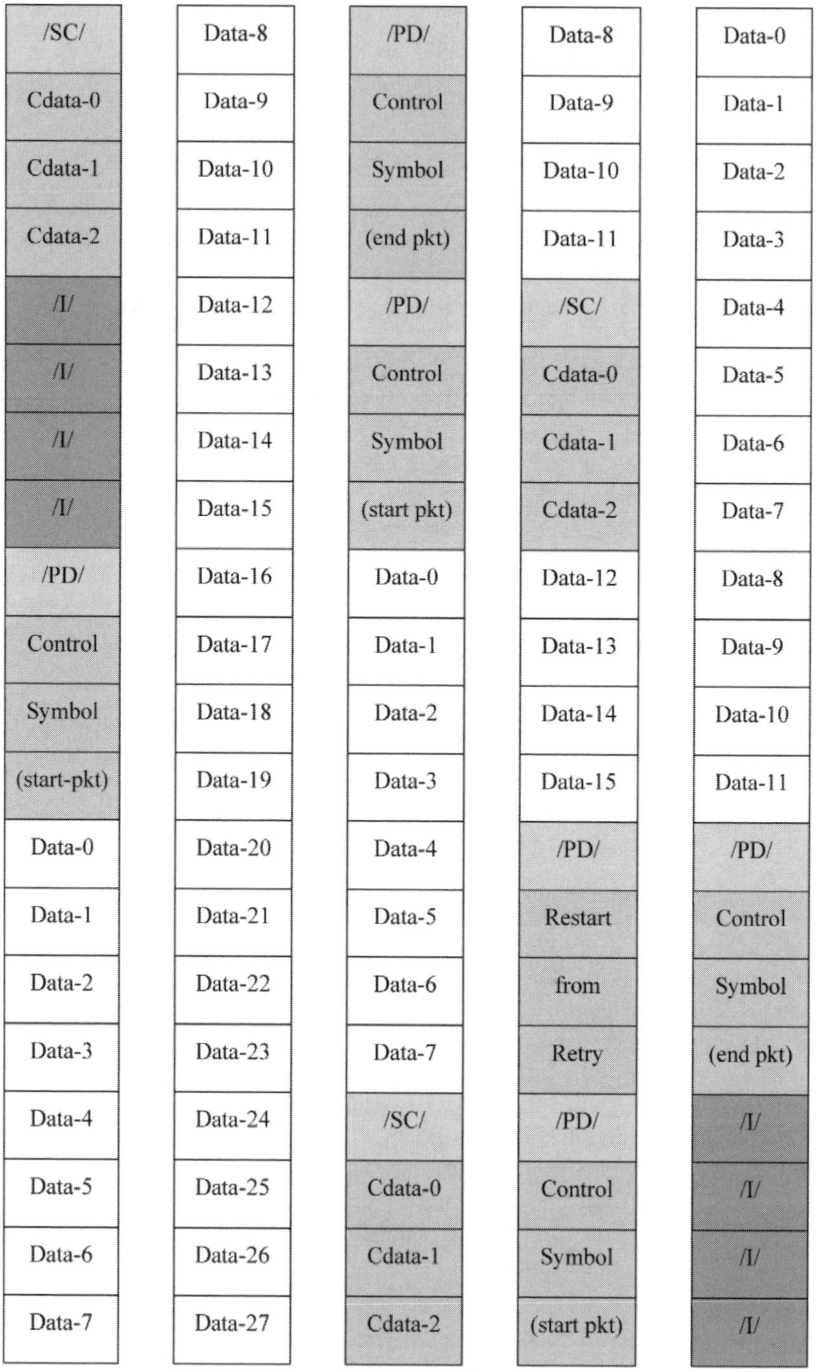

/SC/	Data-8	/PD/	Data-8	Data-0
Cdata-0	Data-9	Control	Data-9	Data-1
Cdata-1	Data-10	Symbol	Data-10	Data-2
Cdata-2	Data-11	(end pkt)	Data-11	Data-3
/I/	Data-12	/PD/	/SC/	Data-4
/I/	Data-13	Control	Cdata-0	Data-5
/I/	Data-14	Symbol	Cdata-1	Data-6
/I/	Data-15	(start pkt)	Cdata-2	Data-7
/PD/	Data-16	Data-0	Data-12	Data-8
Control	Data-17	Data-1	Data-13	Data-9
Symbol	Data-18	Data-2	Data-14	Data-10
(start-pkt)	Data-19	Data-3	Data-15	Data-11
Data-0	Data-20	Data-4	/PD/	/PD/
Data-1	Data-21	Data-5	Restart	Control
Data-2	Data-22	Data-6	from	Symbol
Data-3	Data-23	Data-7	Retry	(end pkt)
Data-4	Data-24	/SC/	/PD/	/I/
Data-5	Data-25	Cdata-0	Control	/I/
Data-6	Data-26	Cdata-1	Symbol	/I/
Data-7	Data-27	Cdata-2	(start pkt)	/I/

Figure 7.20   Data flow across a 1x Serial link

of each packet, or delimited control symbol, is placed in lane 0, the second character is placed in lane 1, the third character is placed in lane 2, and the fourth character is placed in lane 3. The fifth character and subsequent characters wrap around and continue beginning in lane 0.

Because of the fixed 24-bit length of control symbols and the requirement that packets must always be an integer multiple of 32 bits in length, delimiting control symbols will always fall in lane 0 after striping and all packets will form an integer number of contiguous columns. After striping, each of the 4 streams of characters are independently 8B/10B encoded and transmitted.

On reception, each lane is decoded. After decoding, the four lanes are aligned. The ‖A‖ columns transmitted as part of the 4x idle sequence provide the information needed to perform alignment. After alignment, the columns are destriped into a single character stream and then passed to the upper layers.

The lane alignment process eliminates the skew between lanes so that after destriping, the ordering of characters in the received character stream is the same as the ordering of characters before striping and transmission. Since the minimum number of non ‖A‖ columns between ‖A‖ columns is 16, the maximum lane skew that can be unambiguously corrected is the time it takes to transmit 7 code groups on a lane. Figure 7.21 shows an example of idle sequence, packet, and delimited control symbol transmission on a 4x link. This example uses the same sequence of packets as the 1x example shown in Figure 7.20.

7.4.6 Flow Control

This section presents the various Serial RapidIO link level flow control mechanisms. The flow control mechanisms operate between each pair of ports connected by a Serial link. The main purpose of link level flow control is to prevent the loss of packets due to insufficient buffer space in a link receiver.

The Serial protocol defines two methods of flow control. These are named receiver-controlled flow control and transmitter-controlled flow control. Every Serial RapidIO port is required to support receiver-controlled flow control. Transmitter-controlled flow control is optional.

7.4.6.1 Receiver-controlled Flow Control

Receiver-controlled flow control is the simplest and most basic method of flow control. In this method, the input side of a port controls the flow of packets from its link partner by accepting or rejecting (retrying) packets on a packet-by-packet basis. The receiving port provides no information to its link partner about the amount of buffer space it has available for packet reception.

As a result, its link partner transmits packets with no *a priori* expectation as to whether a given packet will be accepted or rejected. A port signals its link partner that it is operating in receiver-controlled flow control mode by setting the buf_status field to all 1s in every control symbol containing the field that the port transmits. Indicating that it has open buffers of packet reception. This method is named receiver-controlled flow control because the receiver makes all the decisions about how buffers in the receiver are allocated for packet reception.

A port operating in receiver-controlled flow control mode accepts or rejects each inbound error-free packet on the basis of whether the receiving port has enough buffer space

Lane-0	Lane-1	Lane-2	Lane-3
/SC/	Cdata-0	Cdata-1	Cdata-2
/I/	/I/	/I/	/I/
/PD/	Control Symbol (Start-of-packet)		
Data-0	Data-1	Data-2	Data-3
Data-4	Data-5	Data-6	Data-7
Data-8	Data-9	Data-10	Data-11
Data-12	Data-13	Data-14	Data-15
Data-16	Data-17	Data-18	Data-19
Data-20	Data-21	Data-22	Data-23
Data-24	Data-25	Data-26	Data-27
/PD/	Control Symbol (End-of-packet)		
/PD/	Control Symbol (Start-of-packet)		
Data-0	Data-1	Data-2	Data-3
Data-4	Data-5	Data-6	Data-7
/SC/	Cdata-0	Cdata-1	Cdata-2
Data-8	Data-9	Data-10	Data-11
/SC/	Cdata-0	Cdata-1	Cdata-2
Data-12	Data-13	Data-14	Data-15
/PD/	Control Symbol (Restart-from-retry)		
/PD/	Control Symbol (Start-of-packet)		
Data-0	Data-1	Data-2	Data-3
Data-4	Data-5	Data-6	Data-7
Data-8	Data-9	Data-10	Data-11
/PD/	Control Symbol (End-of-packet)		
/I/	/I/	/I/	/I/

Figure 7.21   Typical 4x data flow

available at the priority level of the packet. If there is enough buffer space available, the port accepts the packet and transmits a packet-accepted control symbol to its link partner that contains the ackID of the accepted packet in its packet_ackID field. This informs the port's link partner that the packet has been received without detected errors and that the port has accepted it. On receiving the packet-accepted control symbol, the link partner discards its copy of the accepted packet, freeing buffer space in the partner.

If buffer space is not available, the port rejects the packet. When a port rejects (retries) an error-free packet, it immediately enters the Input Retry-stopped state and follows the Input Retry-stopped recovery process. As part of the Input Retry-stopped recovery process, the port sends a packet-retry control symbol to its link partner, indicating that the packet whose ackID is in the packet_ackID field of the control symbol and all packets subsequently transmitted by the port have been discarded by the link partner and must all be retransmitted. The control symbol also indicates that the link partner is temporarily out of buffers for packets of priority less than or equal to the priority of the retried packet.

A port that receives a packet-retry control symbol immediately enters the Output Retry-stopped state and follows the Output Retry-stopped recovery process. As part of the Output Retry-stopped recovery process, the port receiving the packet-retry control symbol sends a restart-from-retry control symbol which causes its link partner to exit the Input Retry-stopped state and resume packet reception. The ackID assigned to that first packet transmitted after the restart-from-retry control symbol is the ackID of the packet that was retried.

Figure 7.22 shows an example of receiver-controlled flow control operation. In this example the transmitter is capable of sending packets faster than the receiver is able to absorb them. Once the transmitter has received a retry for a packet, the transmitter may elect to cancel any packet that is presently being transmitted since it will be discarded anyway. This makes bandwidth available for any higher-priority packets that may be pending transmission.

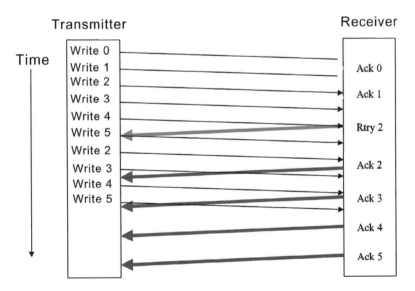

Figure 7.22 Receiver-controlled flow control

7.4.6.2 Transmitter-controlled Flow Control

In transmitter-controlled flow control, the receiving port provides information to its link partner about the amount of buffer space it has available for packet reception. With this information, the sending port can allocate the use of the receiving port's receive buffers according to the number and priority of packets that the sending port has waiting for transmission without concern that one or more of the packets may be forced to retry.

A port signals its link partner that it is operating in transmitter-controlled flow control mode by setting the buf_status field to a value different from all 1s in every control symbol containing the field that the port transmits. This method is named transmitter-controlled flow control because the transmitter makes almost all the decisions about how the buffers in the receiver are allocated for packet reception.

The number of free buffers that a port has available for packet reception is conveyed to its link partner by the value of the buf_status field in the control symbols that the port transmits. The value conveyed by the buf_status field is the number of maximum length packet buffers currently available for packet reception up to the limit that can be reported in the field. If a port has more buffers available than the maximum value that can be reported in the buf_status field, the port sets the field to that maximum value. A port may report a smaller number of buffers than it actually has available, but it can not report a greater number.

A port informs its link partner when the number of free buffers available for packet reception changes. The new value of buf_status is conveyed in the buf_status field of a packet-accepted, packet-retry, or status control symbol. Each change in the number of free buffers a port has available for packet reception need not be conveyed to the link partner.

A port whose link partner is operating in transmitter-control flow control mode should never receive a packet-retry control symbol from its link partner unless the port has transmitted more packets than its link partner has receive buffers, violated the rules that all input buffer may not be filled with low-priority packets or there is some fault condition. If a port whose link partner is operating in transmitter-control flow control mode receives a packet-retry control symbol, the output side of the port immediately enters the Output Retry-stopped state and follows the Output Retry-stopped recovery process.

A port whose link partner is operating in transmitter-controlled flow control mode may send more packets than the number of free buffers indicated by the link partner. Packets transmitted in excess of the free_buffer_count are transmitted on a speculative basis and are subject to retry by the link partner. The link partner accepts or rejects these packets on a packet-by-packet basis in exactly the same way it would if operating in receiver-controlled flow control mode. A port may use such speculative transmission in an attempt to maximize the utilization of the link. However, speculative transmission that results in a significant number of retries and discarded packets can reduce the effective bandwidth of the link.

A simple example of transmitter-controlled flow control is shown in Figure 7.23. Immediately following the initialization of a link, each port begins sending status control symbols to its link partner. The value of the buf_status field in these control symbols indicates to the link partner the flow control mode supported by the sending port.

The flow control mode negotiation rule is as follows. If the port and its link partner both support transmitter-controlled flow control, then both ports shall use transmitter-controlled flow control. Otherwise, both ports shall use receiver-controlled flow control.

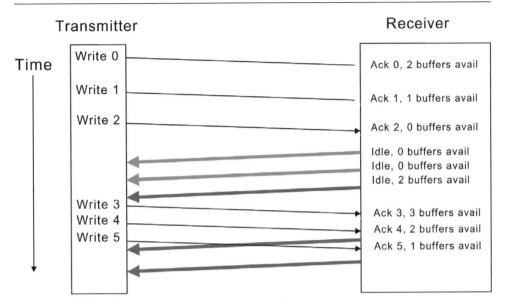

Figure 7.23 Transmitter-controlled flow control

7.4.6.3 Input Retry-stopped Recovery Process

When the input side of a port retries a packet, it immediately enters the Input Retry-stopped state. To recover from this state, the input side of the port takes the following actions.

- Discards the rejected or cancelled packet without reporting a packet error and ignores all subsequently received packets while the port is in the Input Retry-stopped state.
- Causes the output side of the port to issue a packet-retry control symbol containing the ackID value of the retried packet in the packet_ackID field of the control symbol. (The packet-retry control symbol causes the output side of the link partner to enter the Output Retry-stopped state and send a restart-from-retry control symbol.)
- When a restart-from-retry control symbol is received, exit the Input Retry-stopped state and resume packet reception.

7.4.6.4 Output Retry-stopped Recovery Process

To recover from the Output Retry-stopped state, the output side of a port takes the following actions.

- Immediately stops transmitting new packets. Resets the link packet acknowledgment timers for all transmitted but unacknowledged packets. (This prevents the generation of spurious time-out errors.)
- Transmits a restart-from-retry control symbol.
- Backs up to the first unaccepted packet (the retried packet) which is the packet whose ackID value is specified by the packet_ackID value contained in the packet-retry control symbol. (The packet_ackID value is also the value of ackID field the port retrying the packet expects in the first packet it receives after receiving the restart-from-retry control symbol.)

- Exits the Output Retry-stopped state and resumes transmission with either the retried packet or a higher-priority packet, which is assigned the ackID value contained in the packet_ackID field of the packet-retry control symbol.

7.4.7 Link Maintenance Protocol

Beyond the exchange of packets, RapidIO links also must support a link maintenance protocol. This protocol involves request and response pair between ports connected by the link. For software management, the link maintenance request is generated through writes to registers in the configuration space of the sending device. An external host write of a command to the link-request register will cause a link-request control symbol to be issued onto the output port of the device. Only one link-request can be outstanding on a link at a time.

The device that is linked to the sending device will respond with an link-response control symbol if the link-request command requires a response. The external host retrieves the link-response by polling the link-response register through maintenance read transactions. A device with multiple RapidIO interfaces has a link-request and a link-response register pair for each corresponding RapidIO interface.

The automatic error recovery mechanism relies on the hardware automatically generating link-request/input-status control symbols using the corresponding link-response information to attempt recovery.

The retry-stopped state indicates that the port has retried a packet and is waiting to be restarted. This state is cleared when a restart-from-retry (or a link-request/input-status) control symbol is received. The error-stopped state indicates that the port has encountered a transmission error and is waiting to be restarted. This state is cleared when a link-request/input-status control symbol is received.

7.4.8 Canceling Packets

When a port becomes aware of some condition that will require the packet it is currently transmitting to be retransmitted, the port may cancel the packet. This allows the port to avoid wasting bandwidth by not completing the transmission of a packet that the port knows must be retransmitted. Alternatively, the sending port may choose to complete transmission of the packet normally.

A port may cancel a packet if the port detects a problem with the packet as it is being transmitted or if the port receives a packet-retry or packet-not-accepted control symbol for a packet that is still being transmitted or that was previously transmitted. A packet-retry or packet-not-accepted control symbol can be transmitted by a port for a packet at any time after the port begins receiving the packet.

The sending device uses the stomp control symbol, the restart-from-retry control symbol (in response to a packet-retry control symbol), or any link-request control symbol to cancel a packet.

A port receiving a cancelled packet is required to drop the packet. The cancellation of a packet does not result in the generation of any errors. However, if the packet was canceled because the sender received a packet-not-accepted control symbol, the error that caused the packet-not-accepted to be sent is reported in the normal manner.

7.5 TRANSACTION AND PACKET DELIVERY ORDERING RULES

Because RapidIO transactions must support both multiple transaction priorities as well as causality and deadlock avoidance, delivery ordering rules must be established to ensure that the system operates properly at all times. The rules specified in this section are required for the physical layer to support the transaction ordering rules specified by the higher layer RapidIO specifications.

Transaction Delivery Ordering Rules:

1. The physical layer of an end point processing element port shall encapsulate in packets and forward to the RapidIO fabric, transactions comprising a given transaction request flow in the same order that the transactions were received from the transport layer of the processing element. (Packets with the same source ID, destination ID and priority level must remain in order with respect to each other).
2. The physical layer of an end point processing element port shall ensure that a higher-priority request transaction that it receives from the transport layer of the processing element before a lower-priority request transaction with the same sourceID and the same destinationID is forwarded to the fabric before the lower-priority transaction. (Lower-priority packets can't pass higher-priority packets).
3. The physical layer of an end point processing element port shall deliver transactions to the transport layer of the processing element in the same order that the packetized transactions were received by the port. (The receiver hands packets up in the same order that they were received).

Packet Delivery Ordering Rules:

1. A packet initiated by a processing element shall not be considered committed to the RapidIO fabric and does not participate in the packet delivery ordering rules until the packet has been accepted by the device at the other end of the link. (RapidIO does not have the concept of delayed or deferred transactions. Once a packet is accepted into the fabric, it is committed.)
2. A switch shall not alter the priority of a packet. (Packet priorities are fixed and retained end-to-end).
3. Packet forwarding decisions made by a switch processing element shall provide a consistent output port selection which is based solely on the value of the destinationID field carried in the packet. (Paths are fixed based on the destinationID).
4. A switch processing element shall not change the order of packets comprising a transaction request flow (packets with the same sourceID, the same destinationID, the same priority and ftype != 8) as the packets pass through the switch.
5. A switch processing element shall not allow lower-priority non-maintenance packets (ftype != 8) to pass higher-priority non-maintenance packets with the same sourceID and destinationID as the packets pass through the switch.
6. A switch processing element shall not allow a priority N maintenance packet (ftype = 8) to pass another maintenance packet of priority N or greater that takes the same path through the switch (same switch input port and same switch output port).

7.5.1 Deadlock Avoidance

One of the defining attributes of RapidIO is that RapidIO switch devices do not interpret the packets that they are switching. This means that a RapidIO switch does not distinguish or treat differently in any way NREAD and NWRITE or SWRITE transactions, or any other transactions. RapidIO switches make their forwarding and reordering decisions solely on the basis of the device ID bits of the transport header and the priority bits of the physical layer header. The only exception to this rule is for the Type 8 maintenance transactions.

To be more specific, switches are not required to distinguish between packets carrying request transactions and packets carrying response transactions. As a result, it is possible for two end points, A and B to each fill all of their output buffers, the fabric connecting them and the other end point's input buffers with read requests. This would result in an input to output dependency loop in each end point in which there would be no buffer space to hold the responses necessary to complete any of the outstanding read requests.

To break input to output dependencies, end point processing elements must have the ability to issue outbound response packets, even if outbound request packets awaiting transmission are blocked by congestion in the connected device. Two techniques are provided to break input to output dependencies. First, a response packet (a packet carrying a response transaction) is always assigned an initial priority level that is one priority level greater than the priority of the associated request packet (the packet carrying the associated request transaction). Table 7.10 shows how response packet priorities are promoted above request packet priorities.

This process is called promotion. An end point processing element may promote a response packet only to the degree necessary for the packet to be accepted by the connected device.

The following rules define the deadlock prevention mechanism.

Deadlock Prevention Rules:

1. A RapidIO fabric shall be dependency cycle free for all operations that do not require a response. (This rule is necessary as there are no mechanisms provided in the fabric to break dependency cycles for operations not requiring responses.)
2. A packet carrying a request transaction that requires a response shall not be issued at the highest priority.
3. A packet carrying a response shall have a priority at least one priority level higher than the priority of the associated request. (This rule in combination with rule 2 is the basis for the priority assignments in Table 7.10)
4. A switch processing element port shall accept an error-free packet of priority N if there is no packet of priority greater than or equal to N that was previously received by the port and

Table 7.10   Response packet promotion

System priority	Request packet priority	Response packet priority
Highest	2	3
Next	1	2 or 3
Lowest	0	1, 2, or 3

is still waiting in the switch to be forwarded. (This rule has multiple implications which include, but are not limited to, the following. First, a switch processing element port must have at least as many maximum length packet input buffers as there are priority levels. Second, a minimum of one maximum length packet input buffer must be reserved for each priority level. A input buffer reserved for priority N might be restricted to only priority N packets or might be allowed to hold packets of priority greater than or equal to N; either approach complies with the rule.)

5. A switch processing element port that transmits a priority N packet that is forced to retry by the connected device shall select a packet of priority greater than N, if one is available, for transmission. (This guarantees that packets of a given priority will not block higher-priority packets.)

6. An end point processing element port shall accept an error-free packet of priority N if the port has enough space for the packet in the input buffer space of the port allocated for packets of priority N. (Lack of input buffer space is the only reason an end point may retry a packet.)

7. The decision of an end point processing element to accept or retry an error-free packet of priority N shall not be dependent on the ability of the end point to issue request packets of priority less than or equal to N from any of its ports. (This rule works in conjunction with rule 6. It prohibits a device's inability to issue packets of priority less than or equal to N, owing to congestion in the connected device, from resulting in a lack of buffers to receive inbound packets of priority greater than or equal to N which in turn would result in packets of priority greater than or equal to N being forced to retry. The implications and some ways of complying with this rule are presented in the following paragraphs.)

One implication of rule 7 is that a port may not fill all of its buffers that can be used to hold packets awaiting transmission with packets carrying request transactions. If this situation was allowed to occur and the output was blocked because of congestion in the connected device, read transactions could not be processed (no place to put the response packet), input buffer space would become filled and all subsequent inbound request packets would be forced to retry, thus violating rule 7.

Another implication is that a port must have a way of preventing output blockage at priority less than or equal to N, owing to congestion in the connected device, from resulting in a lack of input buffer space for inbound packets of priority greater than or equal to N. There are multiple ways of doing this.

One way is to provide a port with input buffer space for at least four maximum length packets and reserve input buffer space for higher-priority packets in a manner similar to that required by rule 4 for switches. In this case, output port blockage at priority less than or equal to N will not result in blocking inbound packets of priority greater than or equal to N as any responses packets they generate will be of priority greater than N which is not congestion blocked. The port must, however, have the ability to select packets of priority greater than N for transmission from the packets awaiting transmission. This approach does not require the use of response packet priority promotion.

Alternatively, a port that does not have enough input buffer space for at least four maximum length packets or that does not reserve space for higher-priority packets can use the promotion mechanism to increase the priority of response packets until the connected device accepts them. This allows output buffer space containing response packets to be freed, even though all request packets awaiting transmission are congestion blocked.

As an example, suppose an end point processing element has a blocked input port because all available resources are being used for a response packet that the processing element is trying to send. If the response packet is retried by the downstream processing element, raising the priority of the response packet until it is accepted allows the processing element's input port to unblock so the system can make forward progress.

7.6 ERROR DETECTION AND RECOVERY

One of the main applications targets of the RapidIO technology is highly reliable electronic equipment such as carrier class telecommunications equipment and enterprise storage systems. In these systems, it is extremely important to do everything reasonably possible to avoid errors and, when errors occur, to detect and attempt to recover from them. The RapidIO interconnect technology provides extensive error detection and recovery techniques by combining retry protocols with cyclic redundancy codes. RapidIO also employs carefully selected delimiter control characters and response timers to reduce the likelihood of undetected errors in systems.

7.6.1 Lost Packet Detection

Some types of errors, such as a lost request or response packet or a lost acknowledgment, result in a system with hung resources. To detect this type of error time-out counters are used. Because the expiration of one of these timers should indicate to the system that there is a problem, this time interval should be set long enough so that a false time-out is not signaled. The response to this error condition is implementation dependent.

The RapidIO specifications require time-out counters for the physical layer, which are called the port link time-out counters, and counters for the logical layer, which are called the port response time-out counters. The interpretation of the counter values is implementation dependent, based on a number of factors, including link clock rate, the internal clock rate of the device, and the desired system behavior.

The physical layer time-out occurs between the transmission of a packet and the receipt of an acknowledgment control symbol. This time-out interval is likely to be comparatively short because the packet and acknowledgment pair must traverse only a single link.

The logical layer time-out occurs between the issuance of a request packet that requires a response packet and the receipt of that response packet. This time-out is counted from the time that the logical layer issues the packet to the physical layer until the time that the associated response packet is delivered from the physical layer to the logical layer. Should the physical layer fail to complete the delivery of the packet, the logical layer time-out will occur. This time-out interval is likely to be comparatively long because the packet and response pair have to traverse the fabric at least twice and be processed by the target. Error handling for a response time-out is implementation dependent.

7.6.2 Link Behavior Under Error

The Serial RapidIO link uses an error detection and retransmission protocol to protect against and recover from transmission errors. Transmission error detection is done at the input port, and all transmission error recovery is also initiated at the input port.

The protocol requires that the receiving port acknowledge each transmitted packet and that the sender retain a copy of each transmitted packet until either the sender receives a packet-accepted control symbol acknowledgment for the packet or the sender determines that the packet has encountered an unrecoverable error. If the receiving port detects a transmission error in a packet, the port sends a packet-not-accepted control symbol acknowledgment back to the sender, indicating that the packet was corrupted as received. After a link-request/input-status and link-response control symbol exchange, the sender begins retransmission with either the packet that was corrupted during transmission or a higher-priority packet if one is awaiting transmission.

All packets corrupted in transmission are retransmitted. The number of times a packet may be retransmitted before the sending port determines that the packet has encountered an unrecoverable condition is implementation dependent.

7.6.3 Effect of Single-bit Code Group Errors

Single-bit code group errors will be the dominant code group error by many orders of magnitude. It is therefore useful to know the variety of code group corruptions that can be caused by a single-bit error.

Table 7.11 lists all possible code group corruptions that can be caused by a single-bit error. The notation $/X/ \Rightarrow /Y/$ means that the code group for the character X has been corrupted by a single-bit error into the code group for the character Y. If the corruption results in a code group that is invalid for the current receiver running disparity, the notation $/X/ \Rightarrow$ /INVALID/ is used. The table provides the information required to deterministically detect all isolated single-bit transmission errors.

Table 7.11 Code group corruption caused by single-bit errors

Corruption	Detection
$/SC/ \Rightarrow$ /INVALID/	Detectable as an error when decoding the code group. When this error occurs within a packet, it is indistinguishable from a $/Dx.y/ \Rightarrow$ /INVALID/. When this error occurs outside a packet, the type of error can be inferred from whether the /INVALID/ is followed by the three $/Dx.y/$ that comprise the control symbol data
$/PD/ \Rightarrow$ /INVALID/	Detectable as an error when decoding the code group. When this error occurs within a packet, it is indistinguishable from a $/Dx.y/ \Rightarrow$ /INVALID/. When this error occurs outside a packet, the type of error can be inferred from whether the /INVALID/ is followed by the three $/Dx.y/$ that comprise the control symbol data
$/A/, /K/$ or $/R/ \Rightarrow /Dx.y/$	Detectable as an error as $/Dx.y/$ is illegal outside a packet or control symbol and $/A/, /K/$ and $/R/$ are illegal within a packet or control symbol
$/A/, /K/$ or $/R/ \Rightarrow$ /INVALID/	Detectable as an error when decoding the code group
$/Dx.y/ \Rightarrow /A/, /K/$ or $/R/$	Detectable as an error as $/A/, /K/$ and $/R/$ are illegal within a packet or control symbol and $/Dx.y/$ is illegal outside a packet or control symbol
$/Dx.y/ \Rightarrow$ /INVALID/	Detectable as an error when decoding the code group
$/Dx.y/ \Rightarrow /Du.v/$	Detectable as an error by the packet or control symbol CRC. The error will also result in a subsequent unerrored code group being decoded as INVALID, but that resulting INVALID code group may occur an arbitrary number of code groups after the errored code group

7.6.4 Recoverable Errors

A RapidIO port is able to detect and automatically recover from the following four basic types of error:

- an idle sequence error
- a control symbol error
- a packet error
- a time-out waiting for an acknowledgment control symbol

When an error is detected the port will go into an Error-stopped state. After entering the stopped state the port will begin a recovery process to attempt to recover from the error condition. Keep in mind during the following discussion of recovery processes that all the defined mechanisms for recovery expect to be implemented in hardware rather than software. Aside from the possible use of counters to monitor error rates, the system will, in general, not be aware that errors in the link are being detected and recovered from automatically.

7.6.4.1 Input Error-stopped Recovery Process

When the input side of a port detects a transmission error, it immediately enters the Input Error-stopped state. To recover from this state, the input side of the port takes the following actions:

- Record the error(s) that caused the port to enter the Input Error-stopped state.
- If the detected error(s) occurred in a control symbol or packet, discard the control symbol or packet.
- Ignore all subsequently received packets while the port is in the Input Error-stopped state.
- Cause the output side of the port to issue a packet-not-accepted control symbol. If the detected error occurred in a packet, the control symbol packet_ackID field contains the ackID value from the errored packet. Otherwise, the packet_ackID field of the control symbol contains an undefined value. Otherwise, the control symbol packet_ackID field contains an unexpected ackID value. (The packet-not-accepted control symbol causes the output side of the receiving port to enter the Output Error-stopped state and send a link-request/input-status control symbol.)
- When an link-request/input-status control symbol is received, cause the output side of the port to issue a link-response control symbol, exit the Input Error-stopped state and resume packet reception.

7.6.4.2 Output Error-stopped Recovery Process

To recover from the Output Error-stopped state, the output side of a port takes the following actions:

- Immediately stops transmitting new packets.
- Resets the link packet acknowledgment timers for all transmitted but unacknowledged packets. (This prevents the generation of spurious time-out errors.)
- Transmits an input-status link-request/input-status (restart-from-error) control symbol. (The input status link-request/input-status control symbol causes the receiving port to transmit

a link-response control symbol that contains the input_status and ackID_status of the input side of the port. The ackID_status is the ackID value that is expected in the next packet that the port receives.)

When the link-response is received, the port backs up the first unaccepted packet, exits the Output Error-stopped state and resumes transmission with either the first unaccepted packet or a higher-priority packet.

7.6.4.3 Idle Sequence Errors

The idle sequence is comprised of A, K, and R (8B/10B special) characters. If an input port detects an invalid character or any valid character other then A, K, or R in an idle sequence, it shall enter the Input Error-stopped state and follow the Input Error-stopped recovery process.

To limit input port complexity, the port is not required to determine the specific error that resulted in an idle sequence error. Following are several examples of idle sequence errors:

- A single-bit transmission error can change an /A/, /K/, or /R/ code group into a /Dx.y/ (data) code group which is illegal in an idle sequence.
- A single-bit transmission error can change an /A/, /K/, or /R/ code group into an invalid code group.
- A single-bit transmission error can change an /SP/ or /PD/ (control symbol delimiters) into an invalid code group.

7.6.4.4 Control Symbol Errors

There are three types of detectable control symbol error:

- an uncorrupted control symbol with a reserved encoding of the stype0, stype1 or cmd field
- an uncorrupted control symbol that violates the link protocol
- a corrupted control symbol

7.6.4.5 Reserved stype and cmd Field Encodings

For forward compatibility, control symbols received by a port with a reserved encoding of the stype0, stype1 or cmd field are handled as follows:

- If the stype0 field of a control symbol contains a reserved encoding, the stype0 function of the control symbol is ignored and no error is reported.
- If the stype1 field of a control symbol contains a reserved encoding, the stype1 function of the control symbol is ignored and no error is reported.
- If the cmd field of a link-request control symbol (stype1 = 0b100) contains a reserved encoding, the control symbol shall cancel a packet whose transmission is in progress, but the stype1 function of the control symbol is otherwise ignored and no error is reported.

7.6.4.6 Link Protocol Violations

The reception of a control symbol with no detected corruption but that violates the link protocol will cause the receiving port to immediately enter the Input Error-stopped state, if the port is not already in the Input Error-stopped state, and/or the Output Error-stopped state, if the port is not already in the Output Error-stopped state, and follow the appropriate Error-stopped recovery process.

Link protocol violations include the following:

- unexpected packet-accepted, packet-retry, or packet-not-accepted control symbol
- packet acknowledgment control symbol with an unexpected packet_ackID value
- link time-out while waiting for an acknowledgment control symbol

The following is an example of a link protocol violation and recovery. A sender transmits packets labeled ackID 2, 3, 4, and 5. It receives acknowledgments for packets 2, 4, and 5, indicating a probable error associated with ackID 3. The sender then stops transmitting new packets and sends a link-request/input-status (restart-from-error) control symbol to the receiver. The receiver then returns a link-response control symbol indicating which packets it has received properly. These are the possible responses and the sender's resulting behavior:

- expecting ackID = 3 sender must retransmit packets 3, 4, and 5
- expecting ackID = 4 sender must retransmit packets 4 and 5
- expecting ackID = 5 sender must retransmit packet 5
- expecting ackID = 6 receiver got all packets, resume operation
- expecting ackID = anything else, fatal (non-recoverable) error

7.6.4.7 Corrupted Control Symbols

The reception of a control symbol with detected corruption will cause the receiving port to immediately enter the Input Error-stopped state and follow the Input Error-stopped recovery process. For this type of error, the packet-not-accepted control symbol sent by the output side of the port as part of the recovery process will have an unexpected packet_ackID value.

Input ports detect the following types of control symbol corruption:

- a control symbol containing invalid characters or valid but non-data characters
- a control symbol with an incorrect CRC value

7.6.4.8 Packet Errors

The reception of a packet with detected corruption will cause the receiving port to immediately enter the Input Error-stopped state and follow the Input Error-stopped recovery process.

Input ports detect the following types of packet corruption:

- packet with an unexpected ackID value
- packet with an incorrect CRC value
- packet containing invalid characters or valid non-data characters
- packet that overruns some defined boundary such as the maximum data payload.

7.6.4.9 Link Time-out

A link time-out while waiting for an acknowledgment control symbol is handled as a link protocol violation.

7.7 RETIMERS AND REPEATERS

The Serial physical layer allows 'retimers' and 'repeaters.' Retimers amplify a weakened signal, but do not transfer jitter to the next segment. Repeaters also amplify a weakened signal, but transfer jitter to the next segment. Retimers allow greater distances between end points at the cost of additional latency. Repeaters support less distance between end points than retimers, and only add a small amount of latency.

7.8 THE ELECTRICAL INTERFACE

It is beyond the scope of this book to present the details of the Serial RapidIO electrical interface. This section will present a brief overview of the techniques used to connect Serial RapidIO devices together and to achieve the extremely high signaling rates of Serial RapidIO. The 1x and 4x links are electrically identical, except for the use of more lanes in the 4x link. Figure 7.24 shows the signal interface diagram connecting two 1x devices together with the RapidIO serial interconnect.

Figure 7.25 shows the signal interface diagram connecting two 4x devices together with the RapidIO serial interconnect.

Figure 7.26 shows the connections that would be active between a 4x serial device and a 1x serial device.

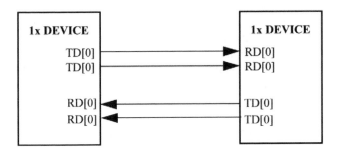

Figure 7.24   RapidIO 1x device to 1x device interface diagram

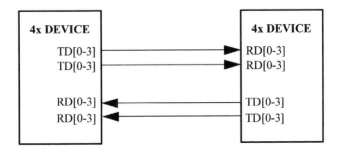

Figure 7.25   RapidIO 4x device to 4x device interface diagram

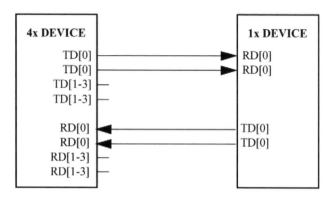

Figure 7.26 RapidIO 4x device to 1x device interface diagram

The RapidIO specification defines two sets of transmitter electrical specifications (short-run and long-run) and a single receiver specification for each of three baud rates, 1.25, 2.50, and 3.125 Giga baud persecond.

Two transmitter specifications allow for solutions ranging from simple board-to-board interconnect to driving through two separate connectors and across a backplane. A single receiver specification is used. The receiver specification will accept signals from both the short-run and long-run transmitter specifications.

The short-run transmitter should be used mainly for chip-to-chip connections on either the same printed circuit board or across a single connector. This covers the case where connections are made to a mezzanine (daughter) card. The minimum swings of the short-run specification can reduce the overall power used.

The long-run transmitter specifications use larger voltage swings that are capable of more reliably driving signals across backplanes. The long-run transmitter is meant to allow a user to drive the RapidIO signals across two connectors and a backplane, with a total distance of at least 50 cm at all the supported baud rates.

To ensure interoperability between drivers and receivers of different vendors and technologies, AC coupling at the receiver input is required.

Serial RapidIO links use differential signaling. This section defines terms used in the description and specification of these differential signals. Figure 7.27 shows how the signals are defined. The figures show waveforms for either a transmitter output (TD and TD) or a receiver input (RD and RD). Each signal swings between A and B volts where $A > B$. Using these waveforms, the definitions are as follows:

1. The transmitter output signals and the receiver input signals TD, TD, RD and RD each have a peak-to-peak swing of $A - B$ V
2. The differential output signal of the transmitter V_{OD} is defined as $V_{TD} - V_{TD}$
3. The differential input signal of the receiver V_{ID} is defined as $V_{RD} - V_{RD}$
4. The differential output signal of the transmitter and the differential input signal of the receiver each range from $A - B$ to $-(A - B)$ V
5. The peak value of the differential transmitter output signal and the differential receiver input signal is $A - B$ V

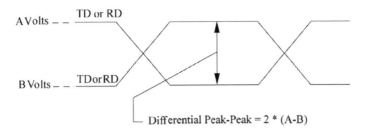

Figure 7.27   Differential signalling

6. The peak-to-peak value of the differential transmitter output signal and the differential
   receiver input signal is $2(A - B)$ V

   To illustrate these definitions using real values, consider the case of a short-run RapidIO
transmitter that has a common mode voltage of 2.25 V and each of its outputs, TD and TD, has
a swing that goes between 2.5 and 2.0V. For these values, the peak-to-peak voltage swing of
the signals TD and TD is 500 mV. The differential output signal ranges between +500 and
−500 mV. The peak differential voltage is 500 mV. The peak-to-peak differential voltage is
then 1000 mV.

   The parameters for the AC electrical specifications are guided by the XAUI electrical
interface specified in Clause 47 of IEEE 802.3ae-2002. XAUI is an IEEE standard for use
in 10-Gigabit Ethernet applications. XAUI has similar application goals to serial RapidIO.
The goal of the RapidIO work is that electrical designs for Serial RapidIO can reuse electrical
designs for XAUI, suitably modified for applications at the baud intervals and reaches as
described.

8

Parallel Physical Layer Protocol

This chapter describes the physical layer protocol for the RapidIO parallel physical layer. The RapidIO parallel physical layer differs from the serial physical layer in the following ways. The packet and control symbol information is striped across an 8- or 16-bit-wide interface, instead of a single- or four-lane interface (Figure 8.1). The clock information is sent on a separate differential pair and is not encoded in the data stream. The clock signal operates at half the frequency of the data; in other words, the data lines operate at double the rate of the clock signal. There is a separate framing signal that is used to distinguish packets from control symbols. The framing signal serves a purpose very similar to the K-codes in the serial physical layer. Control symbols each have a single function on the parallel interface, on the serial interface control symbols can have up to two separate functions. The parallel interface uses

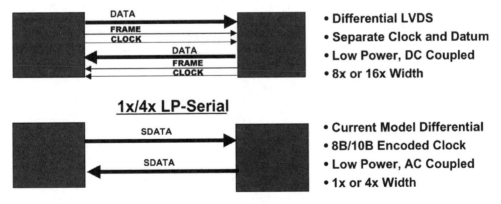

Figure 8.1 RapidIO interface differences

RapidIO® The Embedded System Interconnect. S. Fuller
© 2005 John Wiley & Sons, Ltd   ISBN: 0-470-09291-2

differential electrical signals based on the IEEE LVDS standard, the serial interface uses differential electrical signals based on the IEEE XAUI standard.

The parallel physical layer is simpler to implement than the serial physical layer. The operating frequencies are significantly lower, with specified clock frequencies ranging from 250 MHz to 1 GHz. The parallel physical layer also does not include the 8B/10B encode/ decode layer or the need to generate balanced K- and D-codes. For these reasons, the parallel interface may also offer lower transaction latency than the serial interface layer, making it more attractive as a direct system interface for high-performance microprocessors. Because of its higher pin count, 40 pins for the 8-bit wide interface and 76 pins for the 16-bit wide interface, the parallel physical layer is relatively less attractive for use across backplanes, although this is not prohibited and has been demonstrated to work.

The parallel physical layer functionality includes packet transmission, flow control, error management, and other system functions. In the RapidIO documentation you will often see the parallel end point referred to as 8/16 LP-LVDS. This refers to the width of the interface (8- or 16-bit data lane widths) and the LVDS signaling used. The LP is included to indicate that the Link Protocol is also included as part of the specification. In this book we will use the more common term Parallel RapidIO.

8.1 PACKET FORMATS

Packet formats for the RapidIO parallel physical layer are virtually identical to those used by the serial physical layer. Parallel RapidIO packets differ from Serial RapidIO packets in the following two ways. First, because Parallel RapidIO packets are transmitted on the same data lines as the control symbols and because the parallel interface does not use K-codes to delimit packets and symbols, the first bit of a parallel packet indicates that it is a packet and not a control symbol. This is represented as $S = 0$ in Figure 8.2. This bit is also repeated, inverted in bit position 5. The serial packet does not include an S-bit distinguishing it from a control symbol. Second, where the serial physical layer provides five bits of acknowledgement ID, the parallel physical layer provides only three bits of acknowledgement ID. This means that there is coverage for up to 32 outstanding transactions between two serial link ports. For the parallel interface there is coverage for only eight outstanding transactions between ports. The serial link interface coverage is higher because serial interfaces are more likely to operate over a longer distance and have more unacknowledged transactions active between ports.

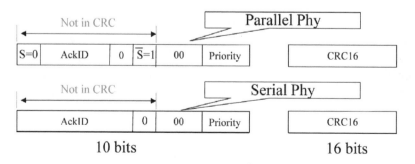

Figure 8.2 Comparison of parallel and serial RapidIO headers

With the exception of these first six bits, there are no differences between the format of packets transmitted over the parallel RapidIO interface compared with packets transmitted over the serial RapidIO interface.

8.2 CONTROL SYMBOL FORMATS

Control symbols are used for packet delineation, transmission, pacing, and other link interface control functions by the RapidIO parallel physical layer.

Control symbols used by the RapidIO parallel interface are 32 bits in length. Although they are 32 bits long, the control symbol is actually a 16-bit quantity that is sent twice. The second 16 bits are an inverted version of the first 16 bits. Figure 8.3 shows the format of the control symbols. From this diagram it is clear that bits 16–31 are inverted representations of bits 0–15. Because of the width of the parallel interface and the fact that data is transmitted at twice the clock rate, the time it takes to transmit a full control symbol is two clock cycles for the 8-bit interface and one clock period for the 16-bit interface. The 32-bit length also ensures that all packets and control symbols are properly aligned to 32-bit boundaries at all times.

This control symbol format is displayed in Figure 8.3. Bits 13–15 and 29–31 contain a field that defines the type of the control symbol. This 3-bit field can encode eight separate types. Six control symbol types are defined with two encodings remaining (Table 8.1). These encodings are reserved for future expansion use.

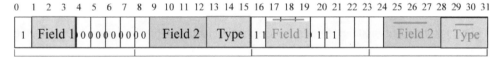

Figure 8.3   Parallel RapidIO control symbol format

Table 8.1   Control symbol types

Type	Type encoding	Name	Description
0	0b000	Packet accepted	Receiver accepted the indicated packet without error
1	0b001	Packet retry	Packet was not accepted and a retry is requested. No error is indicated
2	0b010	Packet-not-accepted	Packet was not accepted due to an error condition. Source should retransmit the packet
3	0b011	Reserved	
4	0b100	Packet control	Used for packet delineation, transmission, pacing and link interface control
5	0b101	Link request	Used to manage the link
6	0b110	Link response	Used for responses to Link requests
7	0b111	Reserved	

8.2.1 Acknowledgment Control Symbol Formats

Control symbol Types 0, 1 and 2 are referred to as acknowledgement control symbols. An acknowledgment control symbol is a transmission status indicator issued by an endpoint when it has received a packet from another endpoint to which it is electrically connected. Acknowledgment control symbols are used for flow control and resource de-allocation between adjacent devices. The following are the different acknowledgment control symbols that can be transmitted back to sending elements from receiving elements:

- packet-accepted (Type 0)
- packet-retry (Type 1)
- packet-not-accepted (Type 2)

Receipt of an acknowledgment control symbol, or any control symbol for that matter, does not imply the end of a packet. A control symbol can be embedded in a packet, as well as sent while an interconnect is idle.

Acknowledgment control symbols contain control and information fields. Three field definitions are provided in Table 8.2.

Table 8.2   Field definitions for acknowledgment control symbols

Field	Definition
packet_ackID	Identifier for acknowledgments back to the request or response packet sender
buf_status	Indicates the number of maximally sized packets that can be received.
cause	Indicates the type of error encountered by an input port.

8.2.1.1 Packet-accepted Control Symbol

The packet-accepted acknowledgment control symbol indicates that the adjacent device in the interconnect fabric has taken responsibility for sending the packet to its final destination and that resources allocated by the sending device can be released. This control symbol is generated only after the entire packet has been received and found to be free of detectable errors.

8.2.1.2 Packet-retry Control Symbol

A packet-retry acknowledgment control symbol indicates that the adjacent device in the interconnect fabric was not able to accept the packet because of some temporary resource conflict, such as insufficient buffering, and the source should retransmit the packet. This control symbol can be generated at any time after the start of a packet, which allows the sender to cancel the packet and try sending a packet with a different priority or destination. This will avoid wasting bandwidth by transmitting all the rejected packet.

8.2.1.3 Packet-not-accepted Control Symbol

A packet-not-accepted acknowledgment control symbol means that the receiving device could not accept the packet because of an error condition, and that the source should retransmit the packet. This control symbol can be generated at any time after the start of a packet, which allows the sender to cancel the packet and try sending a packet with a different priority or destination. Generating this control symbol at any point in packet transmission avoids wasting

Table 8.3 Cause field definition

Encoding	Definition
0b000	Encountered internal error
0b001	Received unexpected ackID on packet
0b010	Received error on control symbol
0b011	Non-maintenance packet reception is stopped
0b100	Received bad CRC on packet
0b101	Received S-bit parity error on packet/control symbol
0b110	Reserved
0b111	General error

bandwidth by not transmitting the entire rejected packet. The packet-not-accepted control symbol contains a field describing the cause of the error condition. If the receiving device is not able to specify the cause for some reason, or the cause is not one of the defined options, the general error encoding is used.

The cause field is used to display informational fields useful for debug. Table 8.3 displays the reasons a packet may not be accepted as indicated by the cause field.

A packet-retry or packet-not-accepted acknowledgment control symbol that is received for a packet that is still being transmitted may result in the sender canceling the packet.

The sending device can use the stomp, restart-from-retry (in response to a packet-retry control symbol), or link-request (in response to a packet-not-accepted control symbol) control symbol to cancel a packet. Because the receiver has already rejected the packet, it will not detect any induced error. Alternatively, the sending device can choose to complete transmission of the packet normally.

8.2.2 Packet Control Symbols

The Type 4 control symbol is referred to as the packet control symbol. There are six subtypes of this control symbol. These subtypes are identified in Table 8.4. The packet control symbols also have a contents field, the meaning of which depends upon the particular control symbol. Of these control symbols, all control symbols that are not defined as terminating a packet may be embedded within a packet.

Table 8.4 shows how the sub_type values function with the values of the contents field. For the idle, eop, and multicast-event control symbols the contents field is used as a buf_status, which is used to provide information on the number of available buffers in the receiver back to the transmitter. For a throttle control symbol, the contents field specifies the number of aligned pacing idle control symbols that the sender should insert in the packet; details of this field are contained in Table 8.5. One of the specified encodings indicates to the sender that it can immediately begin to resume packet transmission. For the stomp and restart-from-retry control symbols, the contents field is unused, tied to all logic 0s and ignored by the receiving device.

The pacing idle count content field for a throttle control symbol is defined in Table 8.5.

8.2.3 Link Maintenance Control Symbol Formats

Link-request/link-response control symbol pairs control maintenance of a link. These control symbols are Types 5 and 6 respectively.

Table 8.4   Sub_type and contents field definitions for type 4 control symbols

sub_type field definition	sub_type encoding	contents field definition
idle	0b000	Used as a buf_status field that indicates the number of maximum-sized packets that can be received.
stomp	0b001	Unused, contents = 0b0000
eop	0b010	Used as a buf_status field that indicates the number of maximum-sized packets that can be received.
restart-from-retry	0b011	Unused, contents = 0b0000
throttle	0b100	Specifies the number of aligned pacing idles that the sender inserts in a packet.
Multicast-event	0b101	Used as a buf_status field that indicates the number of maximally sized packets that can be received.
Reserved	0b110–111	

Table 8.5   Throttle control symbol contents field definition

Encoding	Definition
0b0000	1 aligned pacing idle control symbol
0b0001	2 aligned pacing idle control symbols
0b0010	4 aligned pacing idle control symbols
0b0011	8 aligned pacing idle control symbols
0b0100	16 aligned pacing idle control symbols
0b0101	32 aligned pacing idle control symbols
0b0110	64 aligned pacing idle control symbols
0b0111	128 aligned pacing idle control symbols
0b1000	256 aligned pacing idle control symbols
0b1001	512 aligned pacing idle control symbols
0b1010	1024 aligned pacing idle control symbols
0b1011–1101	Reserved
0b1110	1 aligned pacing idle control symbol for oscillator drift compensation
0b1111	Stop transmitting pacing idles, can immediately resume packet transmission

8.2.3.1 Link Request Control Symbol

A link-request control symbol issues a command to or requests status from the device that is electrically connected, or linked, to the issuing device. Field 1 is called the cmd field, The cmd, or command, field of the link-request control symbol format is defined in Table 8.6. Field 2 is the buf_status field, this field provides information on the status of buffers in the receiver. The link-request control symbol is followed by a complemented version of itself as with the other control symbols. A link-request control symbol cannot be embedded in a packet, but can be used to cancel the packet. Under error conditions a link-request/input-status control symbol acts as a restart-from-error control symbol.

Table 8.6   cmd field definition

cmd encoding	Command name	Description
0b000	Send-training	Send 256 iterations of the training pattern
0b001–010		Reserved
0b011	Reset	Reset the receiving device
0b100	Input-status	Return input port status; functions as a restart-from-error control symbol under error conditions
0b101–111		Reserved

Table 8.7   link_status field definition

link_status encoding	Port status	Description
0b0000–0b0001	Reserved	
0b0010	Error	Unrecoverable error encountered
0b0011	Reserved	
0b0100	Retry-stopped	The port has been stopped by a retry
0b0101	Error-stopped	The port has been stopped by a transmission error; this state is cleared after the link-request/input-status command is completed
0b0110–0b0111	Reserved	
0b1000	OK, ackID0	Working properly, expecting ackID 0
0b1001	OK, ackID1	Working properly, expecting ackID 1
0b1010	OK, ackID2	Working properly, expecting ackID 2
0b1011	OK, ackID3	Working properly, expecting ackID 3
0b1100	OK, ackID4	Working properly, expecting ackID 4
0b1101	OK, ackID5	Working properly, expecting ackID 5
0b1110	OK, ackID6	Working properly, expecting ackID 6
0b1111	OK, ackID7	Working properly, expecting ackID 7

8.2.3.2 Link Response Control Symbol

The link-response control symbol is used by a device to respond to a link-request control symbol as described in the link maintenance protocol described in Section 8.7. The link-response control symbol is the same as all other control symbols in that the second 16 bits are a bit-wise inversion of the first 16 bits. Field 1 is called ackID_status and simply presents a straightforward coding of the expected ackID value. The link_status field definitions are more complicated and are provided in Table 8.7. Note that the ackID information is included in this field as well for additional error coverage if the receiver is working properly (encodings 8–15).

8.3 CONTROL SYMBOL TRANSMISSION ALIGNMENT

This section shows examples of control symbol transmission over the 8-bit and 16-bit interfaces. Figure 8.4 shows the byte transmission ordering on an 8-bit port through time, using an aligned packet-accepted control symbol as an example.

Figure 8.5 shows the same control symbol as it would be transmitted over the 16-bit interface.

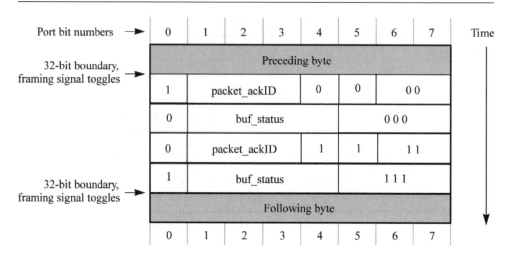

Figure 8.4 Control symbol transmission 8-bit interface

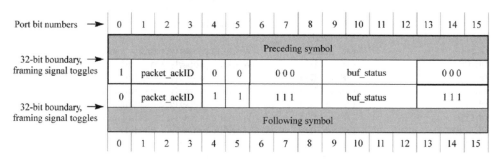

Figure 8.5 Control symbol transmission 16-bit interface

8.4 PACKET START AND CONTROL SYMBOL DELINEATION

The control framing signal used to delineate the start of a packet or a control symbol on the physical port is a no-return-to-zero, or NRZ signal. This frame signal is toggled for the first 8- or 16-bit datum of each packet and for the first 8- or 16-bit datum of each control symbol (depending on the width of the interface). In order for the receiving processing element to sample the data and frame signals, a data reference signal is supplied that toggles on all possible transitions of the interface pins. This type of data reference signal is also known as a double-data-rate clock. These received clocks on devices with multiple RapidIO ports have no required frequency or phase relationship.

The framing signal is toggled once at the beginning of a packet header or at the beginning of a control symbol. However, it is also toggled for all idle control symbols between packets. This means that the maximum toggle rate of the control framing signal is every four bytes, and the framing signal is only allowed to toggle on every fourth byte. Therefore, the framing signal is aligned to a 32-bit boundary as are all of the packets and control symbols.

Figures 8.6 and 8.7 show the relationship between the data, clock and frame signals for the 8- and 16-bit parallel interfaces, respectively. The framing signal should always toggle on the rising edge of the clock signal.

Figure 8.8 shows how consecutive control symbols would be delineated on a 16-bit interface. On the 16-bit interface the control symbol is two beats of data in length and will be transmitted during one clock cycle. The framing signal is an NRZ signal that transitions at the beginning of every control symbol, packet or every fourth Idle packet.

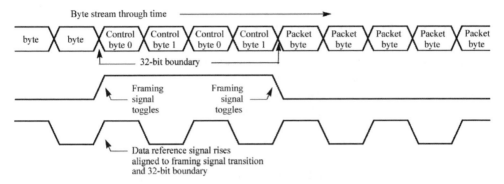

Figure 8.6 Framing signal maximum toggle rate for 8-bit port

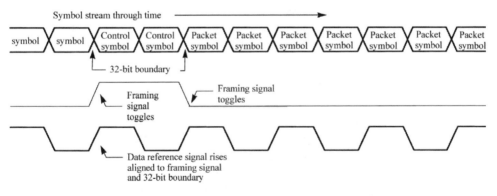

Figure 8.7 Framing signal maximum toggle rate for 16-bit port

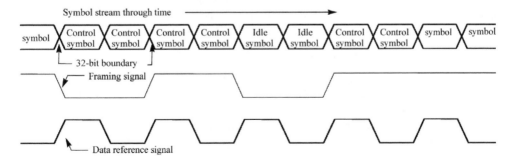

Figure 8.8 Control symbol delineation example for 16-bit port

Errors on the framing and data reference signals can be detected either directly by verifying that the signals transition only when they are allowed and expected to transition, or indirectly by depending upon detection of packet header or CRC or control symbol corruption, etc if these signals behave improperly. Either method of error detection on the framing and data reference signals allows error recovery. For simplicity, the data reference signal will not be included in any additional figures. It is always rising on the 32-bit boundary when it is legal for the frame signal to toggle.

8.5 PACKET EXCHANGE PROTOCOL

The RapidIO parallel physical layer specification defines an exchange of packet and acknowledgment control symbols where the destination or an intermediate switch fabric device explicitly acknowledges receipt of every request or response packet.

If a packet cannot be accepted for any reason, an acknowledgment control symbol indicates that the original packet and any already transmitted subsequent packets should be resent. This behavior provides a flow control and transaction ordering mechanism between processing elements. Link flow control and acknowledgement are performed between adjacent devices. This allows flow control and error handling to be managed between each electrically connected device pair rather than between the original source and the final target of the transaction. This allows the system to optimize the performance of each link and also to pinpoint detected errors to specific links rather than the whole paths between devices (which might contain several links). Devices will always transmit an acknowledge control symbol for a received request. The transaction is always acknowledged before the operation of which it is a component is completed.

8.5.1 Packet and Control Alignment

All packets sent over the RapidIO parallel interface are aligned to 32-bit boundaries. This alignment allows devices to work on packets with a larger internal path width, thus allowing lower core operating frequencies. Packets that are not naturally aligned to a 32-bit boundary are padded. Control symbols are nominally 16-bit quantities, but are defined, in the parallel interface, as a 16-bit control symbol followed by a bit-wise inverted copy of itself to align it to the 32-bit boundary. This, in turn, adds error detection capability to the interface. These 32-bit quantities are referred to as aligned control symbols.

The 16-bit wide port is compatible with an 8-bit wide port. If an 8-bit wide port is properly connected to a 16-bit wide port, the port will function as an 8-bit interface between the devices.

8.5.2 Acknowledge Identification

A packet requires an identifier to uniquely identify its acknowledgment. This identifier, known as the acknowledge ID (or ackID), is three bits, allowing for a range of one to eight outstanding unacknowledged request or response packets between adjacent processing elements; however, only up to seven outstanding unacknowledged packets are allowed at any one time. The ackIDs are assigned sequentially (in increasing order, wrapping back to 0 on overflow) to indicate the order of the packet transmission. The acknowledgments themselves appear as control symbols sent back to the transmitter.

8.6 FIELD PLACEMENT AND DEFINITION

This section contains information on the additional physical layer bit fields and control symbols required to implement the flow control, error management, and other specified system functions in the parallel RapidIO interface.

8.6.1 Flow Control Fields Format

The fields used to control packet flow in the system are described below (Table 8.8).

Figure 8.9 shows the format for the physical layer fields for packets. In order to pad packets to the 16-bit boundary there are three reserved bits in a packet's physical layer fields. These bits are assigned to logic 0 when generated and ignored when received. The figure also shows how the physical layer fields are prefixed to the combined transport and logical layer packet.

The unshaded fields are the physical layer fields defined by the physical layer specification. The shaded fields are the bits associated with the combined transport and logical transaction definitions. The first transport and logical field shown is the two bit tt field described in Chapter 6. The second field is the four bit format type (ftype) defined in Chapters 4, 5, 11 and 12. The third combined field is the remainder of the transport and logical packet. This size and contents of these fields will vary depending on the transaction.

Table 8.8 Fields that control packet flow

Field	Description
S	0b0 RapidIO request or response packet 0b1 Physical layer control symbol
S_bar	Inverse of S-bit for redundancy (odd parity bit)
ackID	Acknowledge ID is the packet identifier for acknowledgments back to the packet sender
prio	Sets packet priority: 0b00 lowest priority 0b01 medium priority 0b10 high priority 0b11 highest priority
buf_status	Specifies the number of available packet buffers in the receiving device
stype	Control symbol type
rsrv	Reserved

S=0	α	rsrv=0	S=1	rsrv=00	prio	:t	ft	Remainder of transport & logical fields
1	3	1	1	2	2	2	4	*n*

Figure 8.9 Flow control fields bit Stream

8.6.2 Packet Priority and Transaction Request Flows

Each packet has a priority that is assigned by the end point processing element that initiates the packet. The priority is carried in the prio field of the packet and has four possible values, 0, 1, 2 or 3, with 0 being the lowest priority and 3 being the highest. Packet priority is used in RapidIO for several purposes including transaction ordering and deadlock prevention.

8.6.3 Transaction and Packet Delivery

Certain physical layer fields and a number of control symbols are used for handling flow control. One physical layer field is the ackID field, this field is assigned by the sending processing element, and expected by the receiving processing element, in a sequential fashion.

Packets are accepted by the receiving processing element only when the ackID values of successive packets occur in the specified sequence. The receiving processing element signals the acceptance of a packet by returning a packet-accepted control symbol to the sender. This order allows a device to detect when a packet has been lost, and also provides a mechanism to maintain ordering.

A device that requests that a packet be resent, what the RapidIO specifications refer to as 'retry,' because of some temporary internal condition, is expected to silently discard all new incoming packets until it receives a restart-from-retry control symbol from the sender. The sender will then retransmit all packets starting from the retried ackID, re-establishing the proper ordering between the devices. The packet sent with the retried ackID may be the original retried packet or a higher priority packet, if one is available, allowing higher-priority packets to bypass lower-priority packets across the link.

Similarly, if a receiving processing element encounters an error condition, it will return a packet-not-accepted control symbol, indicating an error condition, to the sender. It will also silently discard all new incoming packets. If the error condition is due to a transmission error the sender may be able to recover from the effects of the error condition by simply retransmitting the packet.

8.6.4 Control Symbol Protection

The control symbols defined in this specification are protected in two ways:

- The S-bit, distinguishing a control symbol from a packet header, has an odd parity bit to protect a control symbol from being interpreted as a packet.
- The entire aligned control symbol is protected by the bit-wise inversion of the control symbol used to align it to a 32-bit boundary. This allows extensive error detection coverage for control symbols.

A transmission error in the buf_status field, regardless of the control symbol type, may optionally not be treated as an error condition because it is always a reserved or an information-only field that is not critical for proper system behavior. For example, if a corrupt value of buf_status is used, a low value may temporarily prevent a packet from being issued, or a high value may result in a packet being issued when it should not have been, causing a retry. In either case the problems are temporary and will properly resolve themselves through the existing protocol.

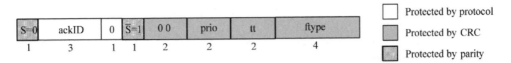

Figure 8.10   Error coverage of first 16 bits of packet header

8.6.5 Packet Protection

All RapidIO packets are protected with a CRC code The CRC code also covers the 2-bit priority field. The S bit is duplicated as it is in the control symbols to protect the packet from being interpreted as a control symbol, the packet is also protected by protocol as described below.

Figure 8.10 shows the error coverage for the first 16 bits of a packet header. The CRC protects the prio, tt, and ftype fields and two of the reserved bits as well as the remainder of the transport and logical fields. Since a new packet has an expected value for the ackID field at the receiver, bit errors on this field are easily detected and the packet is not accepted, because of the unexpected value. An error on the S-bit is detected with the redundant inverted $\bar{S}$-parity bit.

This structure does not require that a packet's CRC value be regenerated when the uncovered physical fields are modified as the packet transits the fabric.

8.6.6 Transactional Boundaries

A system address map usually contains memory boundaries that separate one type of memory space from another. Memory spaces are typically allocated with a preset minimum granularity. These spaces are often called page boundaries. Page boundaries allow the operating system to manage the entire address space through a standard mechanism. These boundaries are often used to mark the start and end of read-only space, peripheral register space, data space, and so forth.

RapidIO allows DMA streaming of data between two processing elements. Typically, in system interfaces that allow streaming, the targeted device of the transaction has a way to disconnect from the master once a transactional boundary has been crossed. The RapidIO specifications do not define a page boundary, nor a mechanism by which a target can disconnect part way through a transaction. Therefore, it is up to the system software and/or hardware implementation to guarantee that a transaction can complete gracefully to the address space requested.

As an example, a RapidIO write transaction does not necessarily have a size associated with it. Given a bus error condition where a packet delimiting control symbol is missed, the target hardware could continue writing data beyond the intended address space, thus possibly corrupting memory. Hardware implementations should set up page boundaries so this condition does not occur. In such an implementation, should a transaction cross the boundary, an error should be indicated and the transaction discarded.

8.7 LINK MAINTENANCE PROTOCOL

To initialize, explore, and recover from errors it is necessary to have a secondary mechanism to communicate between connected system devices. This mechanism is used to establish communications between connected devices, to attempt automatic error recovery and to allow software-managed link maintenance operations.

This protocol involves a request and response pair between electrically connected (linked) devices in the system. For software management, the request is generated through ports in the configuration space of the sending device. An external processing element write of a command to the link-request register with a maintenance write transaction causes a link-request control symbol to be issued onto the output port of the device, but only one link-request can be outstanding on a link at a time. The device that is linked to the sending device responds with a link-response control symbol if the link-request command required it to do so. The external processing element retrieves the link-response by polling the link-response register with I/O logical maintenance read transactions. A device with multiple RapidIO interfaces has a link-request and a link-response register pair for each corresponding RapidIO interface.

The automatic recovery mechanism relies on the hardware generating link-request control symbols under the transmission error conditions and using the corresponding link-response information to attempt recovery. Automatic link initialization also depends upon hardware generation of the appropriate link-requests and link-responses.

8.7.1 Command Descriptions

Table 8.9 contains a summary of the link maintenance commands that are used in the link maintenance protocol described above. Three link-request commands are defined currently. The input-status command generates a paired link-response control symbol; the reset and send-training commands do not.

8.7.1.1 Reset and Safety Lockouts

The reset command causes the receiving device to go through its hard reset or power-up sequence. All state machines and the configuration registers will be reset to the original power on states. The reset command does not generate a link-response control symbol.

Owing to the undefined reliability of various system designs it is necessary to put a safety lockout on the reset function of the link-request control symbol. A device receiving a reset command in a link-request control symbol does not perform the reset function unless it has received four reset commands in a row without any other intervening packets or control symbols, except for idle control symbols. This is meant to prevent spurious reset commands from inadvertently resetting a device.

Table 8.9   Secondary link maintenance command summary

Command	Description
Reset	Resets the device
Input-status	Returns input port status; functions as a restart-from-error control symbol under error conditions
	Generates a paired link-response control symbol
Send-training	Stops normal operation and transmits 256 training pattern iterations

8.7.1.2 Input-status

The input-status command requests the receiving device to return the ackID value it expects to next receive from the sender on its input port. It will also return the current input port operational status. This command causes the receiver to flush its output port of all control symbols generated by packets received before the input-status command. The receiver then responds with a link-response control symbol.

The input-status command is the command used by the hardware to recover from transmission errors. If the input port had stopped because of a transmission error that generated a packet-not-accepted control symbol back to the sender, this input-status command acts as a restart-from-error control symbol, and the receiver is re-enabled to receive new packets after generating the link-response control symbol. This restart-from-error control symbol may also be used to restart the receiving device if it is waiting for a restart-from-retry control symbol after retrying a packet. This situation can occur if transmission errors are encountered while trying to re-synchronize the sending and receiving devices after the retry.

8.7.1.3 Send-training

The send-training command causes the recipient device to suspend normal operation and begin transmitting a special training pattern. The receiving device transmits a total of 256 iterations of the training pattern, followed by at least one idle control symbol, and then resumes operation. The send-training command does not generate a link-response control symbol.

8.7.2 Status Descriptions

The input-status request generates two pieces of information that are returned in the link-response:

- link status
- ackID usage

The first type of data is the current operational status of the interface. These status indicators are described in Table 8.10.

The retry-stopped state indicates that the port has retried a packet and is waiting to be restarted. This state is cleared when a restart-from-retry (or a link-request/input-status) control symbol is received. The error-stopped state indicates that the port has encountered a transmission error and is waiting to be restarted. This state is cleared when a link-request/input-status control symbol is received.

Table 8.10   Link status indicators

Status indicator	Description
OK	The port is working properly
Error	The port has encountered an unrecoverable error and has shut down
Retry-stopped[1]	The port has been stopped by a retry
Error-stopped[1]	The port has been stopped by a transmission error

[1] Valid only with the stopped indicator

The second field returned in the link-response control symbol is state information about the acknowledge identifier usage. The input port returns a value indicating the next ackID expected to be received by the port. The automatic error recovery mechanism uses this information to determine where to begin packet re-transmission after a transmission error condition has been encountered.

8.8 PACKET TERMINATION

A packet is terminated in one of two ways:

- The beginning of a new packet marks the end of a previous packet.
- The end of a packet may be marked with one of the following: an end-of-packet (eop); restart-from-retry; link-request; or stomp control symbol.

The stomp control symbol is used if a transmitting processing element detects a problem with the transmission of a packet. It may choose to cancel the packet by sending the stomp control symbol instead of terminating it in a different, possibly system-fatal, manner such as corrupting the CRC value.

The restart-from-retry control symbol can cancel the current packet. It may also be transmitted on an idle link. This control symbol is used to command the receiver to start accepting packets again after the receiver has retried a packet and entered a stopped state.

The link-request control symbol can cancel the current packet. It may also be transmitted on an idle link. It can be used by software for system observation and maintenance, and it can be used by software or hardware to enable the receiver to start accepting packets after the receiver has refused a packet owing to a transmission error.

A port receiving a cancelled packet is required to drop the packet. The cancellation of a packet does not result in the generation of any errors. If the packet was cancelled because the sender received a packet-not-accepted control symbol, the error that caused the packet-not-accepted control symbol to be sent should be reported in the normal manner.

If a port receiving a cancelled packet has not previously acknowledged the packet and is not in an 'Input-stopped' state (Retry-stopped or Error-stopped), the port should immediately enter the Input Retry-stopped state and follow the Packet Retry mechanism, if the packet was cancelled with a control symbol other than a restart-from-retry or a link-request/input-status control symbol. As part of the Packet Retry mechanism, the port sends a packet-retry control symbol to the sending port, indicating that the cancelled packet was not accepted.

Figure 8.11 shows an example of an end-of-packet control symbol marking the end of a packet. The stomp, link-request, and restart-from-retry control symbol cases would look similar.

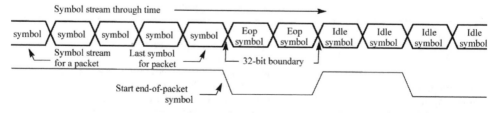

Figure 8.11   End-of-packet control symbol marking end of packet (16-bit port)

8.9 PACKET PACING

If a device cannot transmit a packet in a contiguous stream, perhaps because the internal transfer of data is delayed or cannot keep up with the RapidIO transfer rate, it may insert idle control symbols, called pacing idles, into the packet stream. As with the other control symbols, the pacing idle control symbols are always followed by a bit-wise inverted copy. Any number of pacing idle control symbols can be inserted, up to some implementation-defined limit, at which point the sender should instead send a stomp control symbol and cancel the packet in order to attempt to transmit a different packet. Figure 8.12 shows an example of packet pacing. The receiving device ignores the idle control symbols. More data is sent when it becomes available. Pacing idle control symbols can be embedded anywhere in a packet where they can be legally delineated by the framing signal.

The receiver of a packet may request that the sender insert pacing idle control symbols on its behalf by sending a throttle control symbol specifying the number of pacing idle control symbols to delay. The packet sender then inserts that number of pacing idles into the packet stream. If additional delay is needed, the receiver can send another throttle control symbol.

Note that, when calculating and testing the CRC values for a packet, the pacing idle control symbols that might appear in the symbol stream are not included in the calculation.

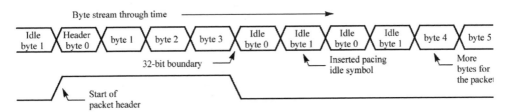

Figure 8.12   Pacing idle insertion in packet (8-bit port)

8.10 EMBEDDED CONTROL SYMBOLS

Control symbols can be embedded anywhere in a packet in the same fashion as pacing idle control symbols, as long as all delineation and alignment rules are followed (Figures 8.13, 8.14). As with the pacing idle control symbols, the embedded control symbols are not included in the CRC value calculation for the packet.

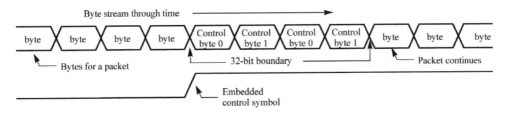

Figure 8.13   Embedded control symbols for 8-bit port

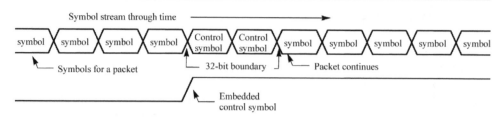

Figure 8.14 Embedded control symbols for 16-bit port

A special error case exists when a corrupt embedded control symbol is detected. In this case a packet-not-accepted control symbol is generated and the embedding packet is discarded.

8.11 PACKET ALIGNMENT

Figure 8.15 shows the byte transmission ordering on a port through time for a small transport format FTYPE 2 packet. Note that for this example the two bytes following the CRC would indicate some form of packet termination such as a new packet or an eop.

Figure 8.16 shows the same packet transmitted over a 16-bit port.

Port bit numbers →	0	1	2	3	4	5	6	7	Time
32-bit boundary, framing signal toggles →	Preceding byte								
	0	ackID			0	1	0 0		
	prio		tt		0 0 1 0				
	destinationID								
	sourceID								
	transaction				rdsize				
	srcTID								
	address[0-7]								
	address[8-15]								
	address[16-23]								
	address[24-28]					wdptr	xamsbs		
	CRC[0Ğ-]								
32-bit boundary, framing signal toggles →	CRC[8-15]								
	Following byte								
	0	1	2	3	4	5	6	7	

Figure 8.15 Request packet transmission, example 1

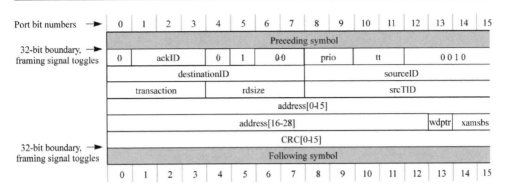

Figure 8.16 Request packet transmission, example 2

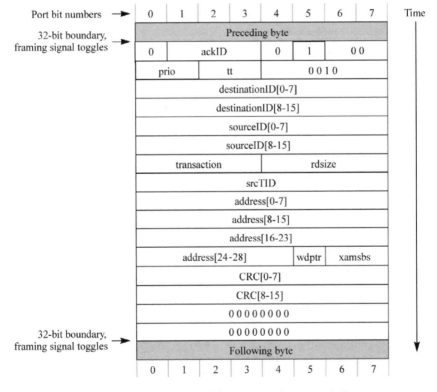

Figure 8.17 Request packet transmission, example 3

Figure 8.17 shows the same example again, but with the large transport format over the 8-bit port. Note that for this example the two bytes following the CRC of the packet are all logic 0 pads.

The packet example in Figure 8.18 is the same as for Figure 8.17, but over the 16-bit port.

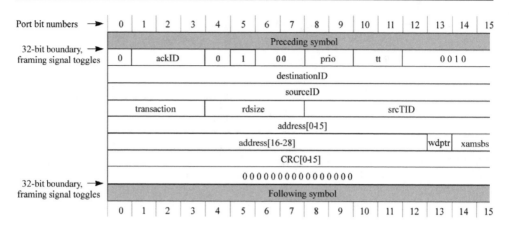

Figure 8.18   Request packet transmission, example 4

8.12 SYSTEM MAINTENANCE

Perhaps of equal importance to the operation of the data link itself are the methods for initializing, configuring, and maintaining the system during operation. For products to be interoperable, the system maintenance should be completely specified. This section describes the system maintenance operations that parallel RapidIO end points are required to support.

8.12.1 Port Initialization

Before a parallel port can exchange packets with another port, the ports and the link connecting the two ports must be initialized. In addition, each 16-bit wide port must decide whether to operate in 8-bit or 16-bit unless it has been statically configured for 8-bit or 16-mode.

8.12.1.1 Port Width Mode Selection

All 16-bit parallel RapidIO ports should be capable of operating in both 8-bit and 16-bit modes. The width mode in which a 16-bit port operates may be either statically or dynamically configured. If the width mode of a port is statically configured, the port will operate correctly only if connected to a port operating in the same width mode. If the width mode of a 16-bit port is not statically configured, the width mode is determined as part of the port initialization process.

When operating in 8-bit mode, only the signal pairs CLK0/$\overline{\text{CLK0}}$, FRAME/$\overline{\text{FRAME}}$ and D[0–7]/$\overline{\text{D[0–7]}}$ are used. The 16-bit mode output signal pairs TCLK1/$\overline{\text{TCLK1}}$, and TD[8–15]/$\overline{\text{TD[8–15]}}$ may be driven as outputs, but the input signal pairs RCLK1/$\overline{\text{RCLK1}}$ and RD[8–15]/RD[8–16] are ignored as inputs.

Dynamic port width selection is based on the presence of valid signals at the inputs of the CLK0, CLK1, FRAME D[0–7] and D[8–15] receivers. If valid signals are present at the inputs of the CLK0, CLK1, FRAME D[0–7] and D[8–15] receivers, a port should operate in 16-bit mode. If valid signals are present at the inputs of the CLK0, FRAME and D[0–7] receivers, but not at the inputs of the CLK1 and D[8–15] receivers, the port should operate in

8-bit mode. If valid signals are not present at the inputs of the CLK0, FRAME and D[0–7] receivers, the width mode is undefined and the port shouldn't exit from the Uninitialized state.

8.12.1.2 Ports Not Requiring Port Initialization

An 8-bit port with fixed input sampling window timing does not require port initialization. Similarly, a 16-bit port with fixed input sampling window timing and with its width mode statically configured does not require port initialization. On device power-up and on device reset, such ports will enter and remain in the Initialized state and will never be in the Uninitialized state. Such ports will transmit idle control symbols until an idle control symbol is received from the connected port. Upon the reception of an idle control symbol from the connected port, the port will transmit an idle control symbol, set the 'port OK' bit in its Port Control and Status Register, enter the normal operation state and may then begin the transmission of packets and non-idle control symbols.

If while waiting to receive an idle control symbol from the connected port, the reception by the port of a link-request/send-training control symbol from the connected port immediately followed by the training pattern indicates that the connected port is not initialized. When this occurs, the port will stop sending idle control symbols and repeatedly send training pattern bursts, each burst followed by an idle control symbol, until an idle control symbol is received from the connected port, indicating that the connected port is now initialized. Upon receiving an idle control symbol from the connected port, the port will complete transmission of the current training pattern burst, transmit an idle control symbol, set the 'port OK' bit in its Port Control and Status Register and enter the normal operation state. The port may then transmit packets and non-idle control symbols.

8.12.1.3 Ports Requiring Port Initialization

Ports that do not have fixed input sampling window timing and 16-bit ports with width mode not statically configured require port initialization. Such ports will enter the Uninitialized state on device power-up and device reset. Such a port will also enter the Uninitialized state if the port loses correct input sampling window timing due to conditions such as excessive system noise or power fluctuations. The algorithm used to determine when a port has lost input sample window alignment is not specified. A port in the Uninitialized state will execute the Port Initialization Process to exit the Uninitialized state.

The output signals of a parallel RapidIO port may be erratic when the device containing the port is powering up or being reset. For example, the output drivers may be temporarily disabled, the signals may have erratic HIGH or LOW times and/or the clock signals may stop toggling. A port must be tolerant of such behavior and should properly initialize after the signals from the connected port return to normal and comply with the electrical specifications.

8.12.1.4 Port Initialization Process

The state of a port is indicated by bits in the associated Port Error and Status Register. The Port Unitialized bit is set when the port is in the Uninitialized state and cleared when the port initialization process is completed and the port is initialized. The Port OK bit is set when the port has received an idle control symbol from the connected port and the port is in the normal operation mode, and is cleared if the port or the connected port is uninitialized or if there is no connected port (the port is not receiving valid input signals).

Any port whose link receiver input sample timing is not fixed and any 16-bit port whose width mode is not statically configured requires port initialization.

Upon entering the Uninitialized state, a port will execute the following Port Initialization Process.

- The port sets the 'Port Uninitialized' bit and clears the 'Port OK' bit in its Port Control and Status Register.
- The port transmits a link-request/send-training control symbol followed by one or more bursts of the training sequence. The port continuously transmits training pattern bursts, each followed by a link-request/send-training or idle control symbol, until the port has achieved input sample timing alignment and has received an idle control symbol from the connected port.
- The port attempts to detect a valid clock signal on its CLK0 input and, if present, on its CLK1 input and to detect the training pattern on its FRAME and D[0–7] inputs and, if present, on its D[8–15] inputs.
- Once valid input signals are detected, a 16-bit port whose width mode is not statically configured determines the width of the connected port and selects the matching width mode.
- Once the width mode of the port is established, either statically or dynamically, the port attempts to achieve input sampling window timing alignment. While attempting to achieve input sampling window timing alignment, the port will transmit a link-request/send-training control symbol after each training pattern burst.
- When the port achieves input sampling window timing alignment, it clears the 'Port Uninitialized' bit in the Port Control and Status Register and transmits an idle control symbol after each training pattern burst instead of a link-request/send-training control symbol. This indicates to the connected port that the port has completed input sampling window alignment.
- Upon receiving an idle control symbol from the connected port, indicating that the connected port has completed input sampling window alignment, the port completes transmitting the current training pattern burst, sends an idle control symbol, sets the 'Port OK' bit in the Port Control and Status Register and enters normal operation.

8.12.2 Link Initialization

After a port is in the Initialized state, the port should not begin transmission of packets and control symbols other than the idle control symbol until it has received an idle control symbol from the connected port. The reception of an idle control symbol indicates that the connected port is in the Initialized state and is ready to receive packets and non-idle control symbols. When both ports connected by a link have received an idle control symbol from the connected port, the link is initialized. Link initialization specifies that each port must receive at least one idle control symbol sequence from its partner before beginning normal operation.

8.12.3 Training

Many ports, especially those operating at higher data rates, must adjust the timing of when they sample input data from the link in order to achieve an acceptable or optimal received bit error rate. This process is called training and is performed while the port is receiving a special training pattern.

Input sampling window alignment is the process in which a port adjusts the timing of when input data is sampled by the link receiver. The timing is adjusted to achieve an acceptable or optimal received bit error rate. This process is also called 'training.' When the process is successfully completed, the port is said to be 'aligned' or 'trained.' The process or algorithm used by a port to align the input sampling window is not specified.

Sampling window alignment is done while the port is receiving a special data pattern called the training pattern. A special data pattern is required to ensure that enough transition timing information is available to the receiver to correctly adjust the input sample timing and to ensure that bytes transmitted by a port operating in 8-bit or that half-words transmitted by a port operating in 16-bit are correctly recovered by the link receiver.

There are two types of training: initialization training and maintenance training. Initialization training is used when a device powers up or is reset or when a port loses input sampling window alignment, owing to events such as excessive system noise or power fluctuations. Maintenance training is used when a port is nominally input sampling window aligned, but in need of some 'fine-tuning' of the input sampling window timing to maintain an acceptable or optimum received bit error rate.

8.12.3.1 Training Pattern

The training pattern used is the bit sequence 0b11110000. The training pattern is transmitted left to right with the left-most bit transmitted first and the right-most bit transmitted last.

When transmitted, the training pattern is transmitted simultaneously on all the data signals, D[0–7] for an 8-bit port or a 16-bit port statically configured to operate in 8-bit mode, D[0–15] for a 16-port not statically configured to operate in 8-bit mode, and the training pattern or its complement is transmitted on the FRAME signal. The training pattern or its complement is selected for transmission on the FRAME signal so that the FRAME signal will toggle at the beginning of training pattern transmission. The training pattern is never transmitted on a CLK signal. The training pattern is never transmitted on signals D[8–15] of a 16-bit port if the port is statically configured to operate in 8-bit mode.

The result of these rules is that during training pattern transmission, FRAME and data signals transmitted by the port have the following properties.

- The FRAME signal toggles at the beginning of training pattern transmission. (Individual data bits may or may not toggle at the beginning of training pattern transmission, depending on their value during the bit time immediately preceding the training pattern.)
- After the first bit of the training pattern, FRAME and all data bits will all toggle at the same nominal time.
- Each byte transmitted by a port transmitting in 8-bit mode is either all ones (0xFF), or all zeros (0x00).
- Each half-word transmitted by a port transmitting in 16-bit mode is either all ones (0xFFFF), or all zeros (0x0000).

The reception of the training pattern by an initialized port is readily identified by looking at RD[0–7] when the frame signal toggles. If the received value of RD[0–7] is either all ones or all zeros, the training pattern is being received.

8.12.3.2 Training Pattern Transmission

When transmitted, the training pattern is transmitted in bursts. Each burst will contain 256 repetitions of the training pattern. Either a link-request/send-training or an idle control symbol will follow each burst.

The training pattern is transmitted by an initialized port only at the request of the connected port. The link-request/send-training control symbol is used to request that the connected port transmit the training pattern. A port that is not initialized and therefore unable to reliably receive control symbols assumes that the connected port is sending link-request/send-training control symbols and therefore continuously transmits training sequence bursts with each burst, followed by a link-request/send-training control symbol as specified by the Port Initialization Process.

The training pattern is never embedded in a packet nor is it used to terminate a packet.

8.12.3.3 Maintenance Training

Depending upon their implementation, some ports may require occasional adjustment of their input sampling window timing while in the Initialized state to maintain an optimal received bit error rate. Such adjustment is called maintenance training. A port requiring maintenance training should do the following.

• The port will transmit a single link-request/send-training control symbol and then resume normal transmit operation.
• If the port is not able to complete maintenance training with one burst of the training pattern, the port may transmit additional link-request/send-training control symbols and shall resume normal transmit operation after transmitting each link-request/send-training control symbol.

A port requiring maintenance training should not transmit the training pattern after transmitting a link-request/send-training control symbol. (The transmission by a port of a link-request/send-training control symbol followed by the training pattern indicates that the port has become uninitialized.)

A port receiving a link-request/send-training control symbol that is not followed by the training pattern should end the transmission of packets and control symbols as quickly as possible without violating the link protocol, transmit one burst of the training pattern followed by an idle control symbol, and then resume normal operation.

8.12.3.4 Unexpected Training Pattern Reception

Once a link has been initialized, the reception of unsolicited training patterns is a protocol violation. It indicates that the sending port has lost input sampling window alignment and has most likely not received some previously sent packets and control symbols. Once the link has been initialized, a port receiving an unsolicited training pattern should enter the Output Error-stopped state. The port should execute the Output Error-stopped recovery process once communication with the connected port has been re-established.

8.13 SYSTEM CLOCKING CONSIDERATIONS

The RapidIO parallel physical interface can be deployed in a variety of system configurations. A fundamental aspect to the successful deployment of RapidIO is clock distribution. This section is provided to point out the issues of distributing clocks in a system.

8.13.1 Example Clock Distribution

Clock distribution in a small system is straightforward (Figure 8.19). In small systems, clocking is typically provided from a single clock source. In this case the timing budget must account for any skew and jitter component between each point. Skew and jitter are introduced by the end point clock regeneration circuitry (PLL or DLL) and by transmission line effects.

Distributing a clock from a central source may not be practical in larger or more robust systems. In these cases it may be more desirable to have multiple clock sources or to distribute the clock through the interconnect. Figure 8.20 displays the distribution of multiple clocks in a larger system.

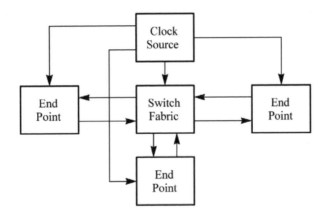

Figure 8.19 Clock distribution in a small system

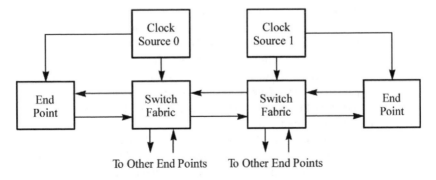

Figure 8.20 Clock distribution in a larger system

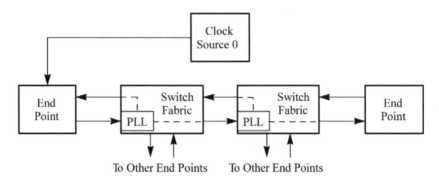

Figure 8.21   Clock distribution through the interconnect

In such a system the clock sources may be of the same relative frequency; however, they are not guaranteed to be always at exact frequency. Clock sources will drift in phase relationship with each other over time. This adds an additional component because it is possible that one device may be slightly faster than its companion device. This requires a packet elasticity mechanism.

If the clock is transported through the interconnect, as shown in Figure 8.21, then additive clock jitter must be taken into account. Assuming that each device gets a clock that was regenerated by its predecessor, and each device adds a certain jitter component to the clock, the resulting clock at the end point may be greatly unstable. This factor must be added to the timing budget.

8.13.2 Elasticity Mechanism

In systems with multiple clock sources, clocks may be of the same relative frequency, but not exact. Their phase will drift over time. An elasticity mechanism is therefore required to keep devices from missing data beats. For example, if the received clock is faster than the internal clock, then it may be necessary to delete an inbound symbol. If the received clock is slower than the internal clock, then it may be necessary to insert an inbound symbol.

The parallel RapidIO interface is source synchronous; meaning that a data element will have an associated clock strobe with which to synchronize. A clock boundary is crossed in the receive logic of the end point as the inbound data is synchronized to the internal clock. The end point should guarantee that the drift between the two clock sources does not cause a setup hold violation resulting in metastability in capturing the data.

To ensure that data is not missed, an end point implements an elasticity buffer. RapidIO uses idle control symbols as the elasticity mechanism. If a receiver needs to skip a symbol during receipt of a large packet, it can issue a throttle control symbol to cause the sender to insert a pacing idle control symbol in to the byte stream.

A data beat is clocked into the elasticity buffer with the external clock. The data beat is pulled out of the elasticity buffer using the internal clock delayed by a number of clocks behind the external clock event. This allows the data to become stable before it is synchronized to the internal clock. If the two clock events drift too close together then it is necessary for the synchronization logic to reset the tap and essentially skip a symbol. By guaranteeing a periodic idle control symbol, it is possible for the receive logic to skip an idle symbol, but not miss a critical packet data bytes.

8.14 BOARD ROUTING GUIDELINES

This section contains board design guidelines for RapidIO-based systems. The information here is presented as a guide for implementing a RapidIO board design. It is noted that the board designer may have constraints such as standard design practices, vendor selection criteria, and design methodology that must be followed. Therefore appropriate diligence must be applied by the designer.

Parallel RapidIO is a source-synchronous differential point-to-point interconnect, so routing considerations are minimal. The very high clock rate places a premium on minimizing skew and discontinuities, such as vias and bends. Generally, layouts should be as straight and free of vias as possible, with controlled impedance differential pairs.

8.14.1 Impedance

Interconnect design should follow standard practice for differential pairs. To minimize reflections from the receiver's 100 Ω termination, the differential pair should have an differential impedance of 50 Ω. The two signals forming the differential pair should be tightly coupled. The differential pairs should be widely spaced, although consistent with skew control and quality routing constraints, so that the crosstalk noise is limited.

8.14.2 Skew

To minimize the skew on a RapidIO channel the total electrical length for each trace within each unidirectional channel should be equal. Several layouts are suggested in Figure 8.22. Because the parallel RapidIO interface is source synchronous, the total length is not critical. Best signal integrity is achieved with a clean layout between opposed parts via routing on a single layer.

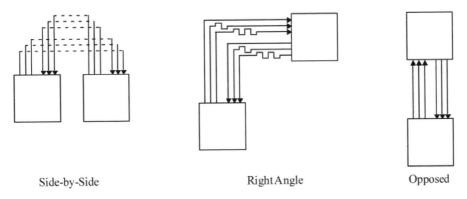

Side-by-Side   Right Angle   Opposed

Figure 8.22   Routing for equalized skew for several placements

The side-by-side layout requires two routing layers and has reduced signal integrity due to the vias between layers. To keep the total electrical length equal, both layers must have the same phase velocity.

Finally, right-angle routing requires meandering to equalize delay, and meandered sections reduce signal integrity while increasing radiation. It may be necessary to place meandered sections on a second routing layer to keep the routing clean. All skew calculations should be taken to the edge of the package.

8.14.3 PCB Stackup

PCB stackup has a significant effect on EMI generated by the relative high frequency of operation of a RapidIO channel, so EMI control must be planned from the start. Several potential stackups are shown in Figure 8.23.

The traditional four-layer stackup provides equal phase velocities on the two routing layers, but the placement of routing on the outside layers allows for easier radiation. This stackup is suitable for very short interconnects or for applications using an add-on shield.

The four-layer stackup can be rearranged to help with EMI control by placing the power and ground layers on the outside. Each routing layer still has equal phase velocities, but orthogonal routing can degrade signal integrity at very high speeds. The power distribution inductance is approximately tripled by the larger spacing between the power and ground planes, so applications using this stackup should plan on using more and higher-quality bypass capacitance.

The six-layer stackup shows one of many possible stackups. High-speed routing is on S1 and S2 in stripline, so signal quality is excellent with EMI control. S3 is for low-speed signals. Both S1 and S2 have equal phase velocities, good impedance control, and excellent isolation. Power distribution inductance is comparable to the four-layer stackup since the extra GND plane makes up for the extra (2×) spacing between PWR and GND. This example stackup is not balanced with respect to metal loading.

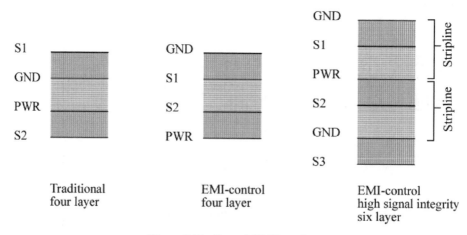

Figure 8.23   Potential PCB stackups

8.14.4 Termination

Depending upon the individual device characteristics and the requirements of the particular application, the board route may be required to encompass external devices such as terminating resistors or networks. The effect of such devices on the board route must be carefully analyzed and controlled.

8.14.5 Additional Considerations

The application environment for a RapidIO channel may place additional constraints on the PCB design.

8.14.5.1 Single Board Environments

A RapidIO channel completely constructed onto a single board offers the highest performance in terms of clock rate and signal integrity. The primary issues are clean routing with minimal skew. Higher clock rates put greater emphasis on the use of quality sockets (in terms of electrical performance) or on eliminating sockets altogether.

8.14.5.2 Single Connector Environments

The high clock rate of the parallel RapidIO physical layer requires the use of an impedance-controlled edge connector. The number of pins dedicated to power should equal the number dedicated to ground, and the distribution of power and ground pins should be comparable. If ground pins greatly outnumber power pins, then bypass capacitors along the length of each side of the connector should be provided. Place the connector as close to one end of the RapidIO interconnect as possible.

8.14.5.3 Backplane Environments

With two connectors, the design considerations from the single connector environment apply, but with greater urgency. The two connectors should either be located as close together or as far apart as possible.

9

Interoperating with PCI Technologies

RapidIO contains a rich set of operations and capabilities. RapidIO can easily support the transport of legacy interconnects such as PCI in a transparent manner. While RapidIO and PCI share similar functionality, the two interconnects have different protocols and require a translation function to move transactions between them. A RapidIO to PCI bridge processing element is required to make the necessary translation between the protocols of the two interconnects. This chapter describes the architectural considerations needed for a RapidIO to PCI bridge. This chapter is not intended as an implementation instruction manual, rather, it is meant to provide direction to the bridge architect and aid in the development of interoperable devices. For this chapter it is assumed that the reader has a thorough understanding of the PCI 2.2 and/or the PCI-X 1.0 specifications.

Figure 9.1 shows a system with devices containing both RapidIO- and PCI-based devices connected by various RapidIO and PCI bus segments. A host bridge is connected to various peripherals via a PCI bus. A RapidIO bridge device is used to translate PCI formatted transactions to the equivalent RapidIO operations to allow access to the rest of the system, including additional subordinate or downstream PCI bus segments.

For transactions that must travel between RapidIO and PCI the following operations are necessary:

- Map address spaces defined on the PCI bus to those of RapidIO interconnect
- Translate PCI transaction types to RapidIO operations
- Maintain the producer/consumer requirements of the PCI bus

This chapter will address each of these considerations for both PCI version 2.2 and PCI-X. The newly defined PCI Express technology would require identical considerations as PCI-X for transport across RapidIO.

RapidIO® The Embedded System Interconnect.   S. Fuller
© 2005 John Wiley & Sons, Ltd   ISBN: 0-470-09291-2

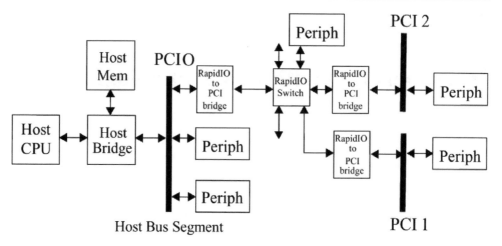

Figure 9.1   Example system with PCI and RapidIO

9.1 ADDRESS MAP CONSIDERATIONS

PCI defines three physical address spaces, specifically, the memory, I/O memory, and configuration spaces. RapidIO, on the other hand, supports only memory and configuration spaces.
I/O memory spaces are treated as memory spaces in RapidIO. Figure 9.2 shows a simple
example of the PCI memory and I/O address spaces for a host bus segment. In order for
devices on the PCI bus to communicate with other devices connected through RapidIO, it is
necessary to provide a memory mapping function. The example PCI host memory map uses a
32-bit physical address space resulting in 4 Gbyte total address space. Host memory is shown
at the bottom of the address map and peripheral devices at the top. The example shows two
RapidIO bridge address windows. One window is located in the memory space and one
window is located in the I/O space.

From the PCI side, any transactions issued to the bus segment with an address that
matches the RapidIO bridge window will be captured by the RapidIO to PCI bridge for

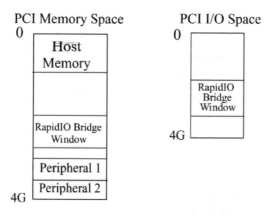

Figure 9.2   Host segment PCI memory map example

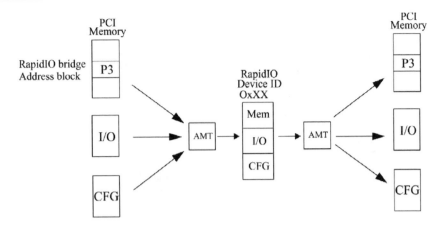

Figure 9.3   AMT and memory mapping

forwarding. Once the transaction has been accepted by the RapidIO to PCI bridge it must be translated to the proper RapidIO context, as shown in Figure 9.3. For the purposes of this discussion this function is called the address mapping and translation function (AMT). The AMT function is responsible for translating PCI addresses to RapidIO addresses. It is also responsible for the translation and assignment of the respective PCI and RapidIO transaction types. The address space defined by the RapidIO bridge window may represent more than one subordinate RapidIO target device. A device on PCI bus segment 0 as shown in Figure 9.1 may require access to a peripheral on PCI bus 1, bus 2, or RapidIO peripheral 5. Because RapidIO uses source addressing (device IDs), the AMT is responsible for translating the PCI address to both a target device ID and an associated offset address. In addition to address translation, RapidIO attributes, transaction types, and other necessary delivery information are established.

Similarly, transactions traveling from a RapidIO bus to a PCI bus must also pass through the AMT function. The address and transaction type are translated back into PCI format, and the AMT selects the appropriate address for the transaction. Memory mapping is relied upon for all transactions bridged between PCI and RapidIO (Figure 9.3).

9.2 TRANSACTION FLOW

In considering the mapping of the PCI bus to RapidIO it is important to understand the transaction flow of PCI transactions through RapidIO.

9.2.1 PCI 2.2 to RapidIO Transaction Flow

9.2.1.1 Posted PCI to RapidIO Transaction

The PCI 2.2 specification defines two classes of transaction types, posted and non-posted. Figure 9.4 shows the route taken by a PCI to RapidIO posted write transaction. Once the request is sent from the PCI master on the bus, it is claimed by the bridge processing element

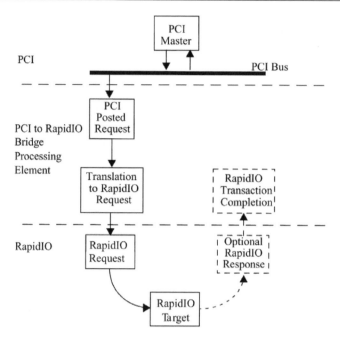

Figure 9.4   PCI mastered posted write transaction flow diagram

which uses the AMT to translate it into a RapidIO request. Only when the transaction is in RapidIO format can it be posted to the RapidIO target. In some cases it may be desirable to guarantee end-to-end delivery of the posted write transaction. For this case the RapidIO NWRITE_R transaction is used which results in a response as shown in the figure.

9.2.1.2 Non-posted PCI to RapidIO Transaction

A non-posted PCI transaction is shown in Figure 9.5. The transaction is mastered by the PCI agent on the PCI bus and accepted by the RapidIO to PCI bridge. The transaction is retried on the PCI bus if the bridge is unable to complete it within the required time-out period. In this case the transaction is completed as a delayed transaction. The transaction is translated to the appropriate RapidIO operation and issued on the RapidIO port. At some time later a RapidIO response is received and the results are translated back to PCI format. When the PCI master subsequently retries the transaction, the delayed results are returned and the operation is completed. This non-posted transaction could be a write or read transaction.

Because PCI allows unbounded transaction data tenures, it may be necessary for the RapidIO to PCI bridge to break the single PCI transaction into multiple RapidIO operations. In addition, RapidIO does not have byte enables and therefore does not support sparse byte transactions. For this case the transaction must be broken into multiple operations as well.

9.2.1.3 RapidIO to PCI Transaction

A RapidIO mastered operation is shown in Figure 9.6. For this case the RapidIO request trans-

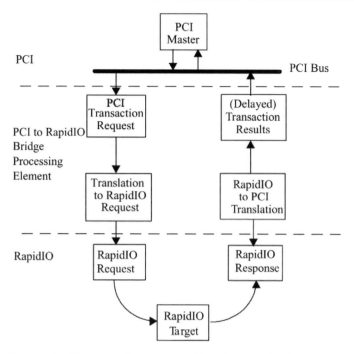

Figure 9.5 PCI mastered non-posted (delayed) transaction flow diagram

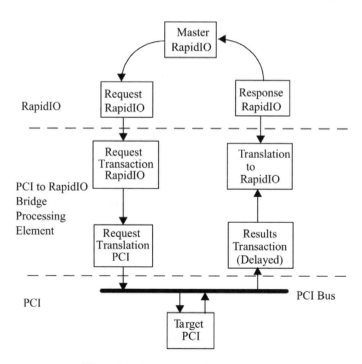

Figure 9.6 RapidIO mastered transaction

action is received at the RapidIO to PCI bridge. The bridge translates the request into the appropriate PCI command which is then issued to the PCI bus. The PCI target may complete the transaction as a posted, non-posted, or delayed non-posted transaction, depending on the command type. Once the command is successfully completed on the PCI bus the results are translated back into the RapidIO format and a response transaction is issued back to the RapidIO master.

9.3 PCI-X TO RAPIDIO TRANSACTION FLOW

The flow of transactions described in the previous section can apply to the PCI-X bus as well. To provide better bus utilization PCI-X also provides support for split transactions. This results in an transaction being broken into a split request and one or more split completions. As a target of a PCI-X split request, the RapidIO to PCI bridge may reply with a split response and complete the request using multiple RapidIO operations. The results of these operations are issued on the PCI-X bus as split completions. If the RapidIO to PCI-X bridge is the initiator of a split request, the target may also indicate that it intends to run the operation as a split transaction with a split response. In this case the target would send the results to the RapidIO to PCI-X bridge as split completions.

The example shown in Figure 9.7 illustrates a transaction completed with a PCI-X split completion. The PCI-X master issues a transaction. The RapidIO to PCI-X bridge determines that it must complete the transaction as a split transaction, and responds with a split response. The transaction is translated to RapidIO and a request is issued on the RapidIO port. The

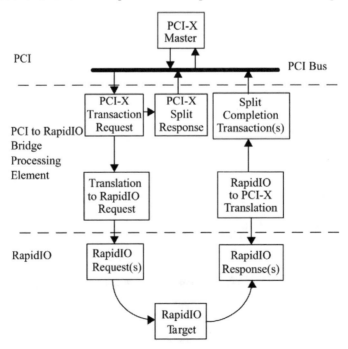

Figure 9.7   PCI-X mastered split response transaction

RapidIO target returns a response transaction which is translated to a PCI-X split completion transaction, completing the operation. PCI-X allows up to a 4 kbyte request. Larger PCI-X requests must be broken into multiple RapidIO operations. The RapidIO to PCI-X bridge may return the results back to the PCI-X master by multiple split completion transactions in a pipelined fashion. Since PCI-X allows devices to disconnect only on 128 byte boundaries it is advantageous to break the large PCI-X request into either 128 or 256 byte RapidIO operations.

9.4 RAPIDIO TO PCI TRANSACTION MAPPING

The RapidIO Logical Input/Output and Globally Shared Memory specifications include the necessary transactions types to map all PCI transactions. Table 9.1 lists the mapping of transactions between PCI 2.2 and RapidIO. A mapping mechanism such as the AMT function described earlier is necessary to assign the proper transaction type based on the address space for which the transaction is targeted.

PCI 2.2 memory transactions do not specify a size. It is possible for a PCI master to read a continuous stream of data from a target or to write a continuous stream of data to a target. Because RapidIO is defined to have a maximum data payload of 256 bytes, PCI transactions that are longer than 256 bytes must be broken into multiple RapidIO operations. Table 9.2 shows the transaction mapping between PCI-X and RapidIO.

Table 9.1   PCI 2.2 to RapidIO transaction mapping

PCI command	RapidIO transaction	Comment
Interrupt-acknowledge	NREAD	
Special-cycle	NWRITE	
I/O-read	NREAD	
I/O-write	NWRITE_R	
Memory-read, memory-read-line, memory-read-multiple	NREAD or IO_READ_HOME	The PCI memory read transactions can be represented by the NREAD operation. If the operation is targeted to hardware-maintained globally coherent memory address space then the I/O Read operation must be used
Memory-write, memory-write-and-invalidate	NWRITE, NWRITE_R, or FLUSH	The PCI Memory Write and Memory-Write-and-Invalidate can be represented by the NWRITE operation. If reliable delivery of an individual write transaction is desired then the NWRITE_R is used. If the operation is targeted to hardware-maintained globally coherent memory address space then the Data Cache Flush operation must be used
Configuration-read	NREAD	
Configuration-write	NWRITE_R	

Table 9.2   PCI-X to RapidIO transaction mapping

PCI-X command	RapidIO transaction	Comment
Interrupt-acknowledge	NREAD	
Special-cycle	NWRITE	
I/O-read	NREAD	
I/O-write	NWRITE_R	
Memory-read DWORD	NREAD or IO_READ_HOME	The PCI-X memory read DWORD transactions can be represented by the NREAD operation. If the operation is targeted to hardware-maintained coherent memory address space then the I/O Read operation must be used. This is indicated in PCI-X using the No Snoop (NS) bit described in Section 2.5 of the PCI-X 1.0 specification
Memory-write	NWRITE, NWRITE_R, or FLUSH	The PCI-X memory write and memory-write-and-invalidate can be represented by the NWRITE operation. If reliable delivery of an individual write transaction is desired then the NWRITE_R is used. If the operation is targeted to hardware-maintained coherent memory address space then the Data Cache Flush operation must be used. This is indicated in PCI-X using the No Snoop (NS) bit described in Section 2.5 of the PCI-X 1.0 specification
Configuration-read	NREAD	
Configuration-write	NWRITE_R	
Split completion		The split completion transaction is the result of a request on the PCI-X, bus that was terminated by the target with a split response. In the case of the RapidIO to PCI-X bridge this would be the artifact of a transaction that either the bridge mastered and received a split response or was the target and issued a split response. This command is equivalent to a RapidIO response transaction and does not traverse the bridge
Memory-read-block	NREAD or IO_READ_HOME	The PCI-X memory read transactions can be represented by the NREAD operation. If the operation is targeted to hardware-maintained globally coherent memory address space then the I/O Read operation must be used. This is indicated in PCI-X using the No Snoop (NS) bit described in Section 2.5 of the PCI-X 1.0 specification
Memory-write-block	NWRITE, NWRITE_R, or FLUSH	The PCI-X memory write and memory-write-and-invalidate can be represented by the NWRITE operation. If reliable delivery of an individual write transaction is desired then the NWRITE_R is used. If the operation is targeted to hardware-maintained globally coherent memory address space then the Data Cache Flush operation must be used. This is indicated in PCI-X using the No Snoop (NS) bit described in Section 2.5 of the PCI-X 1.0 specification

9.5 OPERATION ORDERING AND TRANSACTION DELIVERY

This section discusses what the RapidIO to PCI bridge must do to address the requirements of the ordering rules of the PCI specifications.

9.5.1 Operation Ordering

The RapidIO transaction ordering rules were presented in Chapter 3. The rules guarantee ordered delivery of write data and that results of read operations will contain any data that was previously written to the same location. For bridge devices, the PCI 2.2 specification has the additional requirement that the results of a read command force the completion of posted writes in both directions.

In order for the RapidIO to PCI bridge to be consistent with the PCI 2.2 ordering rules it is necessary to follow the RapidIO transaction ordering rules. In addition, the RapidIO to PCI bridge is required to adhere to the following additional rule:

• Read responses must push ahead all write requests and write responses.

9.5.2 Transaction Delivery Ordering

The RapidIO parallel physical layer specification and RapidIO serial physical layer specification describe the mechanisms by which transaction ordering and delivery occur through the system. When considering the requirements for the RapidIO to PCI bridge it is first necessary to follow the transaction delivery ordering rules in these specifications. Further, it is necessary to add additional constraints to maintain programming model compatibility with PCI.

As described above, PCI has an additional transaction ordering requirement over RapidIO. In order to guarantee interoperability, transaction ordering, and deadlock free operation, it is recommended that devices be restricted to utilizing transaction request flow level 0. In addition, it is recommended that response transactions follow a more strict priority assignment. Table 9.3 illustrates the priority assignment requirements for transactions in the PCI to RapidIO environment.

The PCI transaction ordering model requires that a RapidIO device not issue a read request into the system unless it has sufficient resources available to receive and process a higher-priority write or response packet in order to prevent deadlock. PCI 2.2 states that read responses cannot pass write transactions. The RapidIO specification provides PCI ordering by issuing priority 0 to read requests, and priority 1 to read responses and PCI writes. Since read responses and writes are issued at the same priority, the read responses will not pass writes.

Table 9.3   Packet priority assignments for PCI ordering

RapidIO packet type	Priority	Comment
Read request	0	This will push write requests and responses ahead
Write request	1	Forces writes to complete in order, but allows write requests to bypass read requests
Read response	1	Will force completion of preceding write requests and allows bypass of read requests
Write response	2	Will prevent NWRITE_R request-based deadlocks

9.5.3 PCI-X Relaxed Ordering Considerations

The PCI-X specification defines an additional ordering feature called relaxed ordering. If the PCI-X relaxed ordering attribute is set for a read transaction, the results for the read transaction are allowed to pass posted write transactions. PCI-X read transactions with this bit set allow the PCI-X to RapidIO bridge to ignore the rule described in Section 9.5.1. Table 9.4 shows the results of this additional function.

Table 9.4   Packet priority assignments for PCI-X ordering

RapidIO packet type	Priority	Comment
Read request	0	This will push write requests and responses ahead
Write request	1	Forces writes to complete in order, but allows write requests to bypass of read requests
Read response	1	When PCI-X relaxed ordering attribute is set to 0. Will force completion of preceding write requests and allows bypass of read requests
Read response	2, 3	When PCI-X relaxed ordering attribute is set to 1. The end point may promote the read response to higher priority to allow it to move ahead of posted writes
Write response	2	

9.6 INTERACTIONS WITH GLOBALLY SHARED MEMORY

Traditional systems have two notions of system or subsystem cache coherence. The first, non-coherent, means that memory accesses have no effect on the caches in the system. The memory controller reads and writes memory directly, and any cached data becomes incoherent in the system. This behavior requires that all cache coherence with I/O be managed by software mechanisms, as illustrated in Figure 9.8. In this example the processors and potentially cached contents of local memory are unaware of the request and response transactions to local memory. Software mechanisms must be used to signal changes to local memory so that the caches can be appropriately updated.

The second notion of system cache coherence is that of global coherence. In this scenario, an I/O access to memory will cause a snoop cycle to be issued on the processor bus, keeping all of the system caches coherent with the memory as illustrated in Figure 9.9. Owing to the snoop transaction running on the local interconnect the cache memories are given visibility to the change in the memory contents and may either update their caches with the proper contents or invalidate their copy of the data. Either approach results in correct coherent memory operations.

The example in Figure 9.9 works for systems with traditional bus structures. In RapidIO-based systems there is no common bus that can be used to issue the snoop transaction to. In this type of system global coherence requires special hardware support that goes beyond simply snooping the bus. This leads to a third notion of cache coherence, termed local coherence. For local coherence, a snoop transaction on a processor bus local to the targeted memory controller can be used to keep those caches coherent with that part of memory, but would

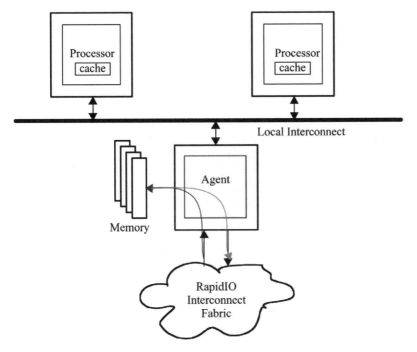

Figure 9.8   Traditional non-coherent I/O access example

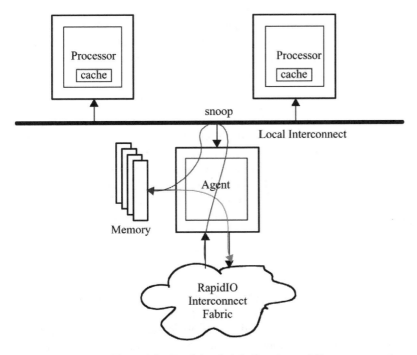

Figure 9.9   Traditional globally coherent I/O access example

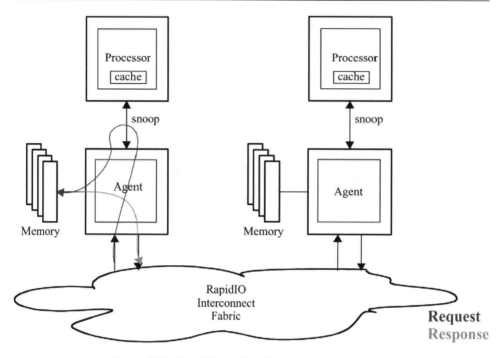

Figure 9.10 RapidIO locally coherent I/O access example

not affect caches associated with other memory controllers, as illustrated in Figure 9.10. What once was regarded in a system as a 'coherent access' is no longer globally coherent, but only locally coherent. Typically, deciding to snoop or not snoop the local processor caches is either determined by design or system architecture policy (always snoop or never snoop), or by an attribute associated with the physical address being accessed. In PCI-X, this attribute is the No Snoop (NS) bit described in Section 2.5 of the PCI-X 1.0 specification.

In order to preserve the concept of global cache coherence for this type of system, the *RapidIO Globally Shared Memory Logical Specification* defines several operations that allow a RapidIO to PCI bridge processing element to access data in the globally shared space without having to implement all of the cache coherence protocol. These operations are the I/O Read and Data Cache Flush operations. For PCI-X bridging, these operations can also be used as a way to encode the No Snoop attribute for locally as well as globally coherent transactions. The targeted memory controller can be designed to understand the required behavior of such a transaction. These encodings are also useful for tunneling PCI-X transactions between PCI-X bridge devices.

The data payload for an I/O Read operation is defined as the size of the coherence granule for the targeted globally shared memory domain. However, the Data Cache Flush operation allows coherence granule, sub-coherence granule, and sub-double-word writes to be performed.

The IO_READ_HOME transaction is used to indicate to the GSM memory controller that the memory access is globally coherent, so the memory controller finds the latest copy of the requested data within the coherence domain (the requesting RapidIO to PCI bridge processing element is, by definition, not in the coherence domain) without changing the state of the participant caches. Therefore, the I/O Read operation allows the RapidIO to PCI bridge to cleanly extract data from a coherent portion of the system with minimal disruption and without having to be a full participant in the coherence domain.

The Data Cache Flush operation has several uses in a coherent part of a system. One such use is to allow a RapidIO to PCI bridge processing element to write to globally shared portions of the system memory. Analogous to the IO_READ_HOME transaction, the FLUSH transaction is used to indicate to the GSM memory controller that the access is globally coherent. The memory controller forces all of the caches in the coherence domain to invalidate the coherence granule if they have a shared copy (or return the data to memory if one had ownership of the data), and then writes memory with the data supplied with the FLUSH request. This behavior allows the I/O device to cleanly write data to the globally shared address space without having to be a full participant in the coherence domain.

Since the RapidIO to PCI bridge processing element is not part of the coherence domain, it is never the target of a coherent operation.

9.7 BYTE LANE AND BYTE ENABLE USAGE

PCI makes use of byte enables and allows combining and merging of transactions. This may result in write transactions with sparse valid bytes. In order to reduce transaction overhead, RapidIO does not support byte enables. RapidIO does, however, support a set of byte encodings that are described in Chapter 4. PCI to RapidIO operations may be issued with sparse bytes. Should a PCI write transaction arrive with byte enables that do not match a specified RapidIO byte encoding, the RapidIO operation that is generated will be broken into multiple separate valid RapidIO operations.

9.8 ERROR MANAGEMENT

Errors that are detected on a PCI bus are delivered to the host by side band signals. The treatment of these signals is left to the system designer and is outside the PCI specifications. Likewise, the delivery of PCI error messages through the RapidIO interconnect is not defined, although they could be implemented as user-defined RapidIO messages.

10

RapidIO Bringup and Initialization Programming

The RapidIO interconnect architecture is promoted as being transparent to software, so a chapter on RapidIO programming would appear to be unnecessary. However, in any embedded system, software is a critical component. RapidIO programming falls into several categories. This book will examine the following categories of RapidIO programming.

- System initialization
- Device enumeration
- Route table configuration
- Memory mapping

In embedded systems there is a much higher degree of interaction between the software and the hardware. In typical computing environments such as a Windows-based desktop computer there is relatively little interaction between the programs running on the main processor and the peripheral devices that, together with the processor, complete the system. In embedded environments there is typically much more interaction between the software and the hardware.

RapidIO systems, once configured, will pass memory and I/O transactions transparently across the RapidIO interconnect. From the point of view of software not specifically interacting with RapidIO, there is no RapidIO hardware. It is invisible to the software. For example, when a processor issues a load instruction to an address space that is configured to reside in or be associated with a device that exists across a RapidIO network, that instruction initiates the generation of a RapidIO READ transaction. The associated RapidIO packet is properly formed and sent across the RapidIO network. The response packet returns the data and the data is forwarded back to the processor to complete the load operation. This process is handled completely in hardware.

This model is very different from that offered for Ethernet; another common interconnect technology. In Ethernet, a processor load or store operation might read and write data to a buffer or it might access an Ethernet controller's configuration or status registers, but it will not automatically generate an Ethernet packet addressed to another device in the system

RapidIO® The Embedded System Interconnect.   S. Fuller
© 2005 John Wiley & Sons, Ltd   ISBN: 0-470-09291-2

configured to provide a reliable request for data from another device that will be automatically forwarded back to the requesting processor. In Ethernet, to accomplish this task, there needs to be cooperating software processes executing on both the sender and the receivers. These software processes communicate with each other through a stack of software-based drivers. For a technology such as RDMA over Ethernet, this operation typically consumes thousands of processor operations on both the sending and receiving device, with transaction latencies measured in milliseconds and tremendous packet inefficiency on the Ethernet channel itself.

The RapidIO operation is handled completely in hardware and is invisible to the software. The software, except through measuring the latency of the load operation, is unable to distinguish loads of data from cache memory, local memory or remote memory.

RapidIO is invisible to software only when it is serving as a memory-mapped I/O bus. This is only one of several possible operating modes. A second important mode of operation is when the message-passing logical layer transactions described in Chapter 5 are being used. Software must be written to explicitly make use of these transactions. They will not be automatically generated by hardware. Software may also play a role in supporting the streaming fabric logical layer transactions. These transactions would be used to support the transport of data streams across RapidIO. Example data streams might include: Ethernet packets, ATM cells, or even technologies such as MPEG-2 digital video streams. The streaming fabric extensions offer mechanisms for data to be segmented and transported with quality of service guarantees across the RapidIO fabric.

Another aspect of RapidIO that is not transparent to software is the configuration of RapidIO itself. What is meant by this is the interaction between the RapidIO hardware and the system software that is needed to initialize the RapidIO end points to ensure that the links are running and properly sending and receiving packets, so that any switches in the system are properly configured and that memory maps are set up to provide the desired system visibility between devices. The rest of this chapter will present the requirements and specified mechanisms for software support of RapidIO in embedded systems.

10.1 OVERVIEW OF THE SYSTEM BRINGUP PROCESS

This section presents a high-level overview of the system bringup process. There are several components to this process. We will look at this process from a high level first and then dive into the details of each component of the process.

1. The host processor fetches the initial boot code (if necessary). If two processors are present, both can fetch the initial boot code.
2. The system exploration and enumeration algorithm is started.
3. All devices have been enumerated and stored in the device database, and routes have been set up between the host device and all end point devices. The enumeration process may optionally choose to do the following:
    (a) Compute and configure optimal routes between the host device and end point devices, and between different end point devices.
    (b) Configure the switch devices with the optimal route information.
    (c) Store the optimal route and alternate route information in the device database.
4. The address space is mapped.

10.1.1 Boot Requirements

The following system state is required for proper system bringup.

After the system is powered on, the state necessary for system enumeration to occur by multiple host processors is automatically initialized as follows:

System devices are initialized with the following base device IDs:

- Non-boot-code and non-host device IDs are set to 0xFF (0xFFFF for 16-bit deviceID systems)
- Boot code device IDs are set to 0xFE (0x00FE for 16-bit deviceID systems)
- Host device IDs are set to 0x00 (0x0000 for 16-bit deviceID systems)

After the system is initialized, the device IDs will all be set to unique values. Before system initialization, the IDs are preconfigured to one of the three values just described.

The RapidIO physical layer links will initialize themselves. This process was described in Chapters 7 and 8 for the serial and physical layers respectively.

The default routing state of all switches between the boot code device and the host devices is set to route all requests for device ID 0xFE (0x00FE for 16-bit device ID systems) to the appropriate boot code device. All response packets are routed back to the requesting host from the boot code device. This will allow the host device to retrieve boot code from the boot device, which will most likely be a flash ROM memory device. RapidIO systems should have only one boot code device and all switches in the system should be configured to forward requests addressed to device ID 0xFE towards the boot code device.

10.1.2 Enumeration

Enumeration is the process of finding all the RapidIO connected devices in a system and assigning them unique device IDs. Although RapidIO networks are quite flexible and provide many features and capabilities, there are a few assumptions and restrictions that can significantly simplify the enumeration process. These assumptions and restrictions are as follows. Only two hosts may simultaneously enumerate a network. Two hosts may be required on a network for fault tolerance purposes. System integrators must determine which hosts should perform this function. Only one host actually completes the network enumeration (this is referred to as the winning host). The second host must retreat and wait for the enumeration to complete or, assuming the winning host has failed, for enumeration to time out. If a time-out occurs, the second host re-enumerates the network.

The enumeration algorithm sets priority on the basis of the value of the power-on device ID. The winning host is the device with the higher power-on host device ID. The losing host has the lower power-on host device ID. The losing host enters a wait state until the winning host completes enumeration or until the wait state times out.

The prioritization mechanism never results in a deadlock if the priorities of both host processors are unique. The enumeration process is initially performed in parallel by both hosts until they meet at a device. When a meeting occurs, prioritization guarantees one winning host—the other host retreats (enters a wait state).

After enumeration, other hosts in the system can passively discover the network to gather topology information such as routing tables and memory maps that may be useful to them.

Any host that participates in the discovery process must change its own device ID to a unique ID value before starting the system initialization process. This value is used by a device's host base device ID lock CSR to ensure only one host can manipulate a device at a time. The allowed ID values for a discovering host are 0x00 (0x0000) and 0x01 (0x0001). A host with an ID of 0x00 (0x0000) has a lower priority than a host with an ID of 0x01 (0x0001).

All host devices, participating in enumeration, should have their master enable bit (port general control CSR) set to 1. Switch devices do not have a master enable bit.

Enumeration is complete when the winning host releases the lock on the losing host. It is the losing host's responsibility to detect that it has been locked by the winning host and to later detect that the lock has been released by the winning host.

As mentioned previously, two hosts can be used to enumerate the RapidIO network. The host with the higher power-on host device ID is generally regarded to have priority over the other host. Because of this pre-defined priority, only one host (the one with higher priority) can win the enumeration task. In this case, the losing host enters a wait state.

If the winning host fails to enumerate the entire network, the losing host's wait state times out. When this occurs, the losing host attempts to enumerate the network. In an open 8-bit device ID system, the losing host must wait 15 Seconds before timing out and restarting the enumeration task. The length of the time-out period in a closed or a 16-bit deviceID system may differ from that of an open system.

10.1.3 Address Space Mapping

Address space mapping would typically occur after device enumeration. Device enumeration builds a database of devices and connectivity in the system. After enumeration it is possible to proceed with address space mapping in the system.

RapidIO devices may offer visibility to zero or more addressable memory regions to other RapidIO devices in a system. Instead of creating a single global address space that all devices must share and map their own regions into, the RapidIO model expects that all devices may have their own independent address regions. Devices can offer a single address space with up to 64 bits of double-word addressability. To make packets more efficient, the address used for any given device may be 34, 50 or 66 bits in length. The target device determines the choice of address length. The address field lengths supported by a target device can be determined by querying the processing element features CAR. The actual length used is established by programming the processing element logical layer control CSR. The CAR tells the system what memory sizes the endpoint can support, the CSR is used to set the size that will be used. A full description of these registers can be found in Chapter 18. Once configured, all memory-addressed transactions to a given end point must use the chosen memory size.

Rather than interact directly with these registers, a set of device access routines (DAR) have been developed. For example the rioDarGetMemorySize() routine may be used to determine the visible memory regions associated with any given RapidIO end point. These routines are described in detail in Section 10.2.4.

The address range that is made visible by a RapidIO end point, in conjunction with the device ID of the end point comprise a unique address in the system. This unique address can be mapped through address translation logic to the local address space of other devices in the system. This means that for any given physical address, composed of a device address and device ID combination, there may be many virtual representations of this address mapped to other devices in the system. Transactions to these apparently local addresses, would be transported across the RapidIO fabric to the target devices physical address.

The RapidIO architecture also anticipates and supports the mapping of PCI-based address ranges into the RapidIO network. PCI devices enter a RapidIO network through a RapidIO to PCI bridge. The address space mapping across this bridge must be done when devices are enumerated and stored in the device database. This allows the address of a found device to be retrieved later and presented to the device access routines during the operating system (OS) initialization. The pseudocode for this process is as follows:

```
FOR every device in the database
    IF the component is a RapidIO device
        ACQUIRE the device's address-space requirement
        MAP the address space into a new host address partition
        EXPAND the partition window to cover the new partition
        UPDATE the device database with the new host address
    ELSE IF the component is a PCI bridge
        ACQUIRE the bridge's PCI bus ID
        ACQUIRE the bridge's address-space requirement
            // All devices that appear behind this PCI bridge must have their address spaces
                mapped within the region specified for this bridge.
        MAP the address space into a new host address partition
        EXPAND the partition window to cover the new partition
        UPDATE the device database with the new host address
    ENDIF
ENDFOR
```

After discovery has been concluded, it is expected that the processing devices in the system will then attempt to load in a software image from a boot device.

10.2 SYSTEM APPLICATION PROGRAMMING INTERFACES

In order to provide a standardized programming interface to RapidIO hardware, the RapidIO Technical Working Group has defined sets of standard applications programming interfaces (API), that should be used by configuration and operating system software to interact with the RapidIO hardware. By working through an API rather than directly interacting with the RapidIO hardware and registers, insulation against low level changes is provided. The most basic level of API provided are the hardware abstraction layer (HAL) functions. The higher-level functions should rely on the support functions provided by the HAL functions to ensure portability between different platforms.

Device-specific (vendor-specific) functions are functions defined by a device manufac-turer that use hardware resources outside of the scope of the RapidIO specifications. The main purpose of these functions is to provide HAL support to the standard RapidIO functions. An important goal of this software API specification is to minimize or eliminate the use of device-specific functions required for enumeration so that the portability of the API across hardware platforms is maximized.

Virtually all of the APIs, when referencing the RapidIO system, will contain at least three parameters. These parameters are: localport, destid and hopcount. Localport refers to the local RapidIO port that the transaction is either targeted at our should be sent out of. Devices can have more that one RapidIO port on them. Localport is used to identify which port is being referred to. Destid is the 8- or 16-bit deviceID of the device that the transaction is targeted at. Although it describes an 8- or 16-bit quantity the destid field is a 32-bit field. The hopcount field is used when the transaction is targeted at a switch. Maintenance transactions contain a hopcount field which is decremented as it is passed through every switch. If the hopcount field reaches zero, it indicates that the transaction is meant for the current switch device. When a transaction is not targeted at a switch the hopcount field should be set to 0xFF.

10.2.1 Basic Hardware Abstraction Layer (HAL) Functions

The hardware abstraction layer (HAL) provides a standard software interface to the device-specific hardware resources needed to support RapidIO system configuration transactions. At the most basic level, configuration read and write operations are used by the HAL functions to access RapidIO device registers. The HAL functions are accessed by the RapidIO enumeration API during system bringup and may be used for ongoing system maintenance.

This section describes the HAL functions and how they can be used to access local and remote RapidIO device registers. These functions must be implemented by every new device-specific host-processing element to support RapidIO system enumeration and initialization. The HAL functions assume the following:

- All configuration read and write operations support only single-word (4-byte) accesses.
- As required by the device, the size of the 8-bit or 16-bit deviceID field is determined by the device implementation.
- An enumerating processor device may have more than one RapidIO end point (local port).

One purpose of the HAL is to provide a unified software interface to configuration registers in both local and remote RapidIO processing elements. This is done by a universal device-addressing scheme. Such a scheme enables HAL functions to distinguish between accesses to local and remote RapidIO end points without requiring an additional parameter. The result is that only one set of HAL functions must be implemented to support local and remote configuration operations.

All HAL functions use the destid and hopcount parameters to address a RapidIO device. The HAL reserves destid=0xFFFFFFFF and hopcount of 0 for addressing configuration registers within the local RapidIO end point. A destid=0xFFFFFFFF and hopcount of 0 value *must* be used to address the local processing end point regardless of the actual destination ID value. This reserved combination does not conflict with the address of other RapidIO devices. The localport parameter is used by the HAL functions to identify a specific local port within RapidIO devices that contain multiple ports.

The most basic functions that form the RapidIO initialization HAL are the rioGetNumLocalPorts, rioConfiguratioRead and rioConfigurationWrite functions described below. These functions are simple APIs that are meant to simplify the task of interaction with the RapidIO CAR and CSR registers. The CAR registers contain the capabilities and attributes of a RapidIO end point or switch and the CSR registers can be used to configure the capabilities of a RapidIO endpoint or switch. Either the device vendor or the RTOS vendor, depending on the product, would provide the code supporting these APIs.

rioGetNumLocalPorts

Prototype:
    INT32 rioGetNumLocalPorts (
      Void
)

Arguments:
    None

Return value:
    0   Error
    n   Number of RapidIO ports supported

Synopsis:
rioGetNumLocalPorts() returns the total number of local RapidIO ports supported by the HAL functions. The number *n* returned by this function should be equal to or greater than 1. A returned value of 0 indicates an error.

rioConfigurationRead

Prototype:
    STATUS rioConfigurationRead (
      UINT8   localport,
      UINT32   destid,
      UINT8   hopcount,
      UINT32   offset,
      UINT32   *readdata
)

Arguments:
    localport   Local port number [IN]
    destid   Destination ID of the target device [IN]
    hopcount   Hop count [IN]
    offset   Word-aligned (four byte boundary) offset—in bytes—of the CAR or CSR [IN]
    *readdata   Pointer to storage for received data [OUT]

Return value:

RIO_SUCCESS	The read operation completed successfully and valid data was placed into the specified location.
RIO_ERR_INVALID_PARAMETER	One or more input parameters had an invalid value.
RIO_ERR_RIO	The RapidIO fabric returned a Response Packet with ERROR status reported. Error status returned by this function may contain additional information from the Response Packet.
RIO_ERR_ACCESS	A device-specific hardware interface was unable to generate a maintenance transaction and reported an error.

Synopsis:

rioConfigurationRead( ) performs a configuration read transaction from CAR and/or CSR register(s) belonging to a local or remote RapidIO device. The function uses a device-specific hardware interface to generate maintenance transactions to remote devices. This hardware sends a configuration read request to the remote device (specified by destid and/or hopcount) and waits for a corresponding configuration read response. After the function receives a configuration read response it returns data and/or status to the caller. The method for accessing registers in a local device is device-specific.

A destid value of HOST_REGS and hopcount of 0 results in accesses to the local hosts RapidIO registers.

rioConfigurationWrite

Prototype:

```
STATUS rioConfigurationWrite (
    UINT8      localport,
    UINT32     destid,
    UINT8      hopcount,
    UINT32     offset,
    UINT32     *writedata
    )
```

Arguments:

localport	Local port number [IN]
destid	Destination ID of the target device [IN]
hopcount	Hop count [IN]
offset	Word-aligned (four-byte boundary) offset, in bytes, of the CAR or CSR [IN]
*writedata	Pointer to storage for data to be written [IN]

Return value:

RIO_SUCCESS	The write operation completed successfully.
RIO_ERR_INVALID_PARAMETER	One or more input parameters had an invalid value.
RIO_ERR_RIO	The RapidIO fabric returned a Response Packet with ERROR status reported. Error status returned by this function may contain additional information from the Response Packet.
RIO_ERR_ACCESS	A device-specific hardware interface was unable to generate a maintenance transaction and reported an error.

Synopsis:

rioConfigurationWrite( ) performs a configuration write transaction to CAR and/or CSR register(s) belonging to a local or remote RapidIO device. The function uses a device-specific hardware interface to generate maintenance transactions to remote devices. This hardware sends a configuration write request to the remote device (specified by destid and/or hopcount) and waits for a corresponding configuration write response. After the function receives a configuration write response it returns status to the caller. The method for accessing registers in a local device is device-specific.

A destid value of HOST_REGS and hopcount of 0 results in accesses to the local hosts RapidIO registers.

10.2.2 RapidIO Bringup HAL Functions

With these three basic HAL functions, we can create a set of more complex functions that can be used to bring up a system. This section describes the RapidIO functions that should be implemented to support system bringup. Functions are defined only for device registers used during the RapidIO enumeration and initialization process, not for all possible RapidIO device registers. Many of the functions may also be implemented as macros that specify predefined parameters for the HAL functions. The standard RapidIO bringup functions can be combined into a library if they are implemented as a set of subroutines.

rioInitLib

Prototype:
  STATUS rioInitLib (
  Void
)

Arguments:
  None

Return value:
  RIO_SUCCESS   Initialization completed successfully.
  RIO_ERROR   Generic error report. Unable to initialize library.

Synopsis:
rioInitLib( ) initializes the RapidIO API library. No routines defined in this chapter may be called unless and untile rioInitLib has been invoked. If rioInitLib returns RIO_ERROR, no routines defined in this chapter may be called.

rioGetFeatures

Prototype:
  STATUS rioGetFeatures (
    UINT8   localport,
    UINT32 destid,
    UINT8   hopcount,
    UINT32 *features
)

Arguments:
  localport   Local port number [IN]
  destid   Destination ID of the processing element [IN]
  hopcount Hop count [IN]
  *features   Pointer to storage containing the received features [OUT]

Return value:

RIO_SUCCESS	The features were retrieved successfully and placed into the location specified by *features.
RIO_ERR_INVALID_PARAMETER	One or more input parameters had an invalid value.
RIO_ERR_RIO	The RapidIO fabric returned a Response Packet with ERROR status reported. Error status returned by this function may contain additional information from the Response Packet.
RIO_ERR_ACCESS	A device-specific hardware interface was unable to generate a maintenance transaction and reported an error.

Synopsis:

rioGetFeatures( ) uses the HAL *rioConfigurationRead*( ) function to read from the processing element features CAR of the specified processing element. Values read are placed into the location referenced by the *features pointer. Reported status is similar to *rioConfigurationRead*( ).

A destid value of HOST_REGS and hopcount of 0 results in accesses to the local hosts RapidIO registers.

rioGetSwitchPortInfo

Prototype:

```
STATUS rioGetSwitchPortInfo (
    UINT8     localport,
    UINT32    destid,
    UINT8     hopcount,
    UINT32    *portinfo
    )
```

Arguments:

localport	Local port number [IN]
destid	Destination ID of the processing element [IN]
hopcount	Hop count [IN]
*portinfo	Pointer to storage containing the received port information [OUT]

Return value:

RIO_SUCCESS	The port information was retrieved successfully and placed into the location specified by *portinfo.
RIO_ERR_INVALID_PARAMETER	One or more input parameters had an invalid value.
RIO_ERR_RIO	The RapidIO fabric returned a Response Packet with ERROR status reported. Error status returned by this function may contain additional information from the Response Packet.
RIO_ERR_ACCESS	A device-specific hardware interface was unable to generate a maintenance transaction and reported an error.

Synopsis:

rioGetSwitchPortInfo( ) uses the HAL *rioConfigurationRead( )* function to read from the Switch Port Information CAR of the specified processing element. Values read are placed into the location referenced by the *portinfo pointer. Reported status is similar to *rioConfigurationRead( )*.

A destid value of HOST_REGS and hopcount of 0 results in accesses to the local hosts RapidIO registers.

rioGetExtFeaturesPtr

Prototype:
```
STATUS rioGetExtFeaturesPtr (
    UINT8     localport,
    UINT32    destid,
    UINT8     hopcount,
    UINT32    *extfptr
    )
```

Arguments:

localport	Local port number [IN]
destid	Destination ID of the processing element [IN]
hopcount	Hop count [IN]
*extfptr	Pointer to storage containing the received extended feature information [OUT]

Return value:

RIO_SUCCESS	The extended feature information was retrieved successfully and placed into the location specified by *extfptr.
RIO_ERR_INVALID_PARAMETER	One or more input parameters had an invalid value.
RIO_ERR_RIO	The RapidIO fabric returned a Response Packet with ERROR status reported. Error status returned by this function may contain additional information from the Response Packet.
RIO_ERR_ACCESS	A device-specific hardware interface was unable to generate a maintenance transaction and reported an error.

Synopsis:

rioGetExtFeaturesPtr( ) uses the HAL *rioConfigurationRead( )* function to read the pointer to the first entry in the extended features list from the assembly information CAR of the specified processing element. That pointer is placed into the location referenced by the *extfptr pointer. Reported status is similar to *rioConfigurationRead( )*.

A destid value of HOST_REGS and hopcount of 0 results in accesses to the local hosts RapidIO registers. Note that if the EF_PTR field of *extfptr is 0, no extended features are available.

rioGetNextExtFeaturesPtr

Prototype:
   STATUS rioGetNextExtFeaturesPtr (
      UINT8   localport,
      UINT32   destid,
      UINT8   hopcount,
      UINT32   currfptr,
      UINT32   *extfptr
)

Arguments:
localport	Local port number [IN]
destid	Destination ID of the processing element [IN]
hopcount	Hop count [IN]
currfptr	Pointer to the last reported extended feature [IN]
*extfptr	Pointer to storage containing the received extended feature information [OUT]

Return value:
RIO_SUCCESS	The extended feature information was retrieved successfully and placed into the location specified by *extfptr.
RIO_ERR_INVALID_PARAMETER	One or more input parameters had an invalid value.
RIO_ERR_RIO	The RapidIO fabric returned a Response Packet with ERROR status reported. Error status returned by this function may contain additional information from the Response Packet.
RIO_ERR_ACCESS	A device-specific hardware interface was unable to generate a maintenance transaction and reported an error.

Synopsis:
rioGetNextExtFeaturesPtr( ) uses the HAL *rioConfigurationRead( )* function to read the pointer to the next entry in the extended features. That pointer is placed into the location referenced by the *extfptr pointer. Reported status is similar to *rioConfigurationRead( )*.

    A destid value of HOST_REGS and hopcount of 0 results in accesses to the local hosts RapidIO registers. Note that if the EF_PTR field of *extfptr is 0, no further extended features are available. Invoking rioGetNextExtFeaturesPtr when currfptr has an EF_PTR field value of 0 will result in a return code of RIO_ERR_INVALID_PARAMETER.

rioGetSourceOps

Prototype:
   STATUS rioGetSourceOps (
      UINT8   localport,
      UINT32   destid,
      UINT8   hopcount,
      UINT32   *srcops
)

Arguments:
    localport   Local port number [IN]
    destid   Destination ID of the processing element [IN]
    hopcount Hop count [IN]
    *srcops   Pointer to storage containing the received source operation information [OUT]

Return value:

RIO_SUCCESS	The source operation information was retrieved successfully and placed into the location specified by *srcops.
RIO_ERR_INVALID_PARAMETER	One or more input parameters had an invalid value.
RIO_ERR_RIO	The RapidIO fabric returned a Response Packet with ERROR status reported. Error status returned by this function may contain additional information from the Response Packet.
RIO_ERR_ACCESS	A device-specific hardware interface was unable to generate a maintenance transaction and reported an error.

Synopsis:
rioGetSourceOps( ) uses the HAL *rioConfigurationRead*( ) function to read from the source operations CAR of the specified processing element. Values read are placed into the location referenced by the *srcops pointer. Reported status is similar to *rioConfigurationRead*( ).

    A destid value of HOST_REGS and hopcount of 0 results in accesses to the local hosts RapidIO registers.

rioGetDestOps

Prototype:
```
STATUS rioGetDestOps (
    UINT8    localport,
    UINT32   destid,
    UINT8    hopcount,
    UINT32   *dstops
    )
```

Arguments:
    localport   Local port number [IN]
    destid   Destination ID of the processing element [IN]
    hopcount Hop count [IN]
    *dstops   Pointer to storage containing the received destination operation information [OUT]

Return value:

RIO_SUCCESS	The destination operation information was retrieved successfully and placed into the location specified by *dstops.
RIO_ERR_INVALID_PARAMETER	One or more input parameters had an invalid value.

RIO_ERR_RIO	The RapidIO fabric returned a Response Packet with ERROR status reported.
	Error status returned by this function may contain additional information from the Response Packet.
RIO_ERR_ACCESS	A device-specific hardware interface was unable to generate a maintenance transaction and reported an error.

Synopsis:

rioGetDestOps( ) uses the HAL *rioConfigurationRead*( ) function to read from the destination operations CAR of the specified processing element. Values read are placed into the location referenced by the *dstops pointer. Reported status is similar to *rioConfigurationRead*( ).

A destid value of HOST_REGS and hopcount of 0 results in accesses to the local hosts RapidIO registers.

rioGetAddressMode

Prototype:

    STATUS rioGetAddressMode (
        UINT8   localport,
        UINT32   destid,
        UINT8   hopcount,
        ADDR_MODE *amode
)

Arguments:

localport	Local port number [IN]
destid	Destination ID of the processing element [IN]
hopcount	Hop count [IN]
*amode	Pointer to storage containing the received address mode (34-bit, 50-bit, or 66-bit address) information [OUT]

Return value:

RIO_SUCCESS	The address mode information was retrieved successfully and placed into the location specified by *amode.
RIO_ERR_INVALID_PARAMETER	One or more input parameters had an invalid value.
RIO_ERR_RIO	The RapidIO fabric returned a Response Packet with ERROR status reported.
	Error status returned by this function may contain additional information from the Response Packet.
RIO_ERR_ACCESS	A device-specific hardware interface was unable to generate a maintenance transaction and reported an error.

Synopsis:

rioGetAddressMode( ) uses the HAL *rioConfigurationRead*( ) function to read from the PE logical layer CSR of the specified processing element. The number of address bits generated by the PE (as the source of an operation) and processed by the PE (as the target of an operation) are placed into the location referenced by the *amode pointer. Reported status is similar to *rioConfigurationRead()*.

A destid value of HOST_REGS and hopcount of 0 results in accesses to the local hosts RapidIO registers.

rioGetBaseDeviceId

Prototype:
```
STATUS rioGetBaseDeviceId (
    UINT8      localport,
    UINT32     *deviceid
    )
```

Arguments:
    localport   Local port number [IN]
    *deviceid   Pointer to storage containing the base device ID [OUT]

Return value:

RIO_SUCCESS	The base device ID information was retrieved successfully and placed into the location specified by *deviceid.
RIO_ERR_INVALID_PARAMETER	One or more input parameters had an invalid value.
RIO_ERR_RIO	The RapidIO fabric returned a Response Packet with ERROR status reported. Error status returned by this function may contain additional information from the Response Packet.
RIO_ERR_ACCESS	A device-specific hardware interface was unable to generate a maintenance transaction and reported an error.

Synopsis:
rioGetBaseDeviceId() uses the HAL *rioConfigurationRead()* function to read from the base device ID CSR of the local processing element (the destid and hopcount parameters used by *rioConfigurationRead()* must be set to HOST_REGS and zero, respectively). Values read are placed into the location referenced by the *deviceid pointer. Reported status is similar to *rioConfigurationRead()*. This function is useful only for local end point devices.

rioSetBaseDeviceId

Prototype:
```
STATUS rioSetBaseDeviceId (
    UINT8      localport,
    UINT32     destid,
    UINT8      hopcount,
    UINT32     newdeviceid
    )
```

Arguments:
   localport   Local port number [IN]
   destid   Destination ID of the processing element [IN]
   hopcount   Hop count [IN]
   newdeviceid   New base device ID to be set [IN]

Return value:
   RIO_SUCCESS   The base device ID was updated successfully.
   RIO_ERR_INVALID_PARAMETER   One or more input parameters had an invalid value.
   RIO_ERR_RIO   The RapidIO fabric returned a Response Packet
                                 with ERROR status reported.
                                 Error status returned by this function may contain
                                 additional information from the Response Packet.
   RIO_ERR_ACCESS   A device-specific hardware interface was unable
                                 to generate a maintenance transaction and reported
                                 an error.

Synopsis:

rioSetBaseDeviceId( ) uses the HAL *rioConfigurationWrite*( ) function to write the base
device ID in the base device ID CSR of the specified processing element (end point devices
only). Reported status is similar to *rioConfigurationWrite*( ).

   A destid value of HOST_REGS and hopcount of 0 results in accesses to the local hosts
RapidIO registers.

rioAcquireDeviceLock

Prototype:
   STATUS rioAcquireDeviceLock (
      UINT8   localport,
      UINT32   destid,
      UINT8   hopcount,
      UINT16   hostdeviceid,
      UINT16   *hostlockid
)

Arguments:
   localport   Local port number [IN]
   destid   Destination ID of the processing element [IN]
   hopcount   Hop count [IN]
   hostdeviceid   Host base device ID for the local processing element [IN]
   *hostlockid   Device ID of the host holding the lock if ERR_LOCK is returned [OUT]

Return value:
   RIO_SUCCESS   The device lock was acquired successfully.
   RIO_ERR_LOCK   Another host already acquired the specified
                                 processor element. ID of the device holding the
                                 lock is contained in the location referenced by the
                                 *hostlockid parameter.

RIO_ERR_INVALID_PARAMETER	One or more input parameters had an invalid value
RIO_ERR_RIO	The RapidIO fabric returned a Response Packet with ERROR status reported. Error status returned by this function may contain additional information from the Response Packet.
RIO_ERR_ACCESS	A device-specific hardware interface was unable to generate a maintenance transaction and reported an error.

Synopsis:

rioAcquireDeviceLock( ) tries to acquire the hardware device lock for the specified processing element on behalf of the requesting host. The function uses the HAL *rioConfigurationWrite*( ) function to write the requesting host device ID into the host base lock device ID CSR of the specified processing element. After the write completes, this function uses the HAL *rioConfigurationRead*( ) function to read the value back from the host base lock device ID CSR. The written and read values are compared. If they are equal, the lock was acquired successfully. Otherwise, another host acquired this lock and the device ID for that host is reported.

This function assumes unique host-based device identifiers are assigned to discovering hosts. For more details, refer to Section 10.3 System Bringup Example. A destid value of HOST_REGS and hopcount of 0 results in accesses to the local hosts RapidIO registers.

rioReleaseDeviceLock

Prototype:
```
STATUS rioReleaseDeviceLock (
    UINT8    localport,
    UINT32   destid,
    UINT8    hopcount,
    UINT16   hostdeviceid,
    UINT16   *hostlockid
)
```

Arguments:

localport	Local port number [IN]
destid	Destination ID of the processing element [IN]
hopcount	Hop count [IN]
hostdeviceid	Host base device ID for the local processing element [IN]
*hostlockid	Device ID of the host holding the lock if ERR_LOCK is returned [OUT]

Return value:

RIO_SUCCESS	The device lock was released successfully.
RIO_ERR_LOCK	Another host already acquired the specified processor element.
RIO_ERR_INVALID_PARAMETER	One or more input parameters had an invalid value

RIO_ERR_RIO	The RapidIO fabric returned a Response Packet with ERROR status reported.
	Error status returned by this function may contain additional information from the Response Packet.
RIO_ERR_ACCESS	A device-specific hardware interface was unable to generate a maintenance transaction and reported an error.

Synopsis:

rioReleaseDeviceLock( ) tries to release the hardware device lock for the specified processing element on behalf of the requesting host. The function uses the HAL *rioConfigurationWrite*( ) function to write the requesting host device ID into the host base lock device ID CSR of the specified processing element. After the write completes, this function uses the HAL *rioConfigurationRead*( ) function to read the value back from the host base lock device ID CSR. The written and read values are compared. If they are equal, the lock was acquired successfully. Otherwise, another host acquired this lock and the device ID for that host is reported.

A destid value of HOST_REGS and hopcount of 0 results in accesses to the local hosts RapidIO registers.

rioGetComponentTag

Prototype:

```
STATUS rioGetComponentTag (
    UINT8     localport,
    UINT32    destid,
    UINT8     hopcount,
    UINT32    *componenttag
    )
```

Arguments:

localport	Local port number [IN]
destid	Destination ID of the processing element [IN]
hopcount	Hop count [IN]
*componenttag	Pointer to storage containing the received component tag information [OUT]

Return value:

RIO_SUCCESS	The component tag information was retrieved successfully and placed into the location specified by *componenttag.
RIO_ERR_INVALID_PARAMETER	One or more input parameters had an invalid value.
RIO_ERR_RIO	The RapidIO fabric returned a Response Packet with ERROR status reported.
	Error status returned by this function may contain additional information from the Response Packet.
RIO_ERR_ACCESS	A device-specific hardware interface was unable to generate a maintenance transaction and reported an error.

Synopsis:

rioGetComponentTag( ) uses the HAL *rioConfigurationRead*( ) function to read from the component tag CSR of the specified processing element. Values read are placed into the location referenced by the *componenttag pointer. Reported status is similar to *rioConfigurationRead*( ).

A destid value of HOST_REGS and hopcount of 0 results in accesses to the local hosts RapidIO registers.

rioSetComponentTag

Prototype:

```
STATUS rioSetComponentTag (
    UINT8     localport,
    UINT32    destid,
    UINT8     hopcount,
    UINT32    componenttag
)
```

Arguments:

localport	Local port number [IN]
destid	Destination ID of the processing element [IN]
hopcount	Hop count [IN]
componenttag	Component tag value to be set [IN]

Return value:

RIO_SUCCESS	The component tag was updated successfully.
RIO_ERR_INVALID_PARAMETER	One or more input parameters had an invalid value.
RIO_ERR_RIO	The RapidIO fabric returned a Response Packet with ERROR status reported. Error status returned by this function may contain additional information from the Response Packet.
RIO_ERR_ACCESS	A device-specific hardware interface was unable to generate a maintenance transaction and reported an error.

Synopsis:

rioSetComponentTag( ) uses the HAL *rioConfigurationWrite*( ) function to write the component tag into the component tag CSR of the specified processing element. Reported status is similar to *rioConfigurationWrite*( ).

A destid value of HOST_REGS and hopcount of 0 results in accesses to the local hosts RapidIO registers.

rioGetPortErrStatus

Prototype:

```
STATUS rioGetPortErrStatus (
    UINT8     localport,
    UINT32    destid,
    UINT8     hopcount,
```

```
UINT16   extfoffset,
UINT8    portnum,
UINT32   *porterrorstatus
)
```

Arguments:

localport	Local port number [IN]
destid	Destination ID of the processing element [IN]
hopcount	Hop count [IN]
extfoffset	Offset from the previously reported extended features pointer [IN]
portnum	Port number to be accessed [IN]
*porterrorstatus	Pointer to storage for the returned value [OUT]

Return value:

RIO_SUCCESS	The read completed successfully and valid data was placed into the location specified by *porterrorstatus.
RIO_ERR_INVALID_PARAMETER	One or more input parameters had an invalid value.
RIO_ERR_RIO	The RapidIO fabric returned a Response Packet with ERROR status reported. Error status returned by this function may contain additional information from the Response Packet.
RIO_ERR_ACCESS	A device-specific hardware interface was unable to generate a maintenance transaction and reported an error.

Synopsis:

rioGetPortErrStatus( ) uses the HAL *rioConfigurationRead( )* function to read the contents of the port *n* error and status CSR of the specified processing element. Reported status is similar to *rioConfigurationRead( )*.

A destid value of HOST_REGS and hopcount of 0 results in accesses to the local hosts RapidIO registers.

10.2.3 Routing-table Manipulation HAL Functions

This section describes the RapidIO functions that must be provided to support routing tables used within the switch fabric. The RapidIO *Common Transport Specification* requires implementing device-identifier-based packet routing. The detailed implementations of routing tables in switches are not described. To simplify the way that software interacts with RapidIO switch devices, the RapidIO HAL defines a couple of routing-table manipulation functions. These functions would be used in conjunction with the standard RapidIO bringup functions to configure the routing tables in a switch.

The routing-table manipulation functions assume the following:

- The destination ID of the device that receives a packet routed by the switch is the *route destination ID*.
- The specific port at the route destination ID that receives a packet routed by the switch is the *route port number*.

- The software paradigm used for routing tables is a linear routing table indexed by the *route destination ID*.
- Switches may implement a global routing table, 'per port' routing tables, or a combination of both.

In order to provide a consistent software view of routing tables the following two functions are defined. RioRouteAddEntry and rioRouteGetEntry. These functions are described in the following pages.

rioRouteAddEntry

Prototype:
```
STATUS rioRouteAddEntry (
    UINT8    localport,
    UINT32   destid,
    UINT8    hopcount,
    UINT8    tableidx,
    UINT16   routedestid,
    UINT8    routeportno
    )
```

Arguments:

localport	Local port number (RapidIO switch) [IN]
destid	Destination ID of the processing element (RapidIO switch) [IN]
hopcount	Hop count [IN]
tableidx	Routing table index for per-port switch implementations [IN]
routedestid	Route destination ID—used to select an entry into the specified routing table [IN]
routeportno	Route port number—value written to the selected routing table entry [IN]

Return value:

RIO_SUCCESS	The routing table entry was added successfully.
RIO_ERR_INVALID_PARAMETER	One or more input parameters had an invalid value.
RIO_ERR_RIO	The RapidIO fabric returned a Response Packet with ERROR status reported. Error status returned by this function may contain additional information from the Response Packet.
RIO_ERR_ACCESS	A device-specific hardware interface was unable to generate a maintenance transaction and reported an error.
RIO_WARN_INCONSISTENT	Used by rioRouteGetEntry—indicates that the routeportno returned is not the same for all ports.

Synopsis:
rioRouteAddEntry( ) adds an entry to a routing table for the RapidIO switch specified by the destid and hopcount parameters. The tableidx parameter is used to select a specific routing table in the case of implementations with 'per-port' routing tables. A value of tableidx=0xFF specifies a global routing table for the RapidIO switch. The routeportno parameter is written to the routing table entry selected by the routedestid parameter.

A destid value of HOST_REGS and hopcount of 0 results in accesses to the local hosts RapidIO registers.

rioRouteGetEntry

Prototype:
```
STATUS rioRouteGetEntry (
    UINT8     localport,
    UINT32    destid,
    UINT8     hopcount,
    UINT8     tableidx,
    UINT16    routedestid,
    UINT8     *routeportno
)
```

Arguments:

localport	Local port number (RapidIO switch) [IN]
destid	Destination ID of the processing element (RapidIO switch) [IN]
hopcount	Hop count [IN]
tableidx	Routing table index for per-port switch implementations [IN]
routedestid	Route destination ID—used to select an entry into the specified routing table [IN]
*routeportno	Route port number—pointer to value read from the selected routing table entry [OUT]

Return value:

RIO_SUCCESS	The routing table entry was added successfully.
RIO_ERR_INVALID_PARAMETER	One or more input parameters had an invalid value.
RIO_ERR_RIO	The RapidIO fabric returned a Response Packet with ERROR status reported. Error status returned by this function may contain additional information from the Response Packet.
RIO_ERR_ACCESS	A device-specific hardware interface was unable to generate a maintenance transaction and reported an error.
RIO_WARN_INCONSISTENT	Used by rioRouteGetEntry—indicates that the routeportno returned is not the same for all ports.

Synopsis:

rioRouteGetEntry() reads an entry from a routing table for the RapidIO switch specified by the destid and hopcount parameters. The tableidx parameter is used to select a specific routing table in the case of implementations with 'per port' routing tables. A value of tableidx=0xFF specifies a global routing table for the RapidIO switch. The value in the routing table entry selected by the routedestid parameter is read from the table and placed into the location referenced by the *routeportno pointer.

Reads from the global routing table may be undefined in the case where per-port routing tables exist. A destid value of HOST_REGS and hopcount of 0 results in accesses to the local hosts RapidIO registers.

10.2.4 Device Access Routine (DAR) Functions

This section defines the device access routine (DAR) interface that must be provided for RapidIO device configuration. The client for this interface is the boot loader responsible for RapidIO network enumeration and initialization. By using a standard DAR interface, the firmware does not need to include knowledge of device-specific configuration operations. Thus, enumeration and initialization firmware can operate transparently with devices from many component vendors.

For each processor type supported by a DAR provider, linkable object files for DARs should be supplied in ELF format. Device-specific configuration DARs should be supplied in C-language source code format.

The functions provided by device-specific configuration DARs must be able to link and execute within a minimal execution context (e.g. a system-boot monitor or firmware). In general, configuration DARs should not call an external function that is not implemented by the DAR, unless the external function is passed to the configuration DAR by the initialization function. Also, configuration DAR functions may not call standard C-language I/O functions (e.g. *printf*) or standard C-language library functions that might manipulate the execution environment (e.g. *malloc* or *exit*). The functions that must be provided for a RapidIO device-specific configuration DAR are described in the following sections.

For the *rioDar_name*GetFunctionTable functions, the *rioDar_name* portion of the function name is replaced by an appropriate name for the implemented driver.

*rioDar_name*GetFunctionTable

Prototype:
    UINT32 *rioDar_name*GetFunctionTable(
        UINT32   specversion,
        RDCDAR_OPS_STRUCT   *darops,
        UINT32   maxdevices,
        UINT32   *darspecificdatabyte
)

Arguments:
specversion	Version number of the DAR interface specification indicating the caller's implementation of the type definition structures [IN]
*darops	Pointer to a structure of DAR functions that are allocated by the caller and filled in by the called function [OUT]
maxdevices	Maximum expected number of RapidIO devices that must be serviced by this configuration DAR [IN]
*darspecificdatabytes	Number of bytes needed by the DAR for the DAR private data storage area [OUT]

Return value:
    RIO_SUCCESS   On successful completion

Synopsis:
rioDar_name*GetFunctionTable*( ) is called by a client to obtain the list of functions implemented by a RapidIO device-specific configuration DAR module. It shall be called once

before enumerating the RapidIO network. The specversion parameter is the version number defined by the revision level of the specification from which the DAR type definition structures are taken. The maxdevices parameter is an estimate of the maximum number of RapidIO devices in the network that this DAR must service. The DAR uses this estimate to determine the size required for the DAR private data storage area. The storage size is returned to the location referenced by the *darspecificdatabytes pointer. After the client calls this function, the client shall allocate a DAR private data storage area of a size no less than that indicated by *darspecificdatabytes. The client shall provide that private data storage area to *rioDarInitialize()*.

rioDarInitialize

Prototype:
```
UINT32 rioDarInitialize (
    UINT32              specversion,
    UINT32              maxdevices,
    RDCDAR_PLAT_OPS     *platops,
    RDCDAR_DATA         *privdata
    )
```

Arguments:

specversion	Version number of the DAR interface specification indicating the caller's implementation of the type definition structures [IN]
maxdevices	Maximum expected number of RapidIO devices that must be serviced by this configuration DAR [IN]
*platops	Pointer to a structure of platform functions for use by the DAR [IN]
*privdata	Pointer to structure containing DAR private data area [IN/OUT]

Return value:
RIO_SUCCESS   On successful completion

Synopsis:
rioDarInitialize() is called by a client to initialize a RapidIO device-specific configuration DAR module. This function shall be called once after calling the rioDar_name*GetFunctionTable()* functions and before enumerating the RapidIO network. The specversion parameter is the version number defined by the revision level of the specification from which the DAR type definition structures are taken. The maxdevices parameter is an estimate of the maximum number of RapidIO devices in the network that this DAR must service. The maxdevices value must be equal to the value used in the corresponding rioDar_name*GetFunctionTable()* function call. The client is responsible for allocating the structure referenced by *privdata. The client is also responsible for allocating a DAR private data storage area at least as large as that specified by the rioDar_name*GetFunctionTable()* call. The client must initialize the structure referenced by *privdata with the number of bytes allocated to the DAR private data storage area and with the pointer to the storage area. After calling *rioDarInitialize()*, the client may not deallocate the DAR private data storage area until after the *rioDarTerminate()* function has been called.

rioDarTerminate

Prototype:
  UINT32 rioDarTerminate (
    RDCDAR_DATA   *privdata
)

Arguments:
  *privdata   Pointer to structure containing DAR private data area [IN/OUT]

Return value:
  RIO_SUCCESS   On successful completion

Synopsis:
rioDarTerminate( ) is invoked by a client to terminate a RapidIO device-specific configuration DAR module. This function shall be called once after all use of the DAR services is completed. After calling this function, the client may de-allocate the DAR private data storage area in the structure referenced by *privdata.

rioDarTestMatch

Prototype:
  UINT32 rioDarTestMatch (
    RDCDAR_DATA   *privdata,
    UINT8   localport,
    UINT32   destid,
    UINT8   hopcount
)

Arguments:
  *privdata   Pointer to structure containing DAR private data area [IN/OUT]
  localport   Local port number used to access the network [IN]
  destid   Destination device ID for the target device [IN]
  hopcount   Number of switch hops needed to reach the target device [IN]

Return value:
  RIO_SUCCESS   Device DAR does provide services for this device
  RIO_ERR_NO_DEVICE_SUPPORT   Device DAR does not provide services for this device.

Synopsis:
rioDarTestMatch( ) is invoked by a client to determine whether or not a RapidIO device-specific configuration DAR module provides services for the device specified by destid. The DAR interrogates the device (using the platform functions supplied during DAR initialization), examines the device identity and any necessary device registers, and determines whether or not the device is handled by the DAR. The DAR does not assume that a positive match (return value of 0) means the DAR will actually provide services for the device. The client must explicitly register the device with *rioDARregister*( ) if the client will be requesting services.

A destid value of HOST_REGS and hopcount of 0 results in accesses to the local hosts RapidIO registers.

rioDarRegister

Prototype:
    UINT32 rioDarRegister (
    RDCDAR_DATA   *privdata,
    UINT8   localport,
    UINT32   destid,
    UINT8   hopcount,
)

Arguments:
  *privdata   Pointer to structure containing DAR private data [IN/OUT]
  localport   Local port number used to access the network [IN]
  destid   Destination device ID for the target device [IN]
  hopcount   Number of switch hops needed to reach the target device [IN]

Return value:
  RIO_SUCCESS   Device DAR successfully registered this
                                                      device.

  RIO_ERR_NO_DEVICE_SUPPORT   Device DAR does not provide services for
                                                      this device.

  RIO_ERR_INSUFFICIENT_RESOURCES   Insufficient storage available in DAR priv-
                                                      ate storage area

Synopsis:
rioDarRegister( ) is invoked by a client to register a target device with a RapidIO device-specific configuration DAR. The client must call this function once for each device serviced by the DAR. The client should first use the *rioDarTestMatch( )* function to verify that the DAR is capable of providing services to the device.

    A destid value of HOST_REGS and hopcount of 0 results in accesses to the local hosts RapidIO registers.

rioDarGetMemorySize

Prototype:
    UINT32 rioDarGetMemorySize (
    RDCDAR_DATA   *privdata,
    UINT8   localport,
    UINT32   destid,
    UINT8   hopcount,
    UINT32   regionix,
    UINT32   *nregions,
    UINT32   *regbytes[2],
    UINT32   *startoffset[2]
)

Arguments:
  *privdata   Pointer to structure containing DAR private data area [IN/OUT]
  localport   Local port number used to access the network [IN]

destid   Destination device ID for the target device [IN]
hopcount   Number of switch hops needed to reach the target device [IN]
regionix   Index of the memory region being queried (0, 1, 2, 3, . . .) [IN]
*nregions   Number of memory regions provided by the target device [OUT]
*regbytes   Size (in bytes) of the queried memory region [OUT]
*startoffset   Starting address offset for the queried memory region [OUT]

Return value:

RIO_SUCCESS	Device DAR successfully returned memory size information for the target device.
RIO_ERR_NO_DEVICE_SUPPORT	Device DAR could not determine memory size information for the target device.

Synopsis:

rioDarGetMemorySize() is invoked by a client to determine the number of, the sizes of, and the offsets for the memory regions supported by a RapidIO target device. The function is intended to support the mapping of PCI or other address windows to RapidIO devices. If the regionix parameter is greater than the number of regions provided by the device (*nregions), the DAR should return a value of zero for the *regbytes and *startoffset parameters, and indicate a 'successful' (0) return code. rioDarGetMemorySize always returns at least one region. The first index, index 0, always refers to the region controlled by the local configuration space base address registers. The client must register the target device with the RapidIO device-specific configuration DAR before calling this function.

rioDarGetSwitchInfo

Prototype:

```
UINT32 rioDarGetSwitchInfo (
    RDCDAR_DATA             *privdata,
    UINT8                   localport,
    UINT32                  destid,
    UINT8                   hopcount,
    RDCDAR_SWITCH_INFO      *info
    )
```

Arguments:

*privdata   Pointer to structure containing DAR private data [IN/OUT]
localport   Local port number to be used to access network [IN]
destid   Destination device ID to reach target switch device [IN]
hopcount   Number of switch hops to reach target switch device [IN]
*info   Pointer to switch information data structure [OUT]

Return value:

RIO_SUCCESS	Device DAR successfully retrieved the information for RDCDAR_PLAT_OPS_STRUCT.
RIO_ERR_NO_DEVICE_SUPPORT	Insufficient switch routing resources available.
RIO_ERR_NO_SWITCH	Target device is not a switch.

Synopsis:

rioDarGetSwitchInfo( ) is invoked by a client to retrieve the data necessary to initialize the RDCDAR_SWITCH_INFO structure. The client must register the target device with the RapidIO device-specific configuration DAR before calling this function.

A destid value of HOST_REGS and hopcount of 0 results in accesses to the local hosts RapidIO registers.

rioDarSetPortRoute

Prototype:

```
UINT32 rioDarSetPortRoute (
    RDCDAR_DATA    *privdata,
    UINT8          localport,
    UINT32         destid,
    UINT8          hopcount,
    UINT8          tableidx,
    UINT16         routedestid,
    UINT8          routeportno
    )
```

Arguments:

*privdata	Pointer to structure containing DAR private data area [IN/OUT]
localport	Local port number to be used to access network [IN]
destid	Destination device ID to reach target switch device [IN]
hopcount	Number of switch hops to reach target switch device [IN]
inport	Target switch device input port [IN]
tableidx	Routing table index for per-port switch implementations [IN]
routedestid	Route destination ID—used to select an entry into the specified routing table [IN]
routeportno	Route port number—value written to the selected routing table entry [IN]

Return value:

RIO_SUCCESS	Device DAR successfully modified the packet routing configuration for the target switch device.
RIO_ERR_NO_DEVICE_SUPPORT	Insufficient switch routing resources available.
RIO_ERR_ROUTE_ERROR	Switch cannot support requested routing.
RIO_ERR_NO_SWITCH	Target device is not a switch.
RIO_ERR_FEATURE_NOT_SUPPORTED	Target device is not capable of per-input-port routing.

Synopsis:

rioDarSetPortRoute( ) is invoked by a client to modify the packet routing configuration for a RapidIO target switch device. The client must register the target device with the RapidIO device-specific configuration DAR before calling this function.

A destid value of HOST_REGS and hopcount of 0 results in accesses to the local hosts RapidIO registers.

rioDarGetPortRoute

Prototype:
```
UINT32 rioDarGetPortRoute (
    RDCDAR_DATA    *privdata,
    UINT8          localport,
    UINT32         destid,
    UINT8          hopcount,
    UINT8          tableidx,
    UINT16         routedestid,
    UINT8          *routeportno
)
```

Arguments:

*privdata	Pointer to structure containing DAR private data area [IN/OUT]
localport	Local port number to be used to access network [IN]
destid	Destination device ID to reach target switch device [IN]
hopcount	Number of switch hops to reach target switch device [IN]
tableidx	Routing table index for per-port switch implementations [IN]
routedestid	Route destination ID—used to select an entry into the specified routing table [IN]
*routeportno	Route port number—pointer to value read from the selected routing table entry [OUT]

Return value:

RIO_SUCCESS	Device DAR successfully modified the packet routing configuration for the target switch device.
RIO_ERR_NO_DEVICE_SUPPORT	Insufficient switch routing resources available.
RIO_ERR_ROUTE_ERROR	Switch cannot support requested routing.
RIO_ERR_NO_SWITCH	Target device is not a switch.

Synopsis:

rioDarGetPortRoute( ) is invoked by a client to read the packet routing configuration for a RapidIO target switch device. The client must register the target device with the RapidIO device-specific configuration DAR before calling this function.

A destid value of HOST_REGS and hopcount of 0 results in accesses to the local hosts RapidIO registers.

10.3 SYSTEM BRINGUP EXAMPLE

This example shows the process employed by a system that offers two host processors, a single switch and two resource board subsystems. The hosts (A and B) are redundant with each other. Host B is assigned a higher priority than Host A. To better demonstrate the enumeration interaction between the host devices, Host A begins the enumeration process first. Figure 10.1 provides a representation of the system connectivity and topology.

The following additional assumptions are made about the example system: system Host A is preloaded with device ID 0x00 and system Host B is preloaded with device ID 0x01. System Bringup advances through a series of time steps. The time steps are not meant to be of

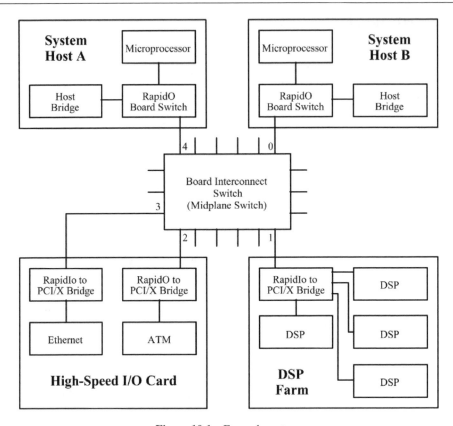

Figure 10.1   Example system

equal length, but they would occur in the specified order. The following time steps would occur as the enumeration process proceeded:

- $T+0$: Host A begins RapidIO enumeration.
- $T+1$: Host B begins RapidIO enumeration and Host A continues RapidIO enumeration.
- $T+2$: Host B discovers another host in the system (Host A) and waits.
- $T+3$: Host A discovers a higher-priority host in the system (Host B) and retreats.
- $T+4$: Host B assumes sole enumeration of the system.
- $T+5$: Host B enumerates the PE on switch port 1.
- $T+6$: Host B enumerates the PEs on switch ports 2, 3 and 4.
- $T+7$: System enumeration is complete.

The following describes the actions taken during each time slot in more detail.

Time T: Host A attempts to acquire the lock from its host base device ID lock CSR by writing 0x00 to the CSR. Host A confirms it has acquired the lock when it reads the value of 0x00 (the host device ID) from the lock CSR. Host A continues by reading the processing element features CAR and adding the information from the CAR to its RapidIO device database. Host A updates its base device ID CSR with the host device ID (0x00).

Time $T+1$: Host B attempts to acquire the lock from its host base device ID lock CSR by writing 0x01 to the CSR. Host B confirms it has acquired the lock when it reads the value of 0x01 (the host device ID) from the lock CSR. Host B continues by reading the processing element features CAR and adding the information from the CAR to its RapidIO device database. Host B updates its base device ID CSR with the host device ID (0x01).

Host A begins neighbor enumeration. It attempts to acquire the lock from the host base device ID lock CSR of the board interconnect switch. A maintenance write of the host device ID (0x00), the destination device ID (0xFF), and the hop count (0) is issued for the lock CSR. Host A confirms it has acquired the lock when it reads the value of 0x00 (the host device ID) from the lock CSR.

Time $T+2$: Host B begins neighbor enumeration. It attempts to acquire the lock from the host base device ID lock CSR of the board interconnect switch. A maintenance write of the host device ID (0x01), the destination device ID (0xFF), and the hop count (0) is issued for the lock CSR. However, after Host B issues a maintenance read from the lock CSR it finds that the device was already locked by host device ID 0x00. Because Host B has a higher priority than the current lock holder (0x01 is greater than 0x00), Host B spins in a delay loop and repeatedly attempts to acquire the lock.

Time $T+3$: Host A continues neighbor enumeration. It issues a maintenance read cycle to the device identity CAR of the board interconnect switch and looks for a matching entry in the device database. Device configuration continues because no match is found (Host A has not enumerated the device). Host A reads the source operations and destination operations CARs for the device. It is determined that the device does not support read/write/atomic operations and does not require a device ID. Host A reads the processing element feature CAR for the device and determines that it is a switch element.

Because the device is a switch, Host A reads the switch port Information CAR and records the device identity in the switch database. Next, Host A adds a set of entries to the switch's routing table. For each previously discovered device ID, an entry is created containing a target ID (0xFF), hop count (0), and the route port number (from the switch port information CAR). The switch database is updated with the same routing information. Host A reads the port error status CSR for switch port 0, verifying that it is possible for the port to have a neighbor PE. An entry is created in the switch's routing table containing target ID (0xFF), hop count (0), and the route port number (0).

Host A continues neighbor enumeration, using a hop count of 1. It attempts to acquire the lock from the host base device ID lock CSR of the neighbor PE on port 0. A maintenance write of the host device ID (0x00), the destination device ID (0xFF), and the hop count (1) is issued for the lock CSR. However, after Host B issues a maintenance read from the lock CSR it finds that the device was already locked by host device ID 0x01. Because Host A has a lower priority than the current lock holder (0x00 is less than 0x01), Host A retreats. It begins the process of backing out all enumeration and configuration changes it has made.

Host A checks its device and switch databases to find all host locks it obtained within the system (system Host A and the board interconnect switch). It issues a maintenance write transaction to their host base device ID lock CSRs to release the locks.

Time $T+4$: As Host B spins in its delay loop, it attempts to acquire the lock from the host base device ID Lock CSR of the board interconnect switch. A maintenance write of the host device ID (0x01), the destination device ID (0xFF), and the hop count (0) is issued for the lock CSR. Because Host A released the lock, Host B is able to confirm it has acquired the lock when it reads the value of 0x01 from the lock CSR.

Host B continues neighbor enumeration. It issues a maintenance read cycle to the device identity CAR of the board interconnect switch and looks for a matching entry in the device database. Device configuration continues because no match is found (Host B has not enumerated the device). Host B reads the source operations and destination operations CARs for the device. It is determined that the device does not support read/write/atomic operations and does not require a device ID. Host B reads the processing element feature CAR for the device and determines that it is a switch element.

Because the device is a switch, Host B reads the switch port information CAR and records the device identity in the switch database. Next, Host B adds a set of entries to the switch's routing table. For each previously discovered device ID, an entry is created containing a target ID (0xFF), hop count (0), and the route port number (from the switch port information CAR). The switch database is updated with the same routing information. Host B reads the port error status CSR for switch port 0, verifying that it is possible for the port to have a neighbor PE. An entry is created in the switch's routing table containing target ID (0xFF), hop count (0), and the route port number (0). Host B detects that it is attached to port 0. Because Host B has already been enumerated, neighbor enumeration continues on the next port.

Time $T+5$: Host B reads the port error status CSR for switch port 1, verifying that it is possible for the port to have a neighbor PE. An entry is created in the switch's routing table containing target ID (0xFF), hop count (0), and the route port number (1).

Host B continues neighbor enumeration using a hop count of 1. It attempts to acquire the lock from the host base device ID lock CSR of the neighbor PE on port 1. A maintenance write of the host device ID (0x01), the destination device ID (0xFF), and the hop count (1) is issued for the lock CSR. Host B confirms it has acquired the lock when it reads the value of 0x01 from the lock CSR.

Host B issues a maintenance read cycle to the device identity CAR of the DSP Farm board as shown in Figure 10.1 and looks for a matching entry in the device database. Device configuration continues because no match is found (Host B has not enumerated the device). Host B reads the source operations and destination operations CARs for the device. It is determined that the device supports read/write/atomic operations. A maintenance write is used to update the base device ID CSR with the value of 0x00 (the first available device ID). Device ID is incremented and compared with the Host B device ID. Because they are equal, device ID is assigned the next available device ID.

Time $T+6$: The process described in the previous step (time $T+5$) is repeated on switch ports 2–4. Device IDs 0x02, 0x03, and 0x04 are assigned to the PEs on switch ports 2, 3 and 4, respectively.

Time $T+7$: Host A detects that its host base device lock CSR has been acquired by another host device, indicating it has been enumerated. Host A can initiate passive discovery to build a local system database.

11

Advanced Features

The RapidIO specification supports several optional advanced features. These features would not be required in all systems or applications of RapidIO. This chapter will present details on the following optional RapidIO features:

- system-level flow control
- error management
- multiprocessor cache coherence
- multicasting transactions
- multicasting symbols

The RapidIO architecture was defined to support optional functionality in a compatible manner. The basis for this compatibility can be found in the separation of responsibility between RapidIO end point and RapidIO switches. End points are responsible for creating and terminating packets. Switches are responsible for forwarding packets towards their final destination. New or optional transaction types must be understood between source and destination, but do not need to be understood by switches. The three optional features described in this chapter offer new logical layer transactions that existing switches will transparently forward through a system. So long as the transactions retain the physical and transport layer header structure they will be compatible with existing RapidIO switch infrastructure.

System-level flow control is an optional feature that would be used in larger-scale systems where traffic patterns might create short-term congestion in the system. The RapidIO link-level flow control is useful for preventing buffer overruns between two adjacent RapidIO devices; the system-level flow control is useful for preventing buffer overruns and general system congestion across many devices. The system-level flow control is optional because it is unlikely to be necessary in small system topologies or in systems with well known and planned for traffic patterns.

Error management extensions are useful in systems that require high levels of reliability, availability and serviceability (RAS). RAS is an important concept in infrastructure

RapidIO® The Embedded System Interconnect. S. Fuller
© 2005 John Wiley & Sons, Ltd ISBN: 0-470-09291-2

equipment such as the telephone networks. It is also important in high-end compute server and enterprise-class storage equipment. RapidIO provides a strong basic level of error detection and recovery. The error management extensions define standard ways for reporting and recovering from errors in a system. Standard approaches to error management can simplify and reduce the cost of deploying highly reliable systems.

The following sections will take a more in-depth look at each of these optional extensions to the base RapidIO protocol.

11.1 SYSTEM-LEVEL FLOW CONTROL

A switch-fabric-based system can encounter several types of congestion, differentiated by the duration of the event. Table 11.1 presents four different classifications of congestion that may occur in systems.

Congestion can be detected inside a switch, at the connections between the switch and other switches and end points. Conceptually, the congestion is detected at an output port that is trying to transmit data to the connected device, but is receiving more information than it is able to transmit. This excess data can then 'pile up' until the switch is out of storage capacity, at this point the congestion may spread to other devices in the system. Therefore, contention for a particular connection in the fabric can affect the ability of the fabric to transmit data unrelated to the contested connection. This type of behavior is highly undesirable for many applications.

The best analogy for this congestion is rush hour traffic congestion. Because of congestion, intersections can back up to the point where they affect the flow of traffic through other intersections; traffic that is not destined for the point of congestion, yet is nevertheless affected by the congestion. As RapidIO is intended to operate in systems with potentially thousands of devices and hundreds of switches it is prudent to anticipate that these congestion scenarios may occur, and to provide mechanisms to reduce the impact of this congestion on the rest of the system. The length of time that the congestion lasts determines the magnitude of the effect the congestion has upon the overall system.

Ultra short-term congestion events are characterized as lasting a very small length of time, perhaps up to 500 ns. In a RapidIO type system these events are adequately handled by a combination of buffering within the devices on either end of a link and the retry-based link layer mechanism defined in the 8/16 LP-LVDS and 1x/4x LP-Serial physical layer specifications. This combination adds 'elasticity' to each link in the system. The impact of ultra-short-term events on the overall system is minor, if noticeable at all.

Table 11.1   Congestion time frames

Congestion classification	Time frame
Ultra-short-term	Hundreds of nanoseconds
Short-term	Tens to hundreds of microseconds
Medium-term	Up to seconds or minutes
Long-term	Ongoing

Short-term congestion events last much longer than ultra-short-term events, lasting up into the dozens or hundreds of microseconds. These events can be highly disruptive to the performance of the fabric (and the system overall), in both aggregate bandwidth and end-to-end latency. Managing this type of congestion requires some means of detecting when an ultra-short-term event has turned into a short-term event, and then using some mechanism to reduce the amount of data being injected by the end points into the congested portion of the fabric. If this can be done in time, the congestion stays localized until it clears, and does not adversely affect other parts of the fabric.

Medium-term congestion is typically a frequent series of short-term congestion events over a long period of time, perhaps seconds or minutes. This type of event is indicative of an unbalanced data load being sent into the fabric. Alleviating this type of congestion event requires some sort of software-based load balancing mechanism to reconfigure the fabric.

Long-term congestion is a situation in which a system does not have the raw capacity to handle the demands placed upon it. This situation is corrected by upgrading (or replacing) the system itself.

11.1.1 Interconnect Fabric Congestion Example

System-level flow control, defined in *Part X: RapidIO Flow Control Extensions* of the RapidIO specifications, is designed to address the problem of short term congestion.

This section will look at how system-level flow control is used to reduce or eliminate short-term congestion in systems. An example switch topology is illustrated in Figure 11.1. In this example the switches are output buffered and a point of congestion is associated with an output buffer of switch 3.

The problem that is to be addressed by system-level flow control is caused by multiple independent transaction request flows, each with burst and spatial locality characteristics that individually would not exceed the bandwidth capacity of the links or end points in the system. However, owing to the statistical combination of these transaction request flows, usually in the middle of multistage topologies, the demand for bandwidth through a particular link may exceed the link's capacity for a long enough period of time so that system performance is compromised.

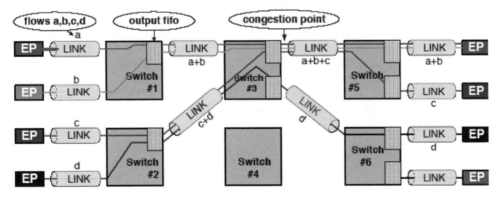

Figure 11.1 Flow control example

In the example, data flows **a**, **b**, and **c** are directed towards an output port of switch 3, as shown. Their peak aggregate bandwidth requirement is higher than the output link can support. As a result, the output buffer for this port can occasionally fill up, causing the link layer flow control to be activated on the links of the preceding switch stages. The output packet buffers for switches 1 and 2 then also fill up as they are throttled by the link layer flow control. Packets for transaction request flows, such as data flow **d**, that are not destined for the output port with the full buffer in switch 3 are now also forced to wait, causing additional system performance loss. This phenomenon is known as higher-order head-of-line blocking.

Past a point of optimal utilization, the aggregate throughput of the fabric will be significantly reduced with increased load if congestion control is not applied. Such non-linear behavior is known as 'performance-collapse.' The purpose of the system-level flow control is to avoid this performance collapse. Research shows that relatively simple 'XON/XOFF' controls on transaction request flows can quite adequately control congestion in fabrics of significant size.

An important design factor for interconnect fabrics is the latency between a congestion control action being initiated and the transaction request flow source acting in response. This latency determines, among other factors, the required buffer sizes for the switches. To keep such buffers small, the latency of a congestion control mechanism must be minimized. For example, if there were 10 data flows contributing to a buffer overflow (forming what is known as a 'hotspot') and it takes 10 packet transmission times for the congestion notification to reach the sources and the last packets sent from the sources to reach the point of congestion after the sources react to the congestion notification, up to 100 packets could be added to the congested buffer.

11.1.2 System-level Flow Control Operations

In devices supporting system-level flow control, the fabric is responsible for detecting congestion and for generating explicit flow control messages referred to as transmit off 'XOFF' and transmit on 'XON' congestion control packets (CCPs). The manner in which the fabric detects congestion is not specified and is left to the manufacturers to determine. The XON and XOFF packets, like any other packet, require link-level packet acknowledgements. The XOFF CCPs are sent to shut off select flows at their source end points. Later, when the congestion event has passed, XON CCPs are sent to the same source end points to restore those flows.

The flow control operation consists of a single FLOW CONTROL transaction. The FLOW CONTROL transaction is issued by a switch or end point back to the original source of congestion with the purpose of controlling the flow of data. This mechanism is backward compatible with RapidIO legacy devices in the same system.

Congestion control packets are regarded as independent traffic flows. They are the most important traffic flow defined by the system. Congestion control packets are always transmitted at the first opportunity and at the expense of all other traffic flows if possible. For the physical layer, this requires marking congestion control packets with a 'PRIO' field value of 0b11, and a 'crf' bit value of 0b1, if supported. These transactions use a normal packet format for purposes of error checking and formatting. The congestion control packet shown in Figure 11.2 is eight bytes in length. If extended system addressing (16-bit device) IDs are used the packet would grow to 12 bytes, including padding.

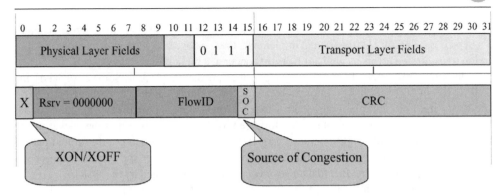

Figure 11.2 Congestion control packet format

11.1.3 End Point Flow Control Rules

There are a number of rules related to flow control that are required of an end point that supports the logical layer flow control extensions.

1. An XOFF flow control transaction stops all transaction request flows of the specified priority and lower, targeted to the specified destination and increments the XON/XOFF counter associated with the specified flowID.
2. A XON flow control transaction decrements the XON/XOFF counter associated with the specified flowID. If the resulting value is zero, the transaction request flows for that flowID and flowIDs of higher priority are restarted.
3. An end point must be able to identify an orphaned XOFF'd flow and restart it. This will typically be accomplished through a time-out mechanism.
4. A destination end point issuing an XOFF flow control transaction must maintain the information necessary to restart the flow with an XON flow control transaction when congestion abates.
5. Upon detection of congestion within one of its ports, the destination end point should send CCP(s) as quickly as possible to reduce latency back to the source end point.

11.1.4 Switch Flow Control Rules

These are the rules related to system-level flow control that a switch is required to support.

1. Upon detection of congestion within a port, the switch shall send a CCP (XOFF) for each congested flow to their respective end points.
2. If a switch runs out of packet buffer space, it is permitted to drop CCPs.
3. A switch issuing an XOFF flow control transaction must maintain the information necessary to restart the flow with an XON flow control transaction when congestion abates.

11.2 ERROR MANAGEMENT EXTENSIONS

To provide for consistent support for the reporting and management of errors that might occur in a system, the RapidIO Trade Association Technical Working Group developed a specification for additional registers and associated functionality to assist in the error manage-ment process. RapidIO hardware that implements these error management extensions are able to log transmission errors as well as errors that occur at a higher level in the system.

In a complex system there are many types of errors that may occur. Some error scenarios will require no software intervention and recovery procedures can be accomplished totally by the hardware. For example a parity error in a packet will be automatically handled by the transmission of a NACK control symbol followed by the retransmission of the packet. If the error was due to a random glitch in the channel, then it is likely that the retransmitted packet will be error free. After proper reception of the retransmitted packet, the system will proceed with normal operation. The fact that an error occurred may be logged by the system.

Other error scenarios will require fault management software for recovery to be suc-cessful. For example, some types of logical layer errors on a Read or Write operation may be recoverable by killing the software process using the affected memory space and removing the memory space from the available system resource pool. For example a Read transaction might be attempted to a device that does not issue a response to the request transaction. A logical layer time-out timer would detect this error. It would be manifested to the request-ing device as an access violation and would most likely invoke fault management software for proper handling.

Another area of fault recovery that requires fault management software to be implemented is that of correcting of system state after an error during an atomic operation. For example, if a request is lost and times out, software can examine the current lock value to determine if the request or the associated response was the transaction that was lost in the switch fabric.

The use of RapidIO message packets relies on the use of higher-layer protocols for error management. Since end points that communicate via messaging are typically running a variety of higher-layer protocols, the message queue management controller handles error reporting of both request and response time-outs locally. Note that side effect errors can occur, for example, ERROR responses or RETRY responses during an active (partially completed) message, which may complicate the recovery procedure.

The error management extensions are intended to provide consistent state information to higher-level software to enable proper diagnosis and resolution of the error condition.

11.2.1 Hot-insertion/extraction

Hot-insertion can be regarded as an error condition where a new part is detected in the system. This section describes an approach for hot insertion. This example is appropriate for high-availability systems, or systems where field replaceable units (FRUs) need to brought into the system in a controlled manner.

At system boot time, the system host identifies all of the unattached links in the machine through the enumeration process and puts them in a locked mode. In locked mode all incom-ing packets are rejected, but the drivers and receivers are enabled. Locked mode is entered by setting the discovered bit in the port general control CSR and the port lockout bit in the port n control CSR. Whenever an FRU is removed, the port lockout bit should be used to ensure that whatever new FRU is inserted cannot access the system until the system host allows it. When a FRU is hot-inserted to the system, the now connected link will automatically start the train-ing sequence. When training is complete (the port OK bit in the port n error and status CSR is set), the locked port generates a maintenance port-write operation to notify the system host of the new connection, and sets the port-write pending bit.

On receipt of the port-write, the system host is responsible for bringing the inserted FRU into the system in a controlled manner. The system host can communicate with the inserted FRU by maintenance operations after clearing all error conditions, if any, clearing the port lockout bit and clearing the output and input port enable bits in the port n control CSR. This procedure allows the system host to access the inserted FRU safely, without exposing itself to incorrect behavior by the inserted FRU.

In order to issue maintenance operations to the inserted FRU, the system host must first make sure that the ackID values for both ends are consistent. Since the inserted FRU has just completed a power-up reset sequence, both its transmit and receive ackID values are the reset value of 0x00. The system host can set the switch device's transmit and receive ackID values to also be 0x00 through the port n local ackID status CSR if they are not already in that state, and can then issue packets normally.

Hot-extraction from a port's point of view behaves identically to a very rapidly failing link and therefore can utilize the error reporting mechanisms. Hot-extraction is ideally done in a controlled fashion by first taking the FRU to be removed out of the system as a usable resource through the system management software so that extraction does not cause switch fabric congestion or result in a loss of data. However, RapidIO provides sufficient capabilities in the error management extensions for software schemes to gracefully manage the unexpected loss of a device in the system.

11.2.2 The Port-write Transaction

The port-write transaction is used to report an error condition to the host. The error management specification provides only one destination for port-write operations, while designers of reliable systems would assume that two is the minimum number. This section explains the rationale for only having one port-write destination.

In the event of an error on a link, assume that both ends of the link will see the error. Thus, there are two parties who can report the error. In the case where the sole link between an end point and a switch fails completely, the switch is expected to detect and report the error. When one of a set of redundant links between an end point and a switch device fails, it is expected that the switch and possibly the end point will report the failure.

When a link between two switches fails, it is assumed that there are multiple paths to the controlling entity available for the port-write to travel. The switches will be able to send at least one, and possibly two, reports to the system host. It is assumed that it is possible to set up a switch's routing parameters such that the traffic to the system host will follow separate paths from each switch.

In some reliable systems, the system host is implemented as multiple redundant subsystems. It is assumed in RapidIO that only one subsystem is actually in control at any one time, and so should be the recipient of all port-writes. If the subsystem that should be in control is detected to be insane, it is the responsibility of the rest of the control subsystem to change the destination for port-writes to be the new subsystem that is in control.

11.2.3 Persistence of Error Management Registers

Under some conditions, a device may be unable to accept any packets because it is in an undefined, 'broken' condition. It is unable to accept maintenance packets to access any of its error detect and capture registers so it cannot be queried by software. Only a device 'reset' is able to bring the device back. The meaning of a link-request/reset condition may be modified for some implementations of the error management extensions to be a 'soft reset' condition. A device that supports a soft reset will still cause a hardware reset however the port n error detect register, the port n error capture registers, the logical/transport error detect register, and the logical/transport error capture registers could retain their previous values. The information contained in these registers could be used to determine the system state prior to the error condition for purposes of logging and recovery processing.

11.2.4 Error Management Extensions

The error management extensions describe added requirements in all physical and logical layers. These extensions add definitions to bits that were previously reserved in the port n control CSR and add new registers that are contained within the error management extended features block. Full descriptions of these registers and their contents can be found in Appendix D. This section describes the behavior of a device when an error is detected and how the new registers and bits would be managed by the system software and hardware.

The occurrence of a transmission error in a device supporting the error management extensions is logged by hardware by setting the appropriate error indication bit in the port n error detect CSR. Transmission errors that are enabled for error capture and error counting will have the corresponding bit set by software in the port n error rate enable CSR. When the capture valid info status bit is not set in the port n error capture attributes CSR, information about the next enabled transmission error is saved to the port n error capture CSRs. The info type and error Type fields are then updated and the capture valid info status bit is to lock the error capture registers. The first 16 bytes of the packet header or the total control symbol that have a detected error are saved in the capture CSRs. Packets smaller than 16 bytes are captured in their entirety. The capture registers are locked after being written to ensure that subsequent errors don't overwrite the capture registers. Software must write a zero to the capture valid info bit to unlock the capture register. This would likely be performed as part of an error handling routine.

The port n error detect CSR does not lock so subsequent error indications would be logged there. However there would only be an indication of a unique error for each unique error type that had possibly occurred. By reading the register, software may see the types of transmission errors that have occurred. Writing it with all logic 0s clears the port n error detect CSR.

Transmission errors are normally hidden from system software since they may be recovered with no loss of data and without software intervention. However this may hide more systemic problems. To help a system monitor the health of its connections, the error management extensions defines two thresholds in the port n error rate threshold CSR. These thresholds can be set to generate a port-write command to system software when the link error rate reaches a level that is deemed by the system to be degraded or unacceptable.

The two thresholds are called respectively the degraded threshold and the failed threshold. When the error rate counter is incremented, the error rate degraded threshold trigger provides a threshold value that, when equal to or exceeded by the value in the error rate counter in the port n error rate register, will cause the error reporting logic to set the output degraded-encountered bit in the port n error and status CSR, and notify the system software.

The error rate failed threshold trigger, if enabled, should be larger than the degraded threshold trigger. It provides a threshold value that, when equal to or exceeded by the value in the error rate counter, will trigger the error reporting logic to set the output failed-encountered bit in the port *n* error and status CSR, and notify system software.

There are other fields in the port *n* error rate CSR that are used to monitor the error rate of the link connected to the associated port. Their use is described in detail in the error management specification.

System software is notified of logical, transport, and physical layer errors in two ways. An interrupt is issued to the local system by an end point device, the method of which is not defined by the RapidIO specifications, or if the error is detected by a switch device, the switch device issues a maintenance port-write operation. Maintenance port-write operations are sent to a predetermined system host. The sending device sets the port-write pending status bit in the port *n* error and status CSR. A 16 byte data payload of the maintenance port-write packet contains the contents of several CSRs, the port on the device that encountered the error condition (for port-based errors), and some optional implementation specific additional information. Software indicates that it has seen the port-write operation by clearing the port-write pending status bit.

In most systems, it is difficult to verify the error handling software. The error management extensions make some registers writable for easier debug.

The logical/transport layer error detect register and the logical/transport layer error capture registers are writable by software to allow software debug of the system error recovery mechanisms. For software debug, software must write the logical/transport layer error capture registers with the desired address and device ID information then write the logical/transport layer error detect register to set an error bit and lock the registers. When an error detect bit is set, the hardware will inform the system software of the error, using its standard error, reporting mechanism. After the error has been reported, the system software may read and clear registers as necessary to complete its error handling protocol testing.

The port *n* error detect register and the port *n* error capture registers are also writable by software to allow software debug of the system error recovery and thresholding mechanism. For debug, software must write the port *n* attributes error capture CSR to set the capture valid info bit and then the packet/control symbol information in the other capture registers. Each write of a non-zero value to the port *n* error detect CSR will cause the error rate counter to increment if the corresponding error bit is enabled in the port *n* error rate enable CSR. When a threshold is reached, the hardware will inform the system software of the error by its standard error reporting mechanism. After the error has been reported, the system software may read and clear registers as necessary to complete its error handling protocol testing.

11.3 MEMORY COHERENCY SUPPORT

This section provides an overview of the *RapidIO Interconnect Globally Shared Memory Logical Specification*, including a description of the relationship between the GSM specifications and the other specifications of the RapidIO interconnect.

The globally shared memory programming model is the preferred programming model for modern general-purpose multiprocessing computer systems, which requires cache coherency support in hardware. This addition of GSM enables both distributed I/O processing and general-purpose multiprocessing to co-exist under the same protocol.

11.3.1 Features of the Globally Shared Memory (GSM) Specification

The RapidIO GSM specification offers the following capabilities, which are designed to satisfy the needs of various applications and systems:

- A cache coherent non-uniform memory access (CC-NUMA) system architecture is supported to provide a globally shared memory model in an environment with point-to-point connectivity between devices as opposed to a traditional bus-based interconnect.
- The size of the supported processor memory requests are either in the cache coherence granularity, or smaller. The coherence granule size may be different for different processor families or implementations.
- The GSM protocols support a variety of cache control and other operations such as block flushes. These functions are provided so that systems can be built offering compatibility with legacy applications and operating systems.

In addition to the capabilities of the GSM protocols, it is also important to point out that these protocols will work in any RapidIO connected system and are not dependent on the use of particular physical layer interfaces, Serial RapidIO or Parallel RapidIO, or system topology. These protocols are not sensitive to the bandwidth or latency of the links.

The GSM protocols are defined to operate with high efficiency and performance. The packets are small to minimize control overhead and are organized for fast and efficient assembly and disassembly. Large address spaces, up to 64 bits in size are fully supported. Moving the coherency support into the memory directories eliminates the need for broadcast transactions. This significantly reduces memory traffic and increases performance.

Under the globally shared distributed memory programming model, memory may be physically located in different places in the machine, yet accessed coherently amongst different processing elements. Coherency means that when a processor reads data from memory it always gets the most up-to-date version of the data that it is requesting. This is not difficult to accomplish when there is a single memory shared by several processors. It becomes more difficult when copies of the data are cached in one or more of the processors cache memories. To support coherency in multiprocessor systems typically requires the use of broadcast transactions. These transactions are visible to all of the processors with caches and are used to all the processor caches to keep track of the state of memory. This approach is sometimes referred to as a bus-based snoopy coherency protocol. A centralized memory controller to which all devices have equal or uniform access usually supports this approach to coherency. Figure 11.3 shows an example of a typical bus-based shared memory system.

While the snoopy bus technique works for small-scale multiprocessor systems, typically with no more that four processors, it does not scale well to higher levels of multiprocessing. Supercomputers, massively parallel, and clustered machines that have distributed memory systems must use a different technique for maintaining memory coherency. Because a broadcast snoopy protocol in these machines is not efficient, given the number of devices that must participate and the latency and transaction overhead involved, coherency mechanisms such as memory directories or distributed linked lists are required to keep track of where the most current copy of data resides. These schemes are often referred to as cache coherent non-uniform memory access (CC-NUMA) protocols. A typical distributed memory system architecture is shown in Figure 11.4. In this system there is no single snoopy bus that is visible to all of the processors. In systems like this another mechanism must be used to provide coherency between the processor groups.

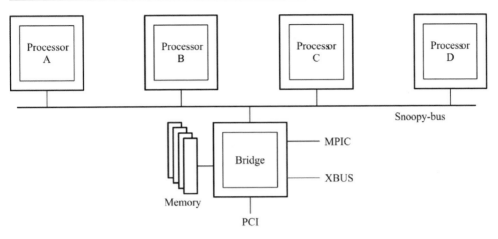

Figure 11.3   A snoopy bus-based multiprocessor system

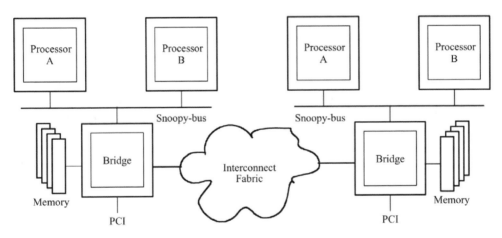

Figure 11.4   A distributed memory system

For RapidIO, a relatively simple directory-based coherency scheme is used. With this approach each memory controller is responsible for tracking where the most current copy of each data element resides in the system. The RapidIO GSM extensions provide a set of cache control and operating system support operations such as block flushes and TLB synchronization mechanisms that together can be used to build a large-scale multiprocessor system that offers hardware-based memory coherency.

To reduce the directory overhead that might otherwise be required for a very large number of processors, the architecture is optimized around small clusters of 16 processors, known as coherency domains. With the concept of domains, it is possible for multiple coherence groupings to coexist within the system as tightly coupled processing clusters.

With a distributed memory system, cache coherence needs to be maintained through a tracking mechanism that keeps records of memory access activity and explicitly notifies processing elements that hold or may hold copies of the data associated with the memory access. For example, if a processing element wishes to write to a memory address, all participant processing elements that have accessed that coherence granule are notified to invalidate that address in their caches. Only when all of the participant processing elements have completed the invalidate operation and replied back to the tracking mechanism is the write allowed to proceed.

11.4 MULTICASTING TRANSACTIONS IN RAPIDIO

The concept of duplicating a single message and sending it to multiple, selected destinations is known as 'multicast.' In many embedded processing systems, it is useful to have the capability to multicast data or messages. This can be accomplished by a variety of means in the system. For example the same data could be individually sent to all the selected targets sequentially. However, the most efficient implementation is to have hardware support present in the switch devices for the duplication of data packets or messages.

Within a RapidIO system, the ability to duplicate data packets or messages should scale with the number of end points in a system. Since the number of end points should scale with the number of switches in the system the multicast implementation for RapidIO mostly affects switches and only minimally impacts end points.

To further simplify the implementation complexity of multicast in RapidIO systems, the multicast functionality is limited to responseless write operations, specifically the NWRITE and SWRITE operations Implementing support for processing RapidIO logical layer response transactions within a switch device, which typically should not be aware of RapidIO logical layer protocols, is problematic and complex.

The ability to send a single message to multiple destinations can be implemented in a variety of ways, depending on system needs. However a standard approach is needed to simplify the task of developing interoperable hardware and software from multiple vendors.

11.4.1 RapidIO Multicast Support

This section describes the added requirements for switch devices that support multicast operations in a RapidIO system. These requirements include the addition of one new standard CAR register and three new standard CSR registers.

A RapidIO multicast operation consists of the replication of a single transaction so that it can be received by multiple end points. Switch devices will typically perform the replication, so that the capability to replicate transactions expands with the number of switches in a system. Each switch is individually programmed to control which ports the replicated packets are sent to, and thus which specific set of end point devices receive the replicated packet.

Only request packets, which do not require a response, are used for multicast operations in a RapidIO system. This is done to simplify multicast support in the end point processing element. Only NWRITE and SWRITE transactions may be multicast. Use of any other transactions results in implementation-defined behavior.

A RapidIO switch, which does not support multicast, may still co-exist in a RapidIO fabric with other switches that do support multicast. The only requirement is that the switch be capable of routing the destination ID(s) used for multicast transactions.

With respect to end point support for multicast, there are a few potential changes needed. If an end point does validate the destination ID of a packet it receives against its own ID prior to processing that packet (this validation is not required by RapidIO specifications) then that end point needs to have a disable bit for the comparison, or have additional ID comparisons available so that the end point will accept a multicast packet.

Because multicast operations are based on NWRITE or SWRITE transactions, which are address-based transactions, it is useful that all end points provide support for the range of addresses used in the multicast transactions. The target addresses are referred to as multicast channels. Multicast channels are address ranges in the RapidIO address space for which an end point may accept a multicast packet. The end point may re-map the RapidIO write address to a valid region of local physical memory. The size and quantity of multicast channels is implementation dependant. All end points participating in a system multicast must be able to accept the address from the packet and direct the write to a valid area of local physical memory. This implies that the multicast channel on each of the end points should be capable of recognizing the multicast address. The switch fabric is prohibited from modifying the multicast address.

11.4.2 Multicast Ordering Requirements

RapidIO packets which are in the same multicast group (same destination ID) with the same priority and received on the same ingress port must be multicast on the egress ports in the same order that the packets were received on the ingress port. Multicast packets in the same transaction request flow share the same source ID, destination ID, and priority fields. There are no ordering requirements between multicast packets and non-multicast packets, or between multicast packets in different multicast groups. Ordering between packets in the same flow allows an application to reliably multicast a completion flag at the end of a potentially large data transfer that was sent to the same multicast group.

11.4.3 Multicast Programming Model

Multicast operations have two control value types multicast masks and multicast groups. The set of target end points which all receive a particular multicast packet is known as a multicast group. Each multicast group is associated with a unique destination ID. The destination ID of a received packet allows a RapidIO switch device to determine that a packet is to be replicated for a multicast.

A multicast mask is the value that controls which egress ports one or more multicast groups (destination IDs) are associated with. Conceptually, a multicast mask is a register with one enable bit for each possible switch egress port. There is one set of multicast masks for the entire switch. All multicast masks in a switch are assigned unique sequential ID numbers, beginning with 0x0000. This assists with the task of managing the configuration of these masks. Software configures switch devices to associate multicast groups (destination IDs) with multicast masks.

Configuring a RapidIO switch to replicate packets for a multicast group is a two-step process. First, a list of egress ports is enabled or disabled by a multicast mask resource. Second, one or more destination IDs, which define multicast groups, are associated with the multicast mask in the switch. During normal system operation, any time a switch receives a

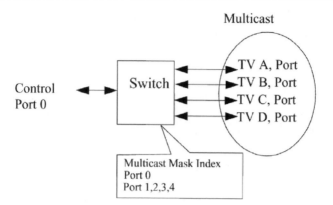

Figure 11.5 Multicast basic concept

packet with a destination ID that has been associated with a multicast mask; it will send that packet to all egress ports enabled by that multicast mask.

Figure 11.5 shows a simple multicast system, where a control unit connected to switch port 0 needs to multicast to televisions (TV) A, B, C and D. TV A, B, C and D are a multicast group. A multicast mask, in this case arbitrarily picked as multicast mask 2, is set up to select which ports in the switch are part of the multicast group. In this example, assume that the system designer assigns 0x80 as the destination ID the control unit should use to multicast to the TVs. In order to multicast, an associate operation must be performed between destination ID 0x80 and multicast mask 2.

The basic approach to multicast is to identify destination IDs that should be multicast and program the switches with this information, and to use the multicast mask to remove selected devices from the group of all devices that otherwise might be multicast to. The association operations are used to program the switch with the information of which device destination IDs should be multicast and for the associated destination IDs which sets of mask should be used to filter egress ports from the multicast pool.

An API for setting up multicast access is being developed to provide for consistent access from multiple device vendors.

11.4.4 Other Multicast Characteristics

Despite the settings of a multicast mask, no packet received on a port will be multicast out of that same port. This feature allows a group of end points which need to multicast to each other to share the same multicast mask.

The default state after a reset for multicast masks is that multicast masks have no ports selected. Additionally, after reset no associations exist between any multicast group/destination ID and a multicast mask. Implementation-specific capabilities may modify the multicast mask values and associations after reset without software intervention.

While each destination ID is associated with a unique multicast group, the programming model allows a destination ID to be mapped to different multicast masks on a per-port basis. It is also possible to map a given destination ID to the same multicast mask for all ports.

11.5 MULTICASTING SYMBOLS

A special characteristic of the RapidIO parallel and serial physical layers is the ability to multicast certain control symbols. The multicast-event control symbol provides a mechanism through which notice that some system-defined event has occurred, can be selectively multicast throughout the system.

When a switch processing element receives a Multicast-Event control symbol, the switch is required to forward the multicast-event by issuing a multicast-event control symbol from each port that is designated in the port's CSR as a multicast-event output port. A switch port should never forward a multicast-event control symbol back to the device from which it received a multicast-event control symbol, regardless of whether the port is designated a multicast-event output or not.

It is intended that at any given time, a single device will source multicast-event control symbols. However, the source device can change (in case of failover, for example). In the event that two or more multicast-event control symbols are received by a switch processing element close enough in time that more than one is present in the switch at the same time, at least one of the multicast-event control symbols shall be forwarded. The others may be forwarded or discarded (device dependent).

The system-defined event whose occurrence is signaled by a multicast-event has no required temporal characteristics. It may occur randomly, periodically, or anything in between. For instance, a multicast-event may be used for a heartbeat function or for a clock synchronization function in a multiprocessor system.

In an application such as clock synchronization in a multiprocessor system, both the propagation time of the notification through the system and the variation in propagation time from multicast-event to multicast-event are of concern. For these reasons control symbols are used to convey multicast-events as control symbols have the highest priority for transmission on a link and can be embedded in packets.

While this specification places no limits on multicast-event forwarding delay or forwarding delay variation, switch functions should be designed to minimize these characteristics. In addition, switch functions shall include in their specifications the maximum value of Multicast-Event forwarding delay (the maximum value of multicast-event forwarding delay through the switch) and the maximum value of multicast-event forwarding delay variation (the maximum value of multicast-event forwarding delay through the switch minus the minimum value of multicast-event forwarding delay through the switch).

11.5.1 Example of Use of Multicast-event Control Symbol

Multi-processor systems frequently have a need for a distributed, low-latency, common real-time or wall-time clock reference for software to use to manage functions such as time-stamps on data, inter-process and inter-processor synchronization, and accurate performance analysis. This clock is conceptually different from the oscillator clock that is used to operate the logic of the processing devices in a system. Because each processing device has its own real-time or wall-time clock, the different copies of the clock may drift in relationship to one, similarly to the way in which watches get out of synchronization with each other after a period of time. If it is important that these clocks stay synchronized to within some tolerable range, it is necessary to provide some means of resynchronizing them. One way to do this is to periodically broad-cast a multicast-event control symbol throughout the system.

The multicast-event control symbol is typically generated at a single point in the system by a central clock master and then distributed to all of the time-of-day clock locations in the system by the interconnect fabric. For example a switch that receives a multicast-event control symbol would transmit a copy of it to a predetermined list of output ports. Eventually, every processing element in the system that has a copy of the real-time or wall-time clock would receive no more than one multicast-event control symbol for each multicast-event control symbol issued by the central clock master device.

Various methods can be used to manage the synchronization of real-time or wall-time clocks in a device:

- A simple, but relatively imprecise and inaccurate distributed real-time clock could be man-aged by software with the receipt of a multicast-event control symbol that generates a clock synchronization interrupt to a local processor. The local processor then manually incre-ments the local clock. For example, a multicast-event control symbol could be distributed through the system every millisecond. At this resolution the multicast-event symbol would arrive once every million clock cycles of a Gigahertz speed processor. The load for process-ing this event would be relatively low although the accuracy would be much lower than the nanosecond range at which the processor operates.
- A slightly more complex scheme (with corresponding better accuracy and precision), might employ a local counter or two to compensate for the interrupt service latency and 'jitter' as well as lost multicast-event control symbols.
- A sophisticated scheme with much better accuracy and precision could employ a number of counters and some associated state machines at the end point processing elements to compensate for control symbol propagation variance between the clock master and the other processing elements in the system, accumulated jitter induced by the fabric, local oscillator drift, and other system-specific and device-specific effects.

The actual approach to distributing the control symbol and compensating for the latency of receiving the control symbol and processing it is left up to the system designer. The multicast-event control symbol is a subtype of the packet control symbol format. More details on these control symbols can be found in Chapters 7 and 8.

12

Data Streaming Logical Layer

Chuck Hill

12.1 INTRODUCTION

The data streaming logical layer is different from the other logical layers in several ways. Data 'streams' are defined as persistent relationships between a source or ingress device, and a destination or egress device. A data stream contains multiple unidirectional transactions separated by discrete intervals of time. The gaps in time between transactions allow many streams to exist simultaneously in a system, and enable the traffic to share a single fabric. A system's ability to manage the quality of service (QoS) offered to the individual streams determines the utilization of the fabric's resources as well as the faithful preservation of each stream's transaction relationships. All of these factors are components of the data streaming logical layer transaction protocol.

The purpose of the RapidIO data streaming logical layer is to provide a standard approach for encapsulation and transport of data streams across a RapidIO fabric. Figure 12.1 shows how RapidIO might be able to interface to a range of data and control elements that might be typically present in a communications system.

The RapidIO Trade Association, to highlight the capabilities of the data streaming logical layer and other newer features of the RapidIO interconnect technology, has coined the term RapidFabric™. RapidFabric is meant to describe the use of RapidIO in dataplane connectivity applications. For a product to claim RapidFabric support, it should at least support the following RapidIO specifications beyond the base RapidIO capabilities:

- Data streaming logical layer (described in this chapter). RapidIO end point devices generate data streaming packets. Legacy RapidIO switches will transport data streaming packets. More advanced switches may use the virtual queue or virtual stream ID to make higher level ordering decisions for traffic management.
- Flow control extensions (described in Chapter 11). Flow control packets are generated by RapidIO switch devices and consumed by RapidIO end point devices.

RapidIO® The Embedded System Interconnect.   S. Fuller
© 2005 John Wiley & Sons, Ltd   ISBN: 0-470-09291-2

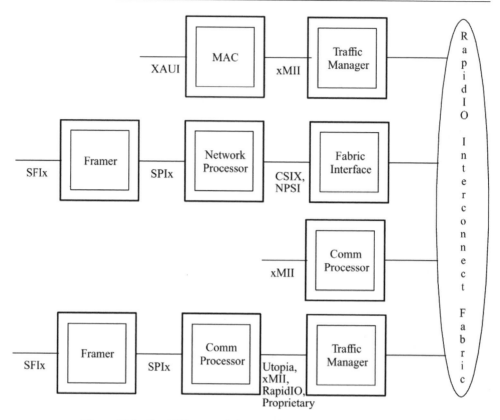

Figure 12.1 RapidIO connectivity to various data-centric interconnects

- Multicast extensions (described in Chapter 11). Multicast support is an attribute of RapidIO switch devices. End points do not generate special multicast packets, but multicast capable switches must be properly configured for multicast to work within the system.

Memory, Input/Output and GSM logical layer operations defined in the I/O logical can be unidirectional or bidirectional operations. In all cases, there is always an implied interaction between both parties. These operations often require some shared knowledge about what is occurring between the two parties. This kind of knowledge may be necessary when a stream is established, but is not needed for each transaction. An example of this would be the mapping of an address range on the target device. In contrast, data streaming devices may be bidirectional in their flow of traffic, but there will rarely be a relationship between the data that is traveling in opposite directions.

Figure 12.2 shows how various streams of data may be prioritized, interleaved and transmitted across a fabric. Streams can be established where the ingress device has no knowledge of the internal workings of the egress device. This ability to construct *virtual* relationships between the ingress and egress devices is a powerful feature of the data streaming logical layer. It uncouples resource dependencies between ingress and egress and allows more diversity in the devices that will interoperate on the same fabric. It is one of the distinctions found between a switched interconnect and a true fabric.

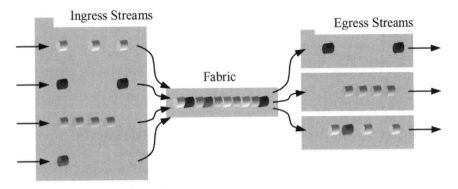

Figure 12.2 Data streams and fabrics

The basic units of transaction in a streaming fabric are self-contained packets of data called *protocol data units* or PDUs. A PDU contains all the information needed by the destination for the continued treatment of the packet. PDUs, as the name implies, are based on a specific protocol. In the communications world, there are many protocols with more being created every year. Offering a technology that can support any protocol without relying on the presence of a specific protocol is another important feature of the data streaming logical layer.

PDUs can come in a variety of sizes. Sizes may range from very small (a few tens of bytes) to very large (many thousands of bytes). The data streaming logical layer offers a *segmentation and reassembly* (SAR) capability to fix the maximum size of the data transfers in the fabric itself. This fixed maximum *segment* size makes latency in the fabric more predictable.

PDUs also have a 'shape.' The shape of a PDU is determined by its timing in its arrival to the ingress, *and* its required timing for delivery to the egress. For example, the transactions within a stream may have hard latency requirements, which if exceeded, render the data useless at its ultimate destination. Voice telephony data is an example of data that has hard latency constraints. This traffic may not only have an upper limit on how long it can take to be delivered, but may also have a lower limit. The data may be useless if it is received too soon. These kinds of data stream may have to share resources with data that can arrive and depart with much more variability.

In addition to fixed packet sizes, the data streaming logical layer contains traffic management semantics for more advanced control of the utilization of system resources. These semantics range from congestion management to congestion avoidance to more sophisticated synchronization of management devices. The traffic management features in data streaming offer increasing sophistication, coupled with increasing complexity. Systems may employ the necessary degree of complexity required to satisfy the needs of the application. The ability to manage traffic that is wide ranging in size, timing, and importance is a critical capability for many systems, and another key feature of the data streaming logical layer.

The features of the RapidIO Data Streaming Logical Layer are:

- Supports independent streaming transactions
- Supports variable size PDU through SAR functionality

- Is independent of the internal PDU protocol
- Supports virtual stream identification
- Supports varying degrees of traffic management

This chapter examines each of these characteristics in more detail and explores how the data streaming logical layer facilitates all these capabilities.

12.2 TYPE 9 PACKET FORMAT (DATA STREAMING CLASS)

The Type 9 packet format is the DATA STREAMING transaction format. Type 9 packets always have a data payload, unless they are terminating the PDU. Unlike other RapidIO logical specifications, the data payload length is defined as a multiple of half-words rather than double-words. A pad bit allows a sender to transmit an odd number of bytes in a packet. An odd bit indicates that the data payload has an odd number of half-words. This bit makes it possible for the destination to determine the end of a data payload if packet padding is done by the underlying transport. An extended header bit allows future expansion of the functionality of the Type 9 packet format. All of these bits are contained in the segmentation control field (SCF). The location and definition of these bits are described in Section 12.3.3. Definitions and encodings of fields specific to Type 9 packets are provided in Table 12.1. Table 12.2 details the various O and P bit combinations.

In addition to these basic fields, there are some specific 'constructs' that are important to data streaming. These constructs are composed of multiple fields. The first is the *virtual stream ID* or VSID. The VSID is a combination of the SourceID or DestinationID + Class + StreamID. The SourceID or DestID is used depending on the frame of reference. The egress, for example sees a stream as identified by the SourceID. VSIDs should be uniquely assigned for all streams in a system. Section 12.3 discusses the use of VSIDs in more detail.

A *virtual queue* contains multiple virtual streams. A virtual queue is identified by the SourceID or DestID + Class. Again, the use of SourceID or DestID is dependent on whether the queue is on the ingress or egress side of the fabric.

A PDU is segmented into one or more fabric segments by the segmentation and reassembly protocol. These segments are designated *start segments, continuation segments, end segments,*

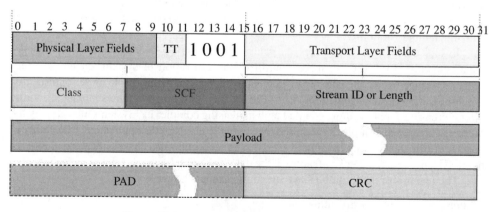

Figure 12.3   Data stream logical layer packet format

Table 12.1 Specific field definitions and encodings for Type 9 packets

Field	Definition
cos	Class of service. This field defines the class of service to be applied by the destination end point (and possibly intervening switch processing elements) to the specified traffic stream
S	Start. If set, this packet is the first segment of a new PDU that is being transmitted. The new PDU is identified by the combination of the source of the packet and the segmentation context field
E	End. If set, this packet is the last segment of a PDU that is being transmitted. Both S and E set indicates that the PDU is fully contained in a single packet
rsrv	Reserved. Assigned to logic 0s by the sender, ignored by the receiver
xh	Extended header. There is an extended header on this packet. Currently there are no defined extended header formats. It is always assigned to 0b0 for Type 9 packets
O	Odd. If set, the data payload has an odd number of half-words
P	Pad. If set, a pad byte was used to pad to a half-word boundary
streamID	Traffic stream identifier. This is an end-to-end (producer-to-consumer) traffic stream identifier
length	PDU length. This is the length in bytes of the segmented PDU 0x0000, 64kbytes 0x0001, 1 byte 0x0002, 2 bytes 0x0003, 3 bytes . . . 0xFFFF – 64kbytes – 1

Table 12.2 Specific field definitions and encodings for Type 9 packets

O bit	P bit	Definition
0b0	0b0	Even number of half-words and no pad byte
0b0	0b1	Even number of half-words and a pad byte
0b1	0b0	Odd number of half-words and no pad byte
0b1	0b1	Odd number of half-words and a pad byte

and *single segments*. A PDU that is equal to or smaller than the *maximum transfer unit* or MTU is transferred as a single segment. PDUs larger than the MTU size are transferred in multiple segments. The segmentation process is described in Section 12.3.3.

12.3 VIRTUAL STREAMS

Virtual Streams are defined as an *ordered set of protocol data units*. The number of streams that might be employed by a system is application specific. A system can have a single stream between an ingress device and an egress device, or it can have a unique stream for every combination of users and traffic types. In the latter case, this could be millions of individual streams. The *virtual stream identifier* supports up to 4 million streams between any source and any destination in a system.

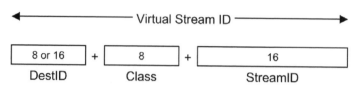

Figure 12.4   Virtual stream ID

As the figure shows, the VSID is a concatenation of the port address (source or destination) the class of service field, and the streamID field. This means that a RapidIO data streaming system can have 64k unique streams per traffic class, per source – destination pair. The VSID is 'virtual' because it can be *overloaded* with any meaning required by the application. The destination defines the way the VSID is allocated. The VSID can be associated with specific buffers, or mapped to the physical address of any downstream device at the destination. VSIDs can be used to segregate incoming traffic by protocol or divert traffic to *subports*. Subports are additional destinations not exposed to the RapidIO fabric address map. The streamID could be used to identify a subport.

The source device is responsible for placing the necessary VSID on a packet so that it will arrive properly at the destination. Usually this process will involve some form of protocol specific *classification*. Classification involves looking at specific fields in the PDU and determining the proper handling of the packet. For most RapidIO traffic, classification would simple determine the destination port and possibly memory address for an operation. In data streaming, the PDU must be labeled with a complete VSID.

The VSID provides a 'tag' for all the handling of the packet through the system. As all RapidIO data streaming packets have this same tag, there is no need for additional protocol-specific handling of the packet. Part of that handling is a process of reducing (potentially millions of) streams to a few flows in the fabric. The VSID is structured as a group of fields so that all the different behaviors of a fabric are contained within this tag.

12.3.1 Class of Service, Virtual Queues

A group of virtual streams share a common treatment in the fabric by sharing a *class of service*. A *virtual queue* is defined as a concatenation of the sourceID or destID and the class of service field.

Virtual queues are an arbitrary interim staging point for traffic entering or exiting the fabric. A simpler fabric may choose not to use this facility. More complex fabric designs would make use of some form of this capability, so RapidIO data streaming offers a standard approach to provide this capability. Systems using virtual queuing may not employ all 256 available classes of service. A smaller subset of virtual queues may be defined by using a subset of the class bits. When this is done the subsets should use the most significant bits of the class field to define the range of the subsets. As an example, systems with two classes of

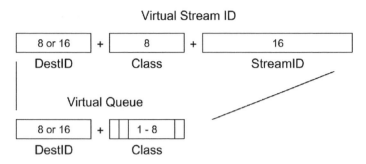

Figure 12.5 Virtual stream ID and virtual queue

service would use a single class bit, systems offering four classes of service would use the two most significant bits, and so forth. Also, if a destination is providing fewer classes of service *per stream* then the remaining class bits may be used to expand the size of the StreamID field and the number of unique streams supported.

Quality of service may be applied to traffic in a variety of different ways. A specific traffic class may offer a *traffic contract* based on priority, latency, bandwidth, or some combination of these factors. Traffic in the fabric will be managed based on the precedence of the most significant bits in the class field. The quality of service *scheduling* that is applied at various interim points in the fabric should not distort the overall desired quality of service between the source and destination. For example, a group of classes providing bandwidth reservation, should share enough bits in common to be distinguished from classes intended to provide strict latency support.

12.3.2 RapidIO Flows

A flow in RapidIO is a logical construct created by the combination of the SourceID, DestID, and the transport 'channel.' RapidIO transactions defined in the 1.2 version of the RapidIO specification rely on physical layer priority bits to complete flow identification. In addition, priority bits enforce specific transaction ordering rules within and between flows and prevent deadlock.

PDUs within a stream are required to remain ordered. The relationship of streams to virtual queues, to flows may vary with implementation, but may not vary dynamically. PDUs are placed into RapidIO flows in *segments*. Segments are also required to be delivered in order. The RapidIO transport and physical layers provide this guarantee of delivery in order.

Figure 12.6 show the relationship between virtual Streams, virtual output queues (VoQ) and flows. A mapping of the data streaming logical layer class bits to the physical layer priority bits is required at the ingress to the fabric to ensure that the fabric will properly handle the ordering of the stream as it transits the fabric.

12.3.3 Segmentation and Reassembly

Fabrics with wide variation in packet sizes can create wide variations in latency. If a packet with a higher quality of service arrives at an ingress point just after a very large packet has

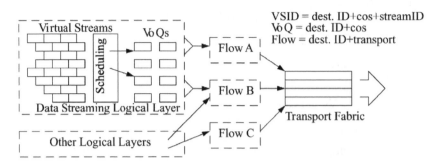

Figure 12.6   Streams, queues and flows hierarchy

begun to be sent, it will have to wait for the entire large packet to be sent. This will create a statistical variation in latency in the traffic. This variation in latency reduces the amount of time available to the applications in systems with hard latency requirements.

To support fixed, smaller, packet sizes, larger PDUs must be segmented. Data streaming includes support for segmentation and reassembly as part of the logical layer protocol. This protocol will break up PDUs that are larger than the maximum payload size of a RapidIO packet in to multiple segments. There are three classes of segments. A typical large PDU will be broken up into a start segment, some number of continuation segments, and an end segment. PDUs at or below the segment size would be transported in a single segment. Bits in the segmentation control field identify the segment type. Figure 12.7 presents the bit definitions for the SCF field of data streaming packets.

The size of the data payload in a segment is determined by a system-wide parameter: *maximum transfer unit* or MTU. All segmentation and reassembly units on a common fabric operate with the same MTU value. MTU size for RapidIO packets may be set between 32 and 256 bytes, however all RapidIO devices must be capable of receiving any MTU payload size up to 256 bytes.

Data is placed in a segment until the MTU size is reached, then a new segment is started. A PDU that is more the 2 MTU in size will be transferred as a start segment, one or more con-

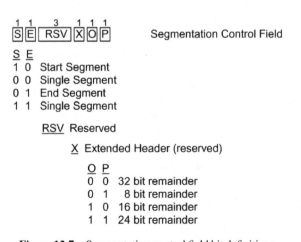

Figure 12.7   Segmentation control field bit definitions

tinuation segments, and completed with a single end segment. If the PDU is greater than one MTU is size, but equal to or less than 2 MTU, it will be transferred as a start segment, and an end segment. If the PDU size is equal to or less than the MTU size, the data is transferred as a single segment.

Start segments and continuation segment payloads are by definition always equal to the MTU size. End segments and single segments contain variable length payloads equal to the portion of the packet left over after any MTU sized segments are filled. RapidIO packets are always a multiple of 32 bits (4 bytes) in length. The O and P bits are used to restore the original payload to byte granularity. The MTU size is set in 4-byte increments. The O and P bits are only used in single or end segments.

12.3.4 Segmentation Contexts

The format for the segment contents varies depending on whether it is a start, continuation or end segment. Start segments contain the information for a complete VSID. When a start segment is received by a destination, a *segmentation context* is created. Additional continuation or end segments are associated with the right context by comparing the source of the packet and the information that defines the channel used by the physical transport; the RapidIO 'flow' it is received on. Continuation and end segments do not contain the StreamID. The VSID from the start segment must be preserved by the destination during reassembly, and then used for further packet handling.

Segmentation contexts are dynamic entities created when a start segment is received and closed when an end segment is received. Destinations will support a finite number of active reassembly contexts. The system must manage the number of contexts a destination will see by limiting the number of flows a source may have with a destination, and the number of sources that may send data to any given destination. The number of segmentation contexts is a 'discovered' parameter that should be available from the register set of a device.

Example 1: a destination port can support 128 segmentation contexts. The system contains 256 ports. In this case only 128 devices may be configured to stream data to the destination, and in this case each source may only use a single flow to that destination.

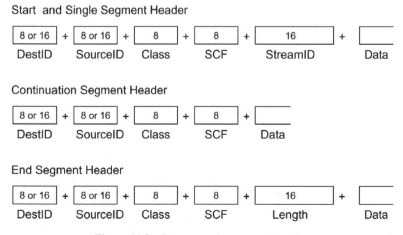

Figure 12.8 Data streaming segment headers

Example 2: A different destination port can also support 128 segmentation contexts. The system has 76 total ports, 4 of these ports are configured to source data through 4 flows, 21 of these ports are configured to use 2 flows, and 50 ports are configured to only use a single flow. The total adds up to 108 contexts. This does not exceed the limit of 128 segmentation contexts for that destination.

With the segmentation capabilities of the RapidIO Data streaming logical layer, a PDU that exceeds the MTU of the RapidIO fabric will be segmented as it is de-queued from a virtual output queue in the source device. The first segment will be a start segment and will contain the 16-bit StreamID field and will be assigned a class. Subsequent segments will be either continuation or end segments. These segments will share the same source, destination and class fields as the start segment. They will not contain the StreamID. Because of the RapidIO ordering rules between sources and destinations at a given priority level, the packets associated with this segmentation context will guarantee that the subsequent segments are received in order at the destination and that the PDU will be able to be correctly reassembled.

12.3.5 Reassembly

As described in Section 12.3.4, when a start segment is received, a segmentation context is opened. When the end bit is received, the context is closed. A single segment (with both bits set) causes the context to be simultaneously opened and closed. In between, the reassembly process can distinguish a number of failures.

- A lost start segment will result in a continuation or end segment arriving without an open context. The additional segments should be discarded until a valid start segment is received.
- For multisegment PDUs, the end segment contains a length field. This permits the reassembled packet to be checked against lost continuation segments. If the total length doesn't match what the end segment presents then an error has occurred. In this situation, the system designer can choose to forward the PDU knowing it has been corrupted, or most likely just discard the whole PDU.
- A lost end segment will result in a start segment arriving for a context that is already open. The previous reassembly should be discarded, and the context reset for the new start segment.
- Continuation and start segments must match the MTU size setting. If either of these packet types is not the correct size, they are considered invalid. With the physical layer error checking (and guaranteed packet delivery) it is very unlikely that partial packets will occur over the links, but intermediate stages of logic could cause packet corruption. It is also possible that software might incorrectly set the MTU register settings.

Discarding corrupted PDUs can reduce overall system overhead. There is no need to burden the remainder of the network by transporting a bad PDU. This frees the RapidIO fabric from needing to provide full support for hardware-based error recovery, as is necessary for the control plane operation. In control plane applications such as processor-to-processor signaling, the fabric must provide a robust connection. In data streaming applications there is often a higher-layer protocol that either tolerates or deals with lost packets. Most data protocols contain provisions for dealing with lost packets. TCP for example will issue retry commands for file segments to avoid corruption. Voice over IP packets will typically use a protocol that can tolerate dropped samples without audible distortion.

12.4 CONFIGURING DATA STREAMING SYSTEMS

The configuration registers for data streaming support 3 parameters:

* maximum transfer unit
* maximum reassembly contexts
* maximum PDU size

Maximum transfer unit is configured to the same value in each data streaming endpoint. Failure to align this value will result in lost data. The value is specified in 4-byte increments from 32 to 256 bytes. This value refers to the size of the payload field. It does not include the fixed overhead of the packet header format. Values other than those identified by the specification may result in unknown behaviors. There is no requirement for the end point designer to enforce certain behaviors.

Maximum reassembly contexts is used by all potential sources to discover this limit in a destination. It is not 'writeable' by other elements in the RapidIO fabric. It may be a locally configurable parameter in end point designs in circumstances where the reassembly process occurs further downstream. An example is a RapidIO to CSIX bridge. (CSIX is a standard interface for network processors defined by the Network Processor Forum.) CSIX uses a segmented interface so a bridge may not need to completely reassemble packets. That may be done by the network processor, so this parameter may need to be configured based on the device to which the bridge is connected.

Another example is a bridge to a general purpose CPU. Reassembly may occur in software and be limited by the software interface. The CPU may preprogram the maximum reassembly contexts based on memory and performance capabilities.

Maximum PDU size tells the sources how big the final PDU buffers are. Data streaming supports up to 64-kbyte PDUs, but allocating 64-kbyte might not be necessary for some applications. An ATM end point, for example may only expect 53-byte PDUs. Some IP protocols are limited to 8-kbyte PDUs. An end point may choose to limit the size of memory used for PDU storage for design reasons. This parameter tells a source not to even try to send a larger PDU. The end point should enforce size limits by discarding PDUs (and subsequent segments) when storage limits are received.

The maximum PDU size parameter is not programmable by the 'system,' but may be locally configurable. Some of the same reasons described above apply. Another consideration is that a system may be able to down configure the buffer size to gain more SAR contexts if the application is known to have size limits. Wireless applications, for example, use UDP/IP packets, but are always size limited because of the handset designs.

12.5 ADVANCED TRAFFIC MANAGEMENT

The data streaming logical layer specification reserves a bit for an 'extended' packet. The extended packet will be used for *advanced traffic management*. At the time of this publication, the format for this packet hasn't been defined. However, the following describes planned features of advanced traffic management.

The existing flow control specification allows end points and intermediate points to control traffic by 'flow.' A flow will have many streams of traffic mapped to it. The existing flow control provides *congestion management* in peak traffic conditions.

The 'class-based' traffic management features, available for data streaming messages, allow traffic to be controlled at the source according to class. This gives a finer grained inter-action between end points, allowing certain classes to be preferred over others and reducing or avoiding flow congestion. This *congestion avoidance* will increase the effectiveness of a fabric in dealing with large variations in traffic. Congestion avoidance can also be applied to streams. The advanced flow control message will support information and management pertaining to specific VSIDs. This creates an even finer grained method for managing traffic.

The extended packet feature in advanced traffic management would provide a bidirec-tional message that allows ingress and egress schedulers to work in conjunction with each other. This extended packet message provides the 'semantics' for very advanced traffic man-agers, but does not specify how the design should to be implemented. An application defined message format for specialized uses is also provided.

Advanced traffic management is optional for data streaming systems. The system may rely on flow control or it may rely on simple class based traffic management. If the applica-tion warrants the use of more sophisticated traffic management functions those may also be employed. The data streaming logical layer supports a wide range of flow, congestion and traffic management capabilities to support the wide variety of applications represented in the embedded communications equipment market.

12.6 USING DATA STREAMING

Data streaming can be applied in a variety of different ways. The critical design issues revolve around the interfaces into and out of the RapidIO fabric. In telecommunications equipment, there are a wide variety of interconnect standards (SPI-3, SPI-4, UTOPIA, GMII, etc.). These standards are primarily focused on interfacing to the external communications networks. Standards for internal communications between chips and boards within a system have not existed up to this point. RapidIO is the first standard to be developed to specifically address this internal chip-to-chip and board-to-board communications requirement. In this role RapidIO can also form a common in-system fabric or interconnect for a wide variety of protocol internetworking applications.

When using the RapidIO data streaming logical layer transactions for protocol inter-networking, the first design decision will be how to use the virtual stream ID. The use of the VSID will vary according to the protocol that is being transported across a RapidIO system. An ATM application, for example, might use a RapidIO to UTOPIA bridge. The UTOPIA bus supports a number of physical devices on the same bus, these physical devices are addressed by the UTOPIA bus. The StreamID may be used as a subport address that translates to the physical UTOPIA address. If multiple UTOPIA ports are supported, the StreamID may be used to select UTOPIA interfaces. When the source examines the VPI/VCI fields for schedul-ing and forwarding the ATM cell, the VSID (port, class, UTOPIA address) is generated. Using the StreamID in this manner keeps the number of RapidIO ports to a minimum. This example is one of the simplest ways to use the VSID.

At the other extreme, a cable modem termination system (CMTS) might manage traffic by user and traffic type. CMTS systems provide streaming voice, streaming video, and Internet access to households. Many cable providers offer up to four computers, three video devices, and multiple telephone connections per household (ten data streams). In the downstream direction, video and voice must be managed for maximum latency for high QoS. Internet traffic must be segregated to avoid interference with the video and voice traffic. Also, the Internet traffic may have bandwidth limits imposed on it according to subscription criteria.

An application of this kind will require sophisticated traffic queuing and scheduling. A CMTS system typically consists of one or more trunk uplinks, and multiple downlink trunks to various cable neighborhood networks. At the trunks, downstream traffic is 'classified' by user and data stream. Each stream type (video1, video2, voice1, computer1, etc.) would be assigned to a traffic class and have appropriate flow or traffic management applied to it. Each user would be assigned a unique StreamID. The scheduling for the trunks places traffic into virtual output queues by class and destination. The ingress scheduling for the fabric takes traffic from the VoQs, prioritizing traffic based on latency and other quality of service metrics.

Using the class and stream features in the VSID allows traffic to be classified a single time at ingress. The queuing at the fabric ingress and egress can be predesigned for various stream types and number of users. Flexibility in the design of the traffic scheduling provides the CMTS with the adaptability to support future services.

The UTOPIA bridge example requires little queuing and no segmentation and reassembly. The traffic from the smaller ports use a single StreamID to send data to centralized scheduling and switching circuitry. The CMTS example uses a hierarchy of queues and scheduling with potentially millions of individually managed streams. The flexibility of the VSID mechanism is at the heart of the ability of the RapidIO data streaming logical layer to meet both sets of requirements with a single approach.

Bridging to other interfaces involves a variety of design decisions. Section 12.6.1 contains a design example of a CSIX to RapidIO bridge. Bridging between protocols involves joining two segmented interfaces, adapting different flow control protocols, and mapping the StreamID.

12.6.1 CSIX to RapidIO Data Streaming Interface

The acronym CSIX stands for 'common switch interface.' The CSIX-L1 interface defines a physical and packet interface as a common mechanism to interface to switch fabrics. The Network Processor Forum, an industry managed standards forum, much like the RapidIO Trade Association, developed CSIX as a way to standardize the fabric interface for network processors that were offered by various vendors. Availability of a CSIX to RapidIO fabric interface would facilitate the interface of most standard network processors in to a RapidIO fabric.

The rest of this section addresses adaptation of the CSIX unicast packet format to a RapidIO data streaming packet format. It also discusses the CSIX flow control features and how they could be mapped to RapidIO flow control constructs.

12.6.1.1 Introduction to CSIX Packet Formats

The CSIX specification describes the packet format as consisting of a base header, optional extended header, payload and vertical parity. A *CFrame* is a CSIX packet defined by the 'base header.' The base header format is shown in Figure 12.9.

The base header type defines the packet as a Unicast, Flow Control, etc. Table 12.3 shows the definitions of the CSIX header fields. For the base header format and for other formats described in this section.

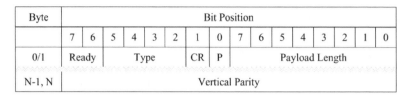

Figure 12.9   CSIX base header format

Table 12.3   Field definitions for CSIX base headers

Field	Encoding	Definition
Ready [7]		Data ready (Unicast, Multicast, Broadcast)
Ready [6]		Control ready (Command Status, Flow Control)
Type		See the CSIX-L1 spec for a complete list of type field definitions
	0b0000	Idle
	0b0001	Unicast
	0b0110	Flow Control
CR		CSIX Reserved
P		Private (may be used for design specific features)
Class		Class of service. 256 Classes are supported
Destination		Destination port address. 4096 ports are supported
FC		Flow Control Type (unicast, multicast, broadcast)
C,P		Class and port wildcards
Speed		Provides 'throttling' of the data from 0b1111 (full on) to 0b0000 (full off)
Payload length		Length in bytes of the payload (0 = 256 bytes). Max payload size is 256 bytes
Vertical parity		Required field, checking of parity is not required

Beyond the base header, all CSIX packets contain a logical 'extended header.' The extended header format varies by packet type. An idle packet has a zero-length extended header and payload. In the absence of data or control packets, idle packets will be continuously sent. A more intelligent bridging function would either delete idle packets or replace them with RapidIO idle symbols as appropriate. A simpler straight protocol mapping function might blindly send the CSIX idle packets as encapsulated RapidIO data streaming packets that would be recovered by the destination port. A CSIX port will require either a continuous stream of data or idle packets to keep the ready bits up to date and the clocks in synchronization.

The Unicast CSIX packet format is the primary packet used for carrying data within a CSIX environment. The basic unicast packet format is shown in Figure 12.10 (with the base header). Bytes 2–5 make up the extended header bytes for the unicast packet. The payload starts with byte 6. The maximum payload size supported by CSIX is 256 bytes. However, devices may not support variable or different fixed frame sizes. The system integrator is responsible for ensuring that all of the CSIX devices used in the system will support the intended frame sizes.

CFrame Unicast Header															
Byte	Bit Position														
	7	6	5	4	3	2	1	0	7	6	5	4	3	2	1	0
0/1	Ready		Type			CR	P	Payload Length							
2/3	Class						P	CSIX Reserved							
4/5	CR		Destination												
6 / N-2	Payload														
N-1, N	Vertical Parity														

Figure 12.10   CSIX Unicast frame format

12.6.1.2 Adapting CSIX to RapidIO Data Streaming Format

This example presumes an ability by the network processor to produce a RapidIO adaptation of the CSIX packet. Essentially the CSIX fields for a unicast frame and other frames are populated such that the adaptation to RapidIO is as straightforward as possible. There are several considerations that must be taken in to account when adapting CSIX to RapidIO:

- Segmentation. CSIX defined an 'L2' interface, but never standardized its use. The interface requires the overlay of the RapidIO data streaming segmentation semantics on the CSIX L1 Frame.
- Source / destination addressing. CSIX supports a 12-bit destination address. RapidIO supports 8- and/or 16-bit device addressing. In addition, RapidIO packets require the placement of a source address in the transport header.
- Payload. The CSIX payload and the RapidIO payload may be different. If all the end points are the same, and the CSIX payload size can be set to match the RapidIO size, then the interface is simpler. If this interface is used in conjunction with other end points, the lack of flexibility here would force all other end points to segment at the fixed size used by this device.
- Data streaming virtual stream ID. The VSID for data streaming requires a length field and a StreamID. The interface might generate the length field, but the StreamID must come from the Network Processor. It has the sophistication to perform the classification step.
- RapidIO 'flow.' The CSIX PDU is mapped to a specific 'flow' on the RapidIO fabric. This implies associating the packets to a specific physical layer priority. One approach would be to use a lookup table on the class field.

An adapted CSIX unicast packet might look something like Figure 12.11. It is assumed in this example that the device attached to the CSIX interface contains a segmentation and reassembly mechanism. Depending on the CSIX payload sizes and the efficiency of the header mapping, this mechanism may not be necessary. In this example, this mechanism must also flag segments as start, continuation, and end segments. Overloading the 'private' bits in the unicast header (byte 3) with the S and E flags, saves space in the overall packet. A CSIX device that is not flexible enough to use this portion of the Unicast header would have to place all this information into the payload.

To support differences in the CSIX payload segment size and the RapidIO maximum transfer unit (MTU) setting, data is temporarily stored in a re-segmentation buffer. The buffer

RapidIO Unicast Header															
Byte	Bit Position														
	7	6	5	4	3	2	1	0	7	6	5	4	3	2	1	0
0 / 1	Ready		Type				CR	P	Payload Length							
2 / 3	Class						S	E	CSIX Reserved							
4 / 5	Dest			Destination												
6 / 7	StreamID/ PDU Length / or PAD														
8 / N-2	Payload														
N-1, N	Vertical Parity														

Figure 12.11   Unicast packet for RapidIO adaptation

is sized to support just a few packets to allow the segment size to be converted up or down, depending on the differences in size. It is not necessary to fully reassemble the PDU before transitioning onto the other interface.

The re-segmentation buffer is composed of storage for the header fields, and storage for the data portion of the payload. The header is stored separately to be able to apply fields as needed to each segment (see header transmogrification). It also identifies a packet with the right context. Matching the destination address, and the class field, with the value in the stored header, associates subsequent packets with the right context. The number of contexts a bridge supports should align with the number of contexts supported by the end device. The *segmentation context capability register* (on the RapidIO maintenance interface) must be programmed with the lesser value of contexts supported by the CSIX device or the bridge interface.

The CSIX information translates almost directly to the RapidIO data streaming header information. Table 12.4 shows how the CSIX fields may be mapped to RapidIO data streaming packet fields.

There are a number of ways to translate the CSIX destination address. An interface supporting only the small transport RapidIO packet format needs to map across only 8 bits of CSIX destination address. If the interface is using the large transport model, the CSIX interface will only be able to offer 12 bits of addressing. If the CSIX device can 'deviate' from the existing specification, some of the reserved bits could allow expanded usefulness of the interface.

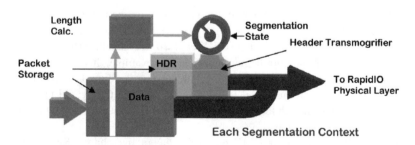

Figure 12.12   CSIX to RapidIO conversion

Table 12.4   CSIX to RapidIO field conversion

CSIX	⇔	RapidIO
Destination ID	If Transport = Small	8 bit DestID
	If Transport = Large	(12) → 16 bit DestID
	DeviceID stored in bridge register	8 or 16 bit SourceID
Class	Maps directly	Class of Service
	256 ⇒ 4 LUT	Physical layer priority
S E flags	Maps directly	S E flags
Payload Length	Used to calculate residual byte count	O P flags
PDU Length	Supplied or calculated	PDU length in end segments
StreamID	Maps directly	StreamID in single and start segments

The PDU length and StreamID fields in the payload portion of the Unicast packet may be handled in multiple ways, depending on the CSIX device. To keep the re-segmentation process simple, the easiest way would be to always exclude these two bytes from any payload content. Otherwise the re-segmentation logic must deal with varying packet formats. The StreamID will be valid on start and single segments, the field would contain the PDU length in end segments, and the two bytes could be a pad in continuation segments.

The payload length must be used to account for odd byte boundary transfers. CSIX uses natural bus widths of 32, 64 or even 96 bits. Some CSIX interfaces will transfer a variable length payload, but the transfer will be a multiple of the bus size. The payload length indicates the actual byte count of the true data payload. In the CSIX ⇒ RapidIO direction, the payload length is used to derive the 'O' and 'P' bits in the segmentation control field. In the RapidIO ⇒ CSIX direction, the payload length is set from the residual size of the end or single segment.

The other aspect of a RapidIO bridge is the use of RapidIO flows. The combination of the SourceID, DestinationID, and physical 'channel' defines a 'flow.' At the time of this writing, RapidIO supports eight priorities in the serial physical layer. When transitioning the packets into the RapidIO transport and physical layers some mapping of class of service must occur to determine the proper priority for each stream. This mapping does not have to be a direct relationship of bits. A 256×3 bit lookup table (LUT) allows any relationship of classes to priorities.

The lack of a standard segmentation process for CSIX creates complexity in developing an interface to RapidIO. Conducting a careful survey of desired CSIX devices to fully understand their capabilities should precede the design of a RapidIO to CSIX interface to ensure that the devices will be properly supported.

12.6.1.3 Flow Control

CSIX supports link-level flow control through the Ready bits. These bits pause traffic flowing into the first level of buffering for the CSIX physical interface. They should not be mapped to any RapidIO interface functions, other than indirectly as a link-level backpressure mechanism.

CSIX also has a flow control packet. The CSIX flow control packet consists of multiple flow control entries and is shown in Figure 12.13. The flow control message allows regulation of specific classes and ports. The wild card bits allow overloading a message with multiple

CFrame Flow Control Header																
Byte	Bit Position															
	7	6	5	4	3	2	1	0	7	6	5	4	3	2	1	0
0/1	Ready		Type				CR	P	Payload Length							
2/3	Class								FC		C	P	Speed			
4/5	P		CR		Destination											
6-9	FC Entry #2															
10-13	FC Entry #3															
N-1, N	Vertical Parity															

Figure 12.13   CSIX flow control header

functions (e.g. turn off all traffic of this class to all ports). The CSIX specification [1] has a more complete description of the valid combinations of fields and wildcards. These commands would map to future data streaming flow control commands.

CSIX flow control packets can be translated to the existing 'Type 7' flow control messages as well. The class field can be 'recovered' from the flow information in the Type 7 packets with a reverse LUT that matches the mapping from class to priority. An alternative is to use flow-based queuing in the bridge and interpret (terminate) the Type 7 flow packets at that interface. Then as that queue starts to back up, create CSIX messages according to the needs of the CSIX device. This is an interworking approach as opposed to direct translation. It is more complex, but may allow the CSIX side to be more programmable, and hide more of the variations from the RapidIO fabric side.

The use of CSIX flow control, like segmentation, is not well specified. To support flow control, the bridge may need some flexibility. One variant, for example, is that some devices only support a CFrame with a single flow control entry. Others will have varying support for wildcards, and may only support explicit messages.

12.6.2 Ethernet and IP Applications

The use of a RapidIO fabric can provide common device connectivity for a variety of uses, including bridging or transporting Ethernet- and IP-based traffic. Ethernet is itself just a basic transport mechanism with a number of physical layers and a link layer based on MAC addresses. Ethernet is most widely used to carry packets in a multitude of formats based on the Internet Protocol (IP) family.

A number of ways exist to utilize RapidIO within a traditional Ethernet infrastructure. These include, but are not limited to:

- Ethernet switching: encapsulation of the Ethernet link layer
- IP switching: emulation or elimination of the Ethernet link layer
- TCP offload: application-to-application layer communication

These are explored individually in each of the following sections:

12.6.3 Ethernet Switching

One way to construct a RapidIO to Ethernet bridge is to interface to one of the intermediate interfaces defined in the IEEE802.3 specification. An interface such as GMII takes traffic from the MAC layer containing the Ethernet transport interface. Bridging GMII to RapidIO replaces the Ethernet physical interface with the RapidIO fabric. This permits the RapidIO fabric to carry the native Ethernet traffic, along with many other traffic types.

The Ethernet to RapidIO process would follow this path:

$$\text{GMII} \Rightarrow \text{MAC Add to VSID} \Rightarrow \text{Segmentation} \Rightarrow \text{RapidIO Physical Layer}$$

Ethernet interfaces such as GMII transfer complete packets, requiring only a single segmentation engine. Segmentation begins as soon as the GMII interface has received enough bytes to meet the maximum transfer unit (MTU) size requirement. A lookup table containing MAC addresses maps the MAC address into a complete virtual stream ID (destinationID, class, and streamID). The number of MAC table entries would depend on the application. A content addressable memory (CAM) may be useful depending on the number of entries and the processing speed required. Conversion to the RapidIO physical layer is similar to the CSIX example where class is mapped onto the channels in the physical layer.

The RapidIO to Ethernet path follows this process:

$$\text{RapidIO Physical Layer} \Rightarrow \text{Reassembly} \Rightarrow \text{Completed packets to GMII}$$

The SourceID and 'channel' identify the RapidIO 'flow' for packets exiting the RapidIO physical layer. Each flow is reassembled maintaining separate contexts. Only a fully assembled protocol data unit (PDU) may be forwarded to the GMII interface. The number of reassembly contexts supported is again application dependent.

12.6.4 IP Switching

Interfacing the various IP protocols directly to RapidIO is similar to any other transport medium. The TCP/IP protocol stack, for example interfaces to a RapidIO driver that maps the TCP/IP packets into RapidIO data streaming transactions.

Internet Protocol transactions are based on IP addresses. A device table usually maps IP addresses to Ethernet MAC addresses when using Ethernet as the underlying transport. For RapidIO, a virtual stream ID is generated to transport the packet. Population of the device table uses the address reply protocol (ARP) when the device with the desired IP address is unknown. ARP emulation would require use of RapidIO's multicast protocol to enable address table learning.

The RapidIO driver segments the PDU, applies the VSID, and introduces the RapidIO data streaming segments into the fabric. The driver interfaces to any standard protocol stack. The RapidIO interface hardware may offer hardware speed up features for segmentation and reassembly and direct memory transfers.

The major advantage of using a RapidIO fabric for IP is quality of service. The VSID may be generated just from the destination IP address, or it may be generated from deeper inspection of the higher-layer protocol fields. Depending on the type of data being transported, different classes of service can be assigned. The VSID offers a 'standard' ID tag to provide more advanced queuing and scheduling of traffic. The classification process occurs once as the IP packet is introduced to the fabric. The RapidIO segments can then receive specific treatment throughout its movement through the fabric.

12.6.5 TCP Offload

TCP is a 'transport protocol' used to reliably move large data files through an IP network. TCP provides for out-of-order delivery of data blocks, and retries for corrupted data. The TCP stack can consume a large amount of processing resources. TCP offload becomes more relevant as data speeds increase.

TCP uses a 'port number' to identify the target that data is streaming to/from. Using components separated by a RapidIO fabric, TCP offload can occur anywhere in the system. The offload engine automates the TCP blocking protocol and delivers intact files directly to an application's buffer. The TCP offload engine does not require direct knowledge of the applications address map or queue structure. The TCP port value maps to a VSID. The receiving driver associates the data stream locally with the buffer.

The virtual nature of the VSID allow local management of resources. When a TCP connection is established, the RapidIO infrastructure remains invisible to the process and facilitates the distribution of resources. The ability to connect streams of traffic directly between applications is highly advantageous compared with using a network such as Ethernet with standard TCP/IP stacks. To move data from one device to another, the data is packaged by the protocol stack and then unpackaged at the other end. RapidIO data streaming offers a much lighter weight mechanism than TCP to transport the data, while still offering data integrity. It also eliminates the need for end points to share knowledge about memory maps, which would be required for DMA approaches to moving data between devices.

12.6.6 RapidIO to ATM

Adaptation of ATM to RapidIO is very straightforward. ATM utilizes fixed sized cells, so a RapidIO fabric carrying ATM is configured such that the MTU size is large enough that cells are transferred as single segments. No segmentation or reassembly is required.

ATM cells are identified by two fields: virtual path ID and virtual channel ID (VPI/VCI). These two IDs map the ATM cell onto a virtual stream. Classification of the ATM cell can occur in a couple of different ways, depending on the location of traffic management in the system. ATM requires management of every cell, as there is no flow control mechanism. A system with a centralized queuing, switching, and traffic management function needs only the cells delivered to a single address.

UTOPIA (Universal Test and Operations PHY Interface for ATM) is a common electrical interface for ATM cells. This bus operates between 25 and 50 MHz and may be 8 or 16 bits wide. It supports data rates of up to 622 Mbps. A UTOPIA to RapidIO bridge, in this example, would simply pack the cell transferred across its UTOPIA to RapidIO interface into a RapidIO single segment and apply a fixed VSID. The ATM switch at the destination will take care of identifying the VPI/VCI and scheduling the traffic. On the RapidIO to UTOPIA interface side, the UTOPIA bus may have multiple physical devices (multi-phy). In this instance, the StreamID field would provide a subport address to identify the destination PHY. The RapidIO to UTOPIA interface does not need to examine any portion of the ATM cell.

A system with a more distributed configuration is required to manage the flow of ATM cells. The traffic manager will be required to 'classify' the ATM cells based on VPI/VCI (and possibly additional fields depending on the ATM adaptation layer used). As traffic enters the RapidIO fabric, the corresponding VSID is used to control the introduction of the ATM cells and the quality of service.

REFERENCE

1. CSIX-L1 Common Switch Interface Specification-L1, August 5, 2000, Revision 1.0, Network Processor Forum.

13

Applications of the RapidIO Interconnect Technology

Brett Niver

13.1 RAPIDIO IN STORAGE SYSTEMS

The challenge of successfully managing more data more easily is central to the professional existence of today's IT manager. A quick Google search of 'data storage management' yields literally thousands of references to storage hardware and software products, consulting services, books, and courses, all offering solutions to help meet this challenge.

Both data storage capacity and data storage creation are growing at tremendous rates. According to a Berkeley Information Study 2003[1], 'Print, film, magnetic, and optical storage media produced about 5 exabytes (2^{60}) of new information in 2002. Ninety-two percent of this new information was stored on magnetic media, mostly in hard disks.'

Another example of the growth of data comes in the form of a quote from Dwight Beeson, a member of the Library Of Congress' Systems Engineering Group [2], 'Since 1993, we have had a compound data storage growth rate of about 120 percent per year.'

13.1.1 Features of Storage

A good example of a typical storage system architecture is the EMC MOSAIC:2000 architecture [3]. This architecture is comprised of a group of front-end processors referred to as channel directors, a large data cache and a set of back-end processors, referred to as disk directors. The front-end processors manage the communications between the host computers and the storage system. The back-end processors manage the communications between the storage system and the physical disk drives. The cache memory is used to retain recently accessed data and to provide a staging area for disassembling data to be sent to multiple physical drives or to reassemble data coming from multiple drives. A system such as this can have dozens of processors, all cooperating in the task of moving data between the hosts and the physical drives. In addition to the transfer of the data, there is also a translation of protocols. The host connections will often be accomplished by a networking interface such as IP operating over

RapidIO® The Embedded System Interconnect.   S. Fuller
© 2005 John Wiley & Sons, Ltd   ISBN: 0-470-09291-2

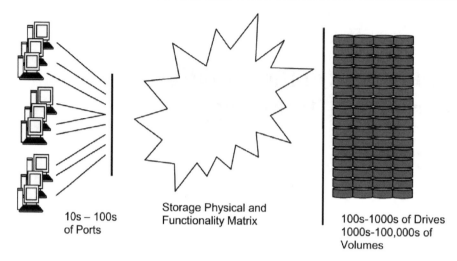

10s – 100s of Ports

Storage Physical and Functionality Matrix

100s-1000s of Drives 1000s-100,000s of Volumes

Figure 13.1 The storage matrix

Ethernet or Fibre Channel. The interface to the physical drives might be Serial ATA or SCSI. Each interface has its own packet or frame size and formatting requirements.

One can see that the 'cloud' shown in Figure 13.1 as the 'storage matrix' is actually a collection of channel directors, cache memory and disk directors. Within the 'storage matrix' is a 'physical matrix' that is the actual collection of physical disks and connectivity in the system (Figure 13.2). However, the 'physical matrix' will not typically be an optimal representation of storage to the host computers. For this reason the storage system will always provide a 'functionality matrix' view to the host computers. This functional view can be quite different in character from the actual physical characteristics of the storage.

13.1.2 Capacity

A data storage system has various important features. The first and most obvious is sheer capacity (Table 13.1). With hard disk drive capacities doubling every year, and data storage requirements growing at similar rates, the offered capacity of a storage system is typically one of the main selling points. Systems offered in 2004 can provide total storage measured in petabytes (2^{50}) with working cache memories reaching the terabyte (2^{40}) range. These are very large systems. Management of systems this large requires many dozens of processors and significant high-speed interconnect technology.

A good analogy for this scale factor is to think of the scale difference between a 2×2 doghouse, a new house built today, the size of the Petagon with office space of over 3.7 million square feet, and the island of Martha's Vineyard which comprises approximately 100 square miles or almost 3 billion square feet (Figure 13.3).

13.1.3 Availability

Another very important aspect of data storage is data integrity. This can be translated into at least two very important criteria, availability and reliability. *Availability* is the real-time

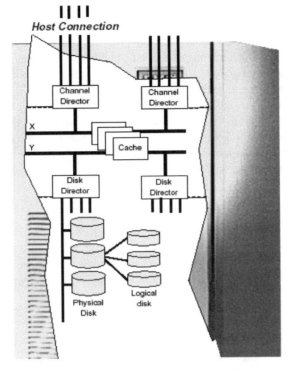

Figure 13.2   Inside a storage system[4]

Table 13.1   Total system resources for three Generations of EMC symmetrix

	Disks	**Cache memory**	**Bandwidth**	**CPU**
Sym4	384, 18–36 GB	4 GB	130 MB/s	> 2k MIPS
Sym5	384, 74 GB, 146 GB, 181 GB	32 GB	330 MB/s	> 8k MIPS
Sym6	576, 146 GB	128 GB	2400 MB/s	32k MIPS

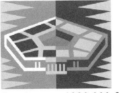
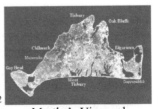

Doghouse 4 ft^2

New House 4000 ft^2

Pentagon > 4000 000 ft^2

Martha's Vineyard
~3000 000 000 ft^2

Gigabytes 2^{30}   **Terabytes 2^{40}**   **Petabytes 2^{50}**   **Exabytes 2^{60}**

Figure 13.3   Analogies of storage growth

accessibility or usefulness of data. With the advent of internet-based commerce and online retailers taking orders at any time of the day or night from anywhere in the world, the unavailability of data translates into lost business. *Reliability* is also critical. How does one ensure that not only is the data, which is being stored, safe, but that it is also accurate? In other words, not only does one need to be able to retrieve the data that was stored, but it must be able to be verified as being exactly the same data that was originally stored.

For Enterprise-class storage systems today, availability means 'highly available.' 'Highly available' means that, even in the event of a failure of a component, the service is uninterrupted. Any component failure and subsequent failover must be completely invisible to the normal user. This also means that during any type of normal service, code or system upgrade or reconfiguration, the service is completely maintained at the normal quality of service.

13.1.4 Density

Enterprise storage systems can contain hundreds of hard disk drives, terabytes of memory and dozens of processors. These systems, to achieve their availability targets are also often fully redundant. These high component counts make it very important that the technologies used in the system, including the interconnect technology, be as efficient and as low power as possible.

13.1.5 Manageability

The cost of managing storage often exceeds the cost of the storage itself. The sheer size of the data being stored is one contributing factor to this cost, but the cost of management is also affected by the growth rate of the use of storage, adding online access methods, performing time-consistent backup, as well as many other aspects of operation.

13.1.6 Storage System Applications of RapidIO

The RapidIO architecture was developed to address the need for high-performance interconnects in embedded systems. This section explores opportunities for the use of RapidIO in enterprise storage systems. Earlier in this chapter the 'physical matrix' of storage was presented. This 'physical matrix' is composed primarily of a high-density matrix of CPUs, cache memories, IO devices and the connectivity to tie all of these devices together.

13.1.6.1 Storage Requirements

In this section an attempt is made to correlate the major connectivity requirements of storage with the characteristics of RapidIO.

- Connectivity/density: we painted a picture in the previous sections of this chapter of the accelerating growth of storage capacity and the requirements this brings for increased density of connectivity, power and performance. RapidIO provides arguably the lowest cost-per-port of any system interconnect available today. The RapidIO protocol was developed to be easily realizable in common process technologies. When compared with technologies such as Ethernet, it can significantly reduce component count and protocol processing overhead. Because RapidIO switching technology can be provided in a single-chip package, the required power-per-port can be quite low, based on currently available circuit fabrication technology.

- Size/scalability: with the extremely dynamic requirements of storage capacity comes the demand for a very scalable system interconnect. Often scalability demands must be met by adding to existing storage systems in the field, often while the system is running. The RapidIO interconnect technology supports up to 64 000 end points in a single fabric, and supports memory addressing modes of up to 66 bits. Sufficient to address exabyte size memory spaces. To help with online scaling demands, RapidIO has built in support for hot-swapping. The destinationID-based routing mechanism provides tremendous topological flexibility to system designers. Topologies can be utilized that would be difficult or impossible to achieve with other interconnect technologies such as PCI Express, Ethernet or HyperTransport. The addressing mechanisms provided by RapidIO also can help with virtualization, another important aspect of storage systems. Virtualization is a mechanism of providing alternate mappings for physical storage. For example a set of five disks can appear as a single virtual disk drive to the host.
- Time-to-market: with the storage needs of information technology continuing to grow so rapidly, it is obviously important for storage vendors to be able to release new products and updates to existing products to address new and growing requirements. Any technology that is being used as an integral piece of a storage product needs to have a very aggressive performance roadmap. RapidIO being a standards-based technology is available from multiple vendors. The fact that RapidIO is not only an industry standard, but an international standard[5] helps ensure its evolution as a ubiquitous technology. The RapidIO Trade Association has also aligned itself with various other high-performance backplane technology groups such as the OIF CEI[6] in a effort to leverage best-available technologies and standards for its future roadmap. These efforts are expected to lead to even higher performance RapidIO technologies becoming available over the next several years.
- Reliability: for enterprise storage, reliability is perhaps the most important feature that a system offers. To reach the reliability targets many techniques are used. These techniques include storing multiple copies of the data to separate disk drives, using redundant array of independent drive (RAID) techniques to spread data across multiple disks, using parity and error-correcting codes (ECC) protected communication paths and memories, and offering redundant processing subsystems and power supplies.

In the design of enterprise storage systems, every component and communication channel must have a backup that can take over in case of component failure. In an environment such as this, the reliability characteristics of RapidIO are very attractive. All RapidIO packets have several inherent protection mechanisms. The 8B/10B coding scheme used for the Serial RapidIO physical layer has an inherent ability to detect channel errors. All RapidIO data transfer, are fully CRC-protected. Long packets contain two CRCs. There is a CRC appended to the packet after 80 bytes and another at the end of the packet, which is a continuation of the first. The acknowledge ID fields are used to ensure that all packets are received in the correct sequence. Packets are not acknowledged until they have been positively confirmed to have been received correctly. Transmission buffers are not released until a positive acknowledgement has been received from the receiver.

Another aspect of RapidIO that aids reliability is the error management extensions specification. The error management extensions provide the ability for switches and other

RapidIO components to proactively provide notification of failure and maintain information associated with the failure event. This is a tremendous aid to the task of architecting a fault-tolerant and high-availability system. The error management extensions also provides high-level support for live insertion. This is a required functionality for creating system architectures that will allow for field replacement of failed components in running systems without any other system disruption.

13.1.7 Conclusion

RapidIO is a well developed system interconnect and fabric standard. It has been designed from the ground up with cost, performance and reliability in mind. The strong correlation between the technology needs of enterprise storage systems and the characteristics of the RapidIO interconnect technology has been demonstrated. RapidIO is an obvious candidate technology for providing the system connectivity for tomorrow's distributed, high-availability, fault-tolerant storage systems.

13.2 RAPIDIO IN CELLULAR WIRELESS INFRASTRUCTURE

Alan Gatherer and Peter Olanders

13.2.1 Introduction to Wireless Infrastructure

The cellular wireless infrastructure network can be divided into three main parts, as shown in Figure 13.4.

Data and calls entering the system from the network arrive via a mobile switching center (MSC) from the public switched telephony network (PSTN) or via a general packet radio system (GPRS) support node (GSN) from the IP network. There are many functions performed on the data at the network interface, including protocol termination and translation, routing, switching, and transcoding and transrating of telephony and video signals. Normally, the number of MSC and GSN nodes in a network is kept to a minimum, so that each node processes a large number of calls. Therefore, the data rates are large, on the order of 10–100 Gigabit/s (Gbps), and high-bandwidth switching and routing occurs. Transcoding and transrating large quantities of voice and video signals requires banks of DSP farms that are connected to the routers by high-speed interconnect.

Once the data has been formatted for the wireless infrastructure, it passes to the radio network controller (RNC) in the radio access layer of the network. The RNC controls access to the basestations and is responsible for setting up and tearing down connections between users and the network. The RNC is responsible for managing the mobility of a user who is moving around between cells and networks. A certain amount of signal processing occurs in the RNC to allow processing of data from multiple basestations.

The basestation is responsible for maintenance of the physical link to the user. Each basestation keeps in contact with the users who are physically located within its cell. A great deal of signal processing occurs in the basestation. Like the MSC and GSN nodes, a basestation will typically house racks of DSP- and ASIC-laden digital baseband boards. RF processing and power amplification also occurs in the basestation. All of the RF and digital

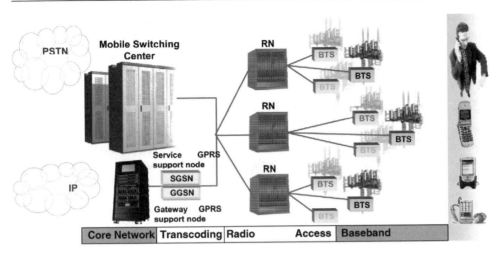

Figure 13.4 Elements of wireless infrastructure

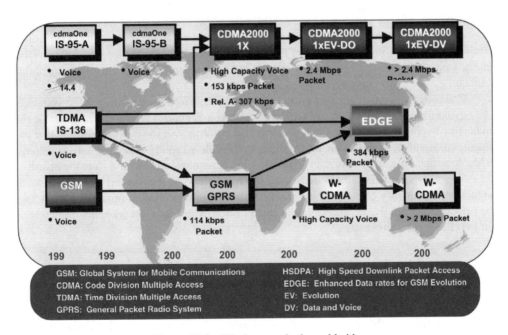

Figure 13.5 Wireless standards worldwide

baseband cards are connected to each other and to miscellaneous control and interface cards via high-speed backplane links. The chip-to-chip data rate on these boards is also quite high and rising as higher data rate technologies such as 3G are deployed.

There are several standards in use today for wireless infrastructure. These include WCDMA[7], GSM, EDGE[8], IS95 and CDMA2000[9]. Figure 13.5 maps out the wireless standards and how they relate to each other. The multitude of standards is due to competition between companies and national organizations as well as the evolution of standards to provide new features and higher data rates. The industry trend is toward higher data rates and more capacity in a single cell, leading directly to increased data rate requirements in the wireless infrastructure equipment itself.

13.2.2 Mobile Switching Center/GPRS Support Node

As was mentioned in Section 13.2.1, wireless infrastructure equipment can be classified into three main parts, the MSC/GSN, the radio network controller and basestation. The challenges for data flow in each of these parts is unique. This section describes how data flow in MSC/GSN equipment has been handled in the past, and highlights some of the problems that need to be solved in future systems.

13.2.3 Core Network Interface

The core network interface can be divided into protocol processing and transcoding. Telephony traffic coming from the telephone service network is coded by pulse coded modulation (PCM) and has a signaling layer, typically SS7, applied so that the calls can be properly routed. Thousands of calls are processed by one MSC. The MSC is made up of racks of DSP cards, with each card containing tens of DSPs and processing hundreds of calls. A local controller deals with the signaling and routing of the calls and the DSP performs the translation of the signal from PCM into the appropriate speech coding standard used by the targeted cellular standard.

Calls are constantly being set up and torn down; therefore the data flow in the system is very dynamic. Microprocessors are typically employed as control processors responsible for managing the flow of data. They will perform load balancing on the boards to ensure that the maximum number of calls can be handled, even with a rapidly changing traffic pattern. Therefore, there is a mixture of data and control information flowing to and from the boards.

Voice calls are very latency-sensitive. Too much delay in the transmission of the voice signal and the quality of the phone call will suffer. Echo cancellation processing must also be performed to remove echoes from the conversation. This leads to very strict latency constraints in the MSC processing.

13.2.4 Transcoding and Transrating

Transcoding refers to the conversion of the PCM voice data from the traditional telephone network to the compressed wireless standard format used by the wireless infrastructure and back again. Transrating is the conversion of a signal stream (usually video in this case) from one data rate to another. An example of transrating is when a multicast video stream is sent to several user devices, with each user device requiring a different display size and resolution. In order to make optimal usage of bandwidth of the wireless channel, the video stream should be converted to the right format for display on each of the target devices before transmission. The data rate requirements for video and picture phones are higher than for voice only phones. As the use of video and pictures increases, the individual data rate to each DSP will increase as well, with an expectation of several hundreds of megabits per second (Mbps) in throughput to each DSP and aggregate data rates reaching in to the gigabits per second range.

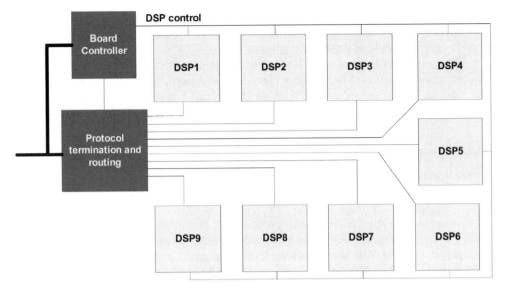

Figure 13.6 Transcoding board architecture

The DSP farm that is used to process this data is a classic embedded processing application. A typical board level architecture is shown in Figure 13.6. In this system, the data enters and leaves the system in packets that are transmitted across a backplane. Each packet is sent to the controller (usually a microprocessor of some kind), which extracts the channel information so that the destination of the packet can be ascertained. The controller then directs the protocol and routing device to route the packet to the DSP that is responsible for processing that channel. The protocol termination and routing chip is usually an FPGA or ASIC designed specifically for the board. Network processors are now seeing some use in this application. The controller must also be able to receive and transmit control information to and from each DSP so that the state of the DSP is known and changes in the channel allocation of a DSP can be communicated. Clearly, there is a lot of routing of both control and data information through a system like this. In addition, because current systems lack a bus that is scalable to the backplane, the protocol on the backplane is often different from that on the board, forcing the board controller and protocol termination devices to bridge between different bus protocols between the board and the backplane.

13.2.5 Radio Network Control

Radio network controllers are responsible for handling mobility management, which keeps track of users, and resource management, which handles radio and network resources. Mobility management localizes where the user is in the network (logically), and carries out handoffs between basestations. This occurs as a cellular user travels within a region. For example while driving at 60 miles per hour handoffs between basestations will occur approximately every ten minutes. The RNC is responsible for coordinating the signals received and transmitted by the basestations and making the decision about when any given basestation should pick up and/or release responsibility for a call.

The resource management function of the RNC refers to the RNC responsibility for keeping track of the available radio spectrum network capacity, and determining which of the available spectra should be used for any given call.

There is a high basestation to RNC ratio, ranging from 100 to 1 to 1000 to 1. Consequently, data rates at the RNC are quite high, they are about the same order as at MSC level. The need for high-speed interconnects at the chip, board and subsystem level have grown quite acute.

13.2.6 Basestation

A basestation architecture is shown in Figure 13.7. In a basestation, there are several cards that need to be connected. In most systems these cards will all reside in the same rack. In larger basestations there can be multiple racks. The network interface card is connected to a transport network based on SONET/SDH, ATM, or possibly IP. This is the communications link to the RNC. The transport protocol is formally terminated in the network card and converted to the internal system protocol. In some cases, the transport network protocol or a derivative is used as the internal protocol. The ATM-like protocol UTOPIA is sometimes used, and recently the OBSAI specification[10] has suggested the use of Ethernet as the internal protocol. RapidIO has been considered by some vendors as a suitable internal system protocol as well.

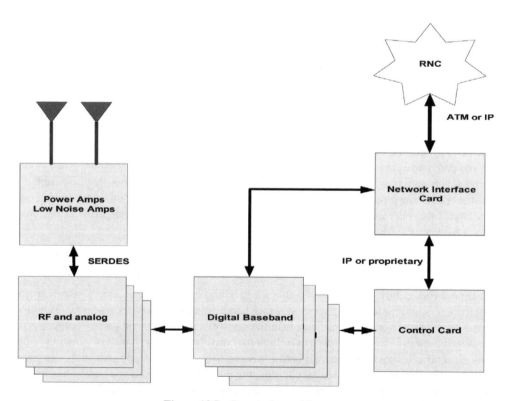

Figure 13.7 Basestation architecture

The controller card decodes the basestation-specific control messages and sets up the baseband modem cards to correctly transmit and receive user data. There may be several baseband modem cards and several analog/RF cards in a single rack. The baseband modem card is sometimes split into transmit and receive cards, which have communication requirements, owing to closed power control loops. The data rate requirements for the network-to-baseband modem cards is in the tens of Mbps. The data rate requirement for the baseband-to-analog interface is several Gbps. Both interfaces require a mixture of control and data.

Basestations come in a variety of sizes, from picostations that serve a single building to large macro cells that can cover hundreds of square kilometers in rural settings. The number of voice users supported by a single basestation varies from tens to thousands. Different basestation rack configurations have been developed to cope with this variety, but since they share many components a flexible, scalable interconnect is required.

The baseband modem card itself presents additional interconnect challenges. A typical third-generation WCDMA modem card architecture is shown in Figure 13.8. The on-chip interconnect for this card poses several challenges:

- Data rate: the data rate from the analog cards can be on the order of several Gbps. The data needs to be routed to all of the front-end ASICs. Since data flow from the ASICs to the DSPs can also amount to Gbps, the ASIC and DSP pin requirements would become quite severe for a traditional parallel bus running close to 100 MHz.
- Interconnect complexity: there are several 'chip rate' DSPs that connect to the ASICs on the board, all of which connect to the 'symbol rate' DSP that does higher-layer processing of the data. Control and status information flowing from the analog boards is usually

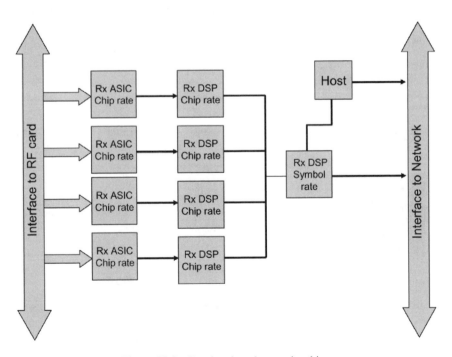

Figure 13.8   Baseband modem card architecture

connected on a separate link to the symbol rate DSP, which needs to process the data and gather statistics as well as monitor for faults.

- Lack of suitable interfaces and a variety of interfaces: typically, a memory interface is used to connect from the chip rate DSPs to the ASICs. This is not an ideal point-to-point interface and may require up to 100% overhead. Different interfaces are used for the DSP-to-DSP connection. Sometimes these are serial ports and sometimes host and memory interfaces. Generally, the lack of high-speed links on the board leads to a large amount of routing and a difficult, expensive board design. Legacy interfaces may not suit new architectures, and new components that have desirable functionality may not have the required legacy interfaces. Software can also be dependent on the interface used, and legacy interfaces often are used just to minimize the software rework that might otherwise be required.

13.2.7 RapidIO For Wireless Infrastructure

As can be seen from the description so far, wireless infrastructure is an excellent target for the use of the RapidIO interconnect technology.

13.2.7.1 Simplification with RapidIO

In the MSC/GSN side there are DSP farms whose cards require low-latency communication from the backplane to a farm of DSPs. There is a need for control for data routing and load sharing, as well as packet data transmission. A switch fabric capable of handling both data and control information provides a cost-effective, scalable solution. As the mobile network moves from mainly voice to higher-bandwidth multimedia applications, DSP farms will be required to process higher data rates in media gateways. A RapidIO-based transcoding board architecture is shown in Figure 13.9. The picture looks very similar to Figure 13.6, but in this case the

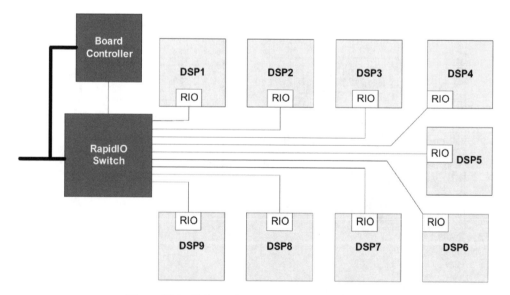

Figure 13.9   Enhanced transcoding board architecture

RapidIO switch is routing both the control and data and is also doing the address translation from the channel address to the DSP. The board controller is used to calculate the load balancing and update the necessary route tables in the switch. RapidIO is also suitable to be extended to the backplane. This significantly simplifies the task of the RapidIO switch, which replaces the protocol termination and routing ASIC or FPGA that a vendor would have had to develop in prior systems.

In the basestation, a variety of protocols are currently used to connect several boards together. The baseband modem board makes use of a variety of mostly proprietary interconnects, some of which require significant bandwidth and low latency. Both control and data information are transmitted through the system. A single protocol that can be used for both the control and data information and for both backplane and chip-to-chip communication minimizes chip count, simplifies the development of different-sized basestations, reduce development cost and can also reduce time to market.

The data rate requirements for basestations continue to increase as the move to third generation (3G) wireless standards brings higher user data rates. The number of antennas supported in a single basestation is also increasing, owing to increased use of techniques such as sectorization and adaptive antenna array processing. A protocol that can support a variety of data rates can be instrumental in reducing the cost and power of the baseband card.

Though the analog card to baseband card interface could be supported by RapidIO, recent moves to standardize this interface in OBSAI and CPRI[11] have focused on frame-based, rather than packet-based protocols. This is because the data transfer is regular and very low jitter is desired. The data is being generated by analog-to-digital converters (ADCs) or is being sent to digital-to-analog converters (DACs). Lower jitter leads to less buffering on the analog card. But there is no reason that streaming write transactions with relatively small packets could not be used in RapidIO to achieve a similar effect. Independent studies at Texas Instruments and Ericsson have shown that the physical area of a RapidIO port that is constrained to a streaming write is about the same as that of an OBSAI or CPRI port. With RapidIO, this interface would allow the use of standard RapidIO switches and routers, leading to less expensive, more flexible solutions. However, as at this point the industry push is to converge on a frame-based protocol, RapidIO switches with OBSAI and CPRI bridges may well be required to serve this market.

13.2.7.2 Architectural Possibilities with RapidIO

As well as providing simplification and cost reduction for today's architectures, we can use the switch fabric nature of RapidIO to enhance the flexibility of the baseband modem card. The architecture shown in Figure 13.8 is restricting in a couple of ways:

- Little support for user mobility: a user generally enters a cell, moves around in the cell, and then leaves. Different antennas in different sectors receive the strongest signals as that user moves. But on the baseband card we do not want to move a user's state information from DSP to DSP. This is time-consuming and can lead to violation of latency constraints for a user that is being moved rather than processed. So a DSP connected to only one ASIC implies that each ASIC must receive all the antenna data. In a world where different cells require different numbers of antennas, this is not a scalable solution. It leads to a needless repetition of antenna interconnect at each ASIC, with implications on ASIC area, board area and power. A more flexible solution is shown in Figure 13.9. RapidIO is used to connect the ASICs to the chip-rate DSPs and a switch. This allows complete access from any DSP to an ASIC. If this flexibility is not needed, then the architecture in Figure 13.9 can be reinvented by simply removing the switch. From the point of view of the DSP and ASIC components, there is no change in the architecture and full software reuse is possible.

- DSP-to-ASIC ratio: the number of DSPs required for every ASIC will vary depending on the type of processing. Architectures that require adaptive antenna array processing or interference cancellation will require more DSP processing. In general, enhancements in signal processing are expected with time, allowing for greater cell capacity and more reliable service. Optimization of DSP code over time and an increase of DSP clock frequency may reduce the number of DSPs required. On the other hand, there is a continuous functional increase, resulting in higher demands on signal processing. It is hard to tell what the trend will be, and a flexible architecture will allow for a varying number of DSPs per ASIC. It is also possible that a heterogeneous system is required. A DSP specializing in matrix processing may be added to enable adaptive antenna arrays in a more cost-effective way than simply adding this processing to each DSP. In Figure 13.10 we show a DSP for specialized processing attached to the fabric. In summary, the switched fabric nature of RapidIO allows for future proofing of the board architecture and the components used.

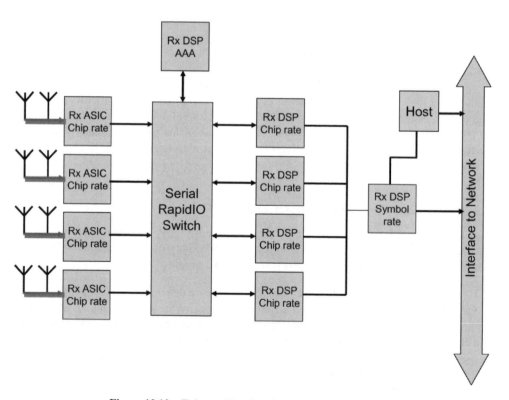

Figure 13.10   Enhanced baseband modem card architecture

13.2.8 Conclusions

The wireless infrastructure space contains two classic embedded processing examples. One is transcoding in the mobile switching center, and the other is the basestation baseband modem card. Today, each suffers from the limitations of presently available interfaces such as serial ports, memory interfaces, host ports, PCI and proprietary interfaces. The introduction of RapidIO will enable a single cost-effective, scalable infrastructure for both backplane and chip-to-chip communication. In addition, the use of a switch fabric allows more flexibility in the design and upgrading of equipment.

13.3 FAULT-TOLERANT SYSTEMS AND RAPIDIO

Victor Menasce

At one time fault tolerance was considered the domain of the lunatic fringe, telecom central office applications, the space shuttle, and so on. Today, the requirement for high availability is mainstream. It's easy to see why. When one adds up all of the network elements that a data packet traverses from source to destination, one commonly crosses 20 or more hardware network elements. While the trend is toward a flatter network, significant hierarchy still exists. The problem with this hierarchy is that it adversely affects reliability. Let's look at a simple example:

- Our example network element has an availability probability of 0.99. A network made up of two of these elements has an availability of $0.99^2 = 0.9801$. A network made up of 20 of these elements has an availability of $0.99^{20} = 0.82$, which most users would consider blatantly unacceptable.

This fact means that each individual network element must achieve sufficiently high availability so that the product of the availabilities yields an acceptable result at the network level. This has become the generally accepted 'five nines', or 0.99999 availability. It is important to note that this availability is measured at the application level. This level of availability equates to 5 min outage downtime per system per year. Let's examine a typical analysis of fault tolerant network element failures shown in Figure 13.11. The systems analyzed in the figure achieve five nines of availability.

The data for this article was obtained from the US Federal Communication Commission (FCC) ARMIS database by the author. The FCC is an independent US government agency

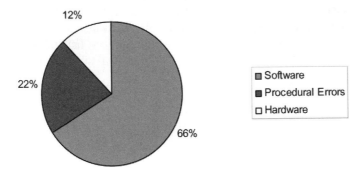

Figure 13.11 Fault proportions for software, hardware and procedures

that regulates interstate and international communications by radio, television, wire, satellite and cable. Access to its ARMIS database is publicly available on the Internet (www.fcc.gov). The raw data used in the article is available in the database, although there are no public reports that match the information presented directly here. The analysis was performed independently by the author who consolidated a range of widely distributed data from the database for the purpose of illustration. It is possible, however, to commission private reports from the FCC that detail data for specific vendors.

The largest proportion of failures are software, for which redundancy is difficult and costly to design. The second largest element is procedural errors. This class of failure is the result of maintenance people making a wrong decision and causing a network outage as a result. The smallest portion of the failure budget is indeed the hardware. Clearly, in order to achieve five nines overall, the hardware needs to be at least an order of magnitude better.

The impact of these outages in terms of outage downtime is affected by the time to repair a given outage. Catastrophic software failures can often be recovered by a system restart. This results in an outage of the order of 2–3 min. However, an outage that requires human intervention to repair the problem can have a significantly longer mean time to repair. Therefore the frequency of hardware outages must be reduced even further to lessen the impact of the statistically longer mean time to repair.

The outage budget can be broken down further into planned and unplanned outages. Some systems are capable of in-service upgrades, but this is rare. This is analogous to replacing the engine on an aircraft while it is in flight. If we allocate 3 min outage downtime to planned outages, then the remaining 2 min can be absorbed by unplanned outages. This means that the hardware needs to deliver better than 24 s of outage downtime per system per year. This equates to 'seven nines', or 99.99994% uptime. This is extremely challenging to achieve. The system designers who deliver this level of product know exactly how hard this is.

Design engineers have traditionally relied on proprietary technology to achieve this kind of performance. However, as companies look to rely increasingly on off-the-shelf technology, industry standards need to incorporate these features. Nowhere is this more important than in the arena of system interconnect. The interconnect architecture can become the weakest link, and also forms the backbone of the system's fault tolerance architecture.

The need for high availability has expanded to include storage systems, servers, network routers and switches.

RapidIO was developed from its inception with high-availability applications in mind. This philosophy permeates the architecture and the specifications. However, there are many legitimate architectures for achieving fault tolerance. The different architectures trade off the level of hardware redundancy, system cost, complexity and availability.

So the interconnect architecture should not assume any one fault tolerance architecture. It should provide the necessary hooks so that any of these fault tolerance architectures can be supported.

13.3.1 Key Elements of Fault Tolerance

There are six key elements to system level fault tolerance:

1. No single point of failure
2. No single point of repair
3. Fault recovery

4. 100% fault detection
5. 100% fault isolation
6. Fault containment

Let's examine what each of these means from a system level.

13.3.1.1 No Single Point of Failure

No single point of failure is quite simple really. It means that you need redundant hardware that can take over in the event of a failure. There are a number of legitimate, but very different sparing architectures.

Examples include:

- redundant hardware, load shared
- duplex hardware, hot/cold standby
- duplex hardware, hot/warm standby
- triplex hardware with voting
- N + M sparing

The RapidIO architecture is capable of supporting all of these different sparing schemes.

13.3.1.2 No Single Point of Repair

No single point of repair is a little more subtle requirement. It means that you shouldn't have to shut down or replace a critical component in order to replace another failed component. An example of a common single point of repair might be a backplane. Live insertion/removal of FRUs is an integral piece of meeting this requirement.

The RapidIO specification supports live insertion and removal through both electrical and logical mechanisms. Live insertion is required to ensure that no single point of repair exists in the system.

The links are point-to-point and as such they do not have the same impact that a shared bus such as PCI may have. Use of extended power pins is common on backplanes, but falls short of the required protocol to bring a new FRU into service in a system.

The logical protocol for managing state information is defined in the RapidIO Error Management Specification. Insertion and removal of a blade is considered an exception condition and the software API is defined in this section of the specification. Bringing a new FRU into the system must be done carefully. It must be throughly tested and configured by the host in the system, prior to being brought into service. New FRUs are assumed to be untrusted until proven trustworthy.

At system boot time, the system host identifies all of the unattached links in the machine, using the system discovery mechanism, and puts them in a locked mode. At this point all incoming packets are rejected, leaving the drivers and receivers enabled. This ensures that whatever new FRU is inserted cannot access the system until the system host permits such access. When a FRU is hot-inserted, connecting to a switch device, the now connected link will automatically start the training sequence. When training is complete, the locked port generates a maintenance port-write operation to notify the system host of the new connection, and sets the port-write pending bit.

On receipt of the port-write, the system host is responsible for bringing the inserted FRU into the system in a controlled manner. The system host can communicate with the inserted FRU using maintenance operations after clearing all error conditions, if any. This procedure allows the system host to access the inserted FRU safely, without exposing itself to incorrect behavior by the inserted FRU.

In order to achieve the first three basic elements, a strong foundation of the last three elements is essential. This is where the features of an interconnect standard can make the task of achieving fault tolerance easy, or nearly impossible. One of the key items is support for live insertion and removal of field-replaceable units.

13.3.1.3 Fault Recovery Support

Fault recovery requires the ability to switch from a failed piece of hardware to a working piece of hardware with little to no interruption of the application. It can also represent the ability to withstand a transient failure at the transaction level. The ability to recover from a fault is entirely dependent on the ability to satisfy the remaining fault tolerance elements. Without this, fault control hardware cannot make informed decisions on the appropriate corrective action.

13.3.1.4 100% Fault Detection

This requirement is essential to determining that a piece of hardware has failed or a transaction has been lost. There can be no datagram transmission of data in the system. This means that all data paths, transactions and storage elements need to be protected by parity or some type of error detection code such as a CRC. This detection mechanism needs to be able to report the error to a control entity when the error occurs. As such it cannot rely upon the transport mechanism that itself may be at fault.

Faulty transactions need to be traceable to the offending transaction. If the exception reporting is imprecise, recovery will be virtually impossible and the only remedy will be a system reset which will result in an outage of the duration of the restart time.

RapidIO provides a rich array of fault detection mechanisms. First, all transactions are protected by a CRC code. All handshake control symbols are protected by a 5-bit CRC, or are transmitted twice. In addition, all transactions must be positively handshaked and strictly ordered. This ensures that dropped transactions do not permit propagation of logical layer failures.

The Serial RapidIO protocol is 8B/10B encoded. This provides an additional layer of fault detection in addition to the CRC protecting the packet. A single bit error can manifest in illegal codes, which are detected at the PCS layer of the PHY. In addition, the K-codes that were chosen for use in RapidIO were chosen specifically so that no single-bit error could be misinterpreted by the PHY logic as a special use character. Other interconnect standards such as PCI Express, Fibre Channel and Ethernet have this vulnerability.

Some system failures are visible as degradation in reliability, prior to a catastrophic failure. For example, the bit error rate on a link may degrade gradually over time. Detection of this trend can provide an early warning of a bad link before the failure becomes a hard failure. RapidIO supports a rich array of performance and reliability monitors. These monitors have multiple thresholds. The traffic light analogy works well here. Working hardware is denoted by a green light. A hard failure is denoted by a red light. A degraded link falls into the yellow light category. This can give maintenance software plenty of time to take corrective action prior to the degradation becoming a hard failure. Thresholds can be set by software to determine what level of protection is optimum for a given system.

13.3.1.5 100% Fault Isolation

Fault isolation usually requires a strong partnership between hardware and maintenance software. The only exception to this is in the case of triplicated hardware where a simple vote determines which of the three sets of hardware is at fault. This is a costly method. Most fault tolerance architects prefer duplicated hardware or an N+M redundancy approach. Let's examine how this might work.

In a duplex system, two identical hardware components compare against one another executing the same application. As long as they agree, the hardware is assumed to be good. At some point, in the presence of a failure, the two hardware components diverge. This is an excellent way to detect an error. However, how do you know which of the two pieces of hardware to trust? This is where fault isolation enters the picture.

At the point of detecting the failure, maintenance software must take over and look for signs of trouble in the recent history. If one of the two pieces of hardware handled an exception, and the other did not, then there may be something for software to investigate. The only way that software can isolate the failure is if the system maintains a history of what was happening in the recent past. The shorter the time between a failure occurrence and the detection of the failure, the better. A shorter interval minimizes the amount of transaction history that needs to be maintained.

Isolating the error is key to understanding the severity of the error, and hence the severity of the recovery process that must be initiated. This can determine whether it is necessary to restart a transaction, a software process, or the entire system. By limiting the scope of recovery, the time impact of the failure can be minimized. This will aid in the uptime statistics.

Most fault-tolerant applications have a restart progression that starts from the lowest downtime impact and grows to the highest downtime impact.

1. Restart the thread of execution
2. Restart the process
3. Restart the operating system from a known point in memory
4. Reload the operating system and application from disk
5. Reload the operating system and application from backup media

Restarting a software process is much faster than a heavy-handed reset of the entire system.

13.3.1.6 Fault Containment

All RapidIO links have full protection at the physical layer. This ensures that a failure cannot propagate beyond a single RapidIO link. However, sometimes the failure is in the end point and is not associated with the link.

In this instance, the failure can propagate beyond the failed component. A rogue transmitter can resist attempts to cease transmission or reset the component. The resulting surge in

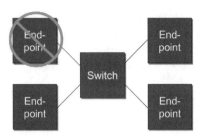

Figure 13.12   How a switch can isolate failures

traffic can cause congestion, buffer overflows, and ultimately data loss. The best protection for this class of failure is to ensure that propagation of the failure does not occur. Figure 13.12 shows the case in question.

RapidIO switches use a table-based routing algorithm. The source and destination addresses in a switch get mapped to specific port numbers. By replicating the mapping table at each port, it is possible to map the ports in an asymmetric way. Doing this will isolate the traffic from a rogue transmitter away from harming the system. This is done by mapping all entries in the table associated with the source end point to an undefined port in the system. All traffic from this source is then discarded and the system is protected. Switches from most RapidIO vendors are able to support this kind of asymmetric mapping of address ranges to ports.

13.3.2 Live Lock and Head of Queue Blocking

In addition to these mechanisms, failures such as live lock and head of queue blocking should be detected and handled in an appropriate manner. These conditions can occur in a system in the presence of either hardware or software failures.

In some systems, it is either desirable or necessary to bound the length of time a packet can remain in a switch. To enable this functionality, a switch will monitor the length of time each packet accepted by one of its ports has been in the switch. The acceptance of a packet by a port is signaled by the port issuing a packet-accepted control symbol for the packet. The timing begins when the port accepts the packet. If a packet remains in a switch longer than the time-to-live specified by the time-to-live field of the packet time-to-live CSR, the packet is discarded rather than forwarded, the output packet-dropped bit is set in the port *n* error and status CSR and the system is notified by a port write to the location of the system host.

Transactions in RapidIO are handshaked at the physical layer. The transmitter of a transaction on a RapidIO link is required to maintain a copy of the transaction in its output buffer until it receives positive acknowledgement of receipt from the receiver. If that acknowledgement indicates an error, or the acknowledgement is not received in time, the packet is retransmitted. In this case, the RapidIO link provides detection, isolation and recovery, all at the hardware level with no software intervention.

13.3.3 A System Example

Let's examine a failure of a switch in a dual-star system to see how this can be made into a fault-tolerant application. In this example, we have four end points and two switches. The

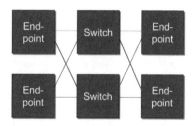

Figure 13.13   Spared system topology

switches are 100% redundant and the end-points are N+M spared. So there are two different fault tolerance architectures being used in conjunction with one another. The system is configured in a warm standby manner with only one switch handling traffic at a time. This method of managing switch resources means that the system must be programmed to address only one switch at a time. The end points are spared as N+M, which in this example is 1+1. The hosts in the system are also spared as N+M, which again in this example is 1+1. So the system requires only one switch, one I/O end point and one host to function correctly. Figure 13.13 illustrate the topology.

Every end point has a connection to each switch. In this case there is only one logical RapidIO end point implementation and two physical layer implementations. The end point design uses a multiplexer allowing connectivity to each PHY. The end point will attempt to train on both the active link and the inactive link as though they were both active. However, the logical mapping of ports at the end point will ensure than only one link is chosen to be active. The other will be disabled.

The end point which contains the system host must have two full end point implementations. This is because the paths are logically different from a maintenance perspective, even though they are logically equivalent from an application perspective. This logical separation needs to be maintained so that the state of the redundant switches can be known to the host. Both switches are active so that, in the event of a failure, the redundant switch is capable of carrying maintenance traffic to and from the host in the system. The host must listen to both the active and inactive switches.

In the event of a failure, a sequence of events will take place to deal with the failure. Figure 13.14 shows the scenario.

There is a failure on an active link. The switch on the bottom fails to acknowledge a transaction initiated by the end point on the lower left. This will result in a transaction time-out

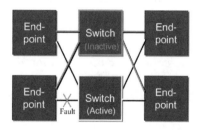

Figure 13.14   System failure example

on the originating end point on the lower left. The end point will initiate an attempt to retrain the link. Eventually, all attempts to retrain the link will have been exhausted. At this point, only the link is assumed to have failed. This is beyond the scope of what the hardware state machine can handle, so an exception is raised to the maintenance software in the system. However, the end point can't get to the host on the failed link. The end point on the lower left then switches physical activity to the redundant link. No reprogramming of the mapping tables is required. A port-write is initiated to the host through the redundant switch. The host then interrogates the failure conditions by looking at the control and status registers in the originating end point. The warm standby switch is used for this transaction.

Maintenance software determines that the link between the end point and the switch on the bottom has failed. There was enough information sent in the original port-write from the end point to the host. However, the host will still want to interrogate the end point and other parts of the system to ensure the problem is correctly identified. So the entire switch on the bottom is assumed to be at fault and must be isolated. This is a catastrophic hardware failure and some data could be lost in the buffers of the failed switch. However, the most important thing is to restore operation quickly. Maintenance software will reprogram the state map of the system hardware by tagging the lower switch as failed and inactive. The system must then be reset. This will be a software-initiated reset. This is the most orderly kind of reset in a system and will ensure the best retention of software state in the system. The reset controller in the system is obliged to hold failed hardware in a reset state. This will ensure that the failed switch remains in the reset state. While emerging from reset, all of the end points will attempt to retrain with the switches. However, only one of the two links will succeed because the failed switch is being held in reset. This ensures that all end points will choose the good switch on their links.

The I/O end points cannot be assumed to have intelligence. So the end points will notify the host again of the failure on the inactive links. Maintenance software will again query the control and status registers. These failures will appear to have grown in scope. However software will note that the bad switch is being held in an isolated state and will ignore the newly reported failures.

A software-initiated reset out of memory without reloading the application and data can occur in seconds in a small system, or could take up to 2 minutes in a large system. This is still significant. However, the frequency of these failures is generally low enough that, statistically, the impact on overall outage downtime metrics falls within the outage downtime budget.

13.3.4 Summary

Fault tolerance is now a mainstream requirement of many network elements, wireline, wireless, voice and data. The features required of fault-tolerant systems must be embedded into the underlying hardware technologies that are used in those systems.

It is important to note that RapidIO by itself does not make a system fault tolerant. A proper fault-tolerant architecture must be developed for the system, and the failure case analysis must be completed for each potential failure scenario. This requires an exhaustive effort to model failures in the system. Design solutions for each potential failure case must be developed. This will require specialized hardware in some way. Generally, the combination of specialized reset controllers and fault management hardware when combined with commercially available silicon devices can yield excellent results.

REFERENCES

1. Berkeley Information Study 2003, 'How Much Information', School of Information Management and Systems, University of California at Berkeley.
2. Washington Technology, As Data Flourishes, So Do Agency Storage Efforts, 04/17/00, Vol. 15, No 2.
3. EMC White Paper: Architecture and Computer Storage, July 2002.
4. EMC White Paper: What's Going Inside The Box *ISV Access to Symmetrix Performance and Utilization Metrics*, January 2000.
5. ECMA International—ECMA-342 and ISO/IEC JTC1.
6. Optical Internetworking Forum Common Electrical Interface.
7. http://www.3GPP.org
8. http://www.gsmworld.com/technology/edge/index.shtml
9. http://www.3gpp2.org/
10. http://www.obsai.org/
11. www.**cpri**.info

14

Developing RapidIO Hardware

Richard O'Connor

14.1 INTRODUCTION

The RapidIO interconnect technology enables a broad range of system applications by providing a feature rich and layered architecture. As has already been discussed in this book, the RapidIO architecture is segmented into logical, transport and physical layers.

The logical layer supports a variety of programming models, enabling devices to choose a transaction model suitable for the specific application. Examples include DMA transactions, direct processor-to-processor messages and streaming data flows.

The transport layer supports both small and large networks, allowing devices a flexible network topology. Hundreds to thousands of devices are easily supported in a single RapidIO network.

The defined physical layers support latency-tolerant backplane applications as well as latency-sensitive memory applications. RapidIO's layered architecture allows different device implementations to co-exist under a unified hardware/software base. This reduces or eliminates the need for bridging technology and additional protocol conversion overhead.

Device implementations of RapidIO can be broadly classified as end points and switches. End points source and sink transactions to and from the RapidIO network. Switches route packets across the RapidIO network (Figure 14.1 from source to destination, using the information in the packet header, without modifying the logical layer or transport layer of the packet. Switches determine the network topology and play an important role in overall RapidIO system performance. Switches and end points both support maintenance transactions for access to architectural registers. Some device implementations can act as both endpoints and switches.

This chapter explores the necessary considerations for a designer developing RapidIO end points and switches.

RapidIO® The Embedded System Interconnect.   S. Fuller
© 2005 John Wiley & Sons, Ltd   ISBN: 0-470-09291-2

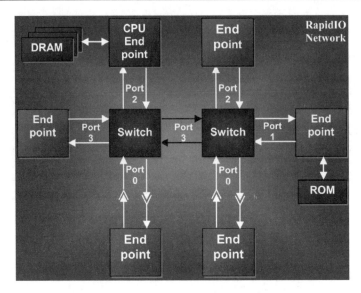

Figure 14.1   A simple RapidIO network

14.1.1 Implementation Performance–Cost Analysis

Based on actual implementations, the cost of implementing a RapidIO end point with a parallel 8/16 bit PHY (~1Ghz data rate) and support for all programming models is less than 145 kgates. A less aggressive implementation supporting only the I/O programming model is approximately equivalent to 100 kgates, this is comparable to a PCI (not PCI Express) implementation.

The cost of implementing a RapidIO end point with a serial PHY (1x/4x) is roughly equivalent to that of a parallel PHY implementation, with the addition of the SERDES logic.

14.1.2 Logical Layer Transaction Considerations

When implementing RapidIO, the designer must first select the correct programming model for their particular application. The RapidIO interconnect specification provides four different data transfer mechanisms with appropriate programming models for each.

Depending on the application, one of the following mechanisms is used:

- globally shared memory (GSM)
- input/output
- message passing
- data streaming

The globally shared memory (GSM) programming model is used for applications that require coherent data movement in a distributed computing environment. It provides support for ccNUMA (cache coherent non-uniform memory access)-based systems, and provides high-performance data sharing among participating devices over a RapidIO network. For example, ccNUMA is used to support data movement among a cluster of processing elements operating coherently from a set of distributed memories. Typical applications are general-purpose computing systems, such as desktops, servers and high-end embedded systems (routers or storage area networks). Global data coherency is achieved using a directory-based scheme across the system. Local data coherency is achieved by a snoop-based scheme. Further details on the GSM logical layer are found in Chapter 11.

The input-output (I/O) programming model is used for applications that do not require coherent data movement. Examples include a non-coherent DMA transfer over RapidIO or bridging from PCI to RapidIO. Typical applications are high-end embedded systems such as DSP farms, media gateways, and 3G base station controllers. The I/O programming model supports a maximum transaction size of 256 bytes, which is the most commonly used mode of operation in RapidIO. The I/O logical layer is described in Chapter 4.

Message passing is used in applications that require data movement from a source that does not have visibility into the target's address space, for example, a data transfer from one processor (running one operating system) to another monolithic system (running a separate operating system), with no visibility between the system address spaces. This is similar to data transfer over Ethernet. The maximum transaction size supported by this programming model is 4096 bytes. This is achieved by chaining 16 independent segments of 256-byte transactions automatically in hardware. Chaining eliminates the need for large buffers to hold the transaction data in temporary storage before it is committed to memory. Message passing also supports very short messages called 'doorbells' which can be used as in-band interrupts among participants in a RapidIO network. Message passing transactions are described in Chapter 5.

The RapidIO Trade Association has also developed a data streaming logical layer optimized for the encapsulation of communications protocols such as IP, Ethernet and ATM. The data streaming logical layer is described in Chapter 12.

The RapidIO specification also provides maintenance transactions that allow endpoints to communicate with other end points and RapidIO switches. Maintenance transactions are also used to initialize and manage the RapidIO network.

The choice of logical layer transaction support is largely a function of the device that is being developed. A network processor device might choose to support only the data streaming logical layer. A general-purpose processor might support all of the logical layers. A PCI bridge device might only need to support the I/O logical layer. While any pair of devices that must communicate with each other through a RapidIO network need to support the same transactions, all devices do not need to support all logical layer transactions.

Implementing only the required software support based on a programming model eliminates extra hardware and overhead. The choice of programming model plays an important role in determining the size of a RapidIO implementation.

14.1.3 RapidIO Transport Layer Considerations

The RapidIO transport layer specification supports small and large networks. Small networks require only 8-bit source and target IDs (addresses for 256 unique endpoints). Large networks, of up to 64 k unique end points, require 16-bit source and target IDs, thereby increasing the size of the packet and changing the controller alignment and storage requirements. When selecting an implementation, the designer must carefully weigh the benefits of supporting large networks against the additional software overhead and hardware cost required.

14.1.4 RapidIO Physical Layer Considerations

The physical layer provides a robust error detection and recovery capability. The choice of physical layer (PHY) is a critical factor in determining the performance and overall cost of an implementation. The RapidIO physical layer specification supports both serial and parallel PHY interfaces, and provides a well-defined software interface for error reporting, handling and statistics generation.

The parallel implementation has either an 8- or 16-bit bidirectional PHY interface offering low latency and extremely high bandwidth. It requires more PCB real estate than the serial implementation (for the parallel bus), and has a relatively short reach (maximum permitted trace lengths are 36 inches). This makes parallel PHY implementations especially suitable for processor memory applications.

The serial PHY interface specification supports 1x/4x bidirectional serial lanes with maximum trace lengths measured in multiple meters. The longer reach and smaller number of traces makes serial RapidIO an excellent candidate for backplane applications and for clustering a large number of end points, such as in a DSP farm.

The cost of the implementation is directly proportional to the performance needs of the application. The following table summarizes the performance, distance, power and board real estate for the RapidIO physical interfaces.

Table 14.1   Physical layer comparisons

	1x-Serial	4x-Serial	8b-Parallel	16b-Parallel
Performance (unidirectional) (Gbps)	2.5	10	16	32
Distance	Meters	Meters	3 Feet	3 Feet
Power	Lowest	Low	High	Highest
Board area	Lowest	Low	High	Highest

14.1.5 Protocol Bridging Considerations

Bridging between RapidIO and a different interface increases the complexity and cost of an implementation. For example, bridging from PCI to RapidIO requires special hardware to split non-contiguous PCI transaction data, represented by byte enables, into multiple contiguous RapidIO transactions. Bridging from RapidIO to PCI requires special hardware to split 256-byte RapidIO transactions into smaller PCI target transactions. There are commercially available bridge ICs that can handle these tasks, but the increase in controller complexity and cost is proportional to the difference in functionality between RapidIO and the interface being bridged. In the end, the designer must decide whether it is necessary or cost-effective to provide connectivity between an older I/O standard and RapidIO.

14.1.6 Transaction Buffering Considerations

Transaction buffering is another significant factor in determining the cost of a RapidIO implementation. The choice of an optimal number of buffers is critical in determining the overall

RapidIO system performance. The RapidIO specification provides flow control both for link-to-link and end-to-end transactions. It also provides multiple transaction flows, which are used to implement transaction ordering among different flow levels as well as within a single flow level. Transaction flows are also used for deadlock avoidance. A more indepth discussion of transaction flows can be found in Chapter 3.

There is a relationship between the number of buffers supported and the utilization of the link bandwidth. It is possible to starve a link by not offering enough transaction buffers to support the number of outstanding packets that might be transiting the link. Optimal buffering is determined by the round trip latency of a transaction for each flow level. Round trip latency at the transmitting end includes the link ACK latency for all transaction types and endpoint RESPONSE latency for GSM and Message transactions (since they support a RETRY response). Link ACK retries can be 'expensive' in terms of latency since all transactions in the latency loop sent after the retried transaction need to be reissued.

The buffering scheme must also allow reordering of higher flow priorities over lower priorities for deadlock avoidance, as specified in the ordering rules section of the RapidIO interconnect specification. In the end, the designer must make trade-offs for a buffering scheme that balances maximum performance against cost.

14.1.7 Clocking Considerations

The RapidIO interface is source synchronous. This means that data received from the RapidIO interface is asynchronous to the clock local to the receiving RapidIO endpoint. Data synchronization can be achieved by a circular queue mechanism at the transmitter and receiver. The circular queues at the transmitter must not enter an under-run situation, to avoid data starvation and the resulting reduction in interface efficiency. Circular queues at the receiver from an over-run situation must also be prevented, as they cause dropped packet data and require a costly re-transmit.

Over-run and under-run can be prevented by choosing an optimal RapidIO clock to local clock relationship and an optimal internal data-path width. The RapidIO endpoint latency is dependent on the frequency and the performance requirement of the given application. Here over-run and under-run are investigated purely from a clocking perspective for guaranteeing error free link operation. The RapidIO interconnect specification provides link-to-link and end-to-end flow control mechanisms for preventing over-run and under-run when the link/ network is functional and encountering very high traffic.

14.1.8 Packet Data Alignment Considerations

A RapidIO end point's internal data-path width contributes to the performance and cost of the implementation. As the internal data-path increases in width, the performance and complexity increases. RapidIO transaction data is word aligned and packets end on word boundaries. RapidIO packets and control symbols are multiples of 32 bits in length. If the end point internal data-path width is 32 bit, only one control symbol or one packet needs to be handled at a time. If an internal data-path is 64 bit, there will be additional complexity for handling packet alignment, since the following combinations need to be handled: two packets, two control symbols, a packet and a control symbol (in any order). Similar considerations have to be applied when implementing the serial physical layer interface SERDES because there are additional alignment requirements included the serial PHY interface specifications.

14.1.9 Data Recovery Considerations

The parallel RapidIO physical layer specifications provide a clock line for every data byte to serve as a timing reference. Clock and data re-timing circuits are required to help the system deal with skew on the incoming clock and data lines so the data can be reliably recovered. As the frequency of the interface increases, the cost and complexity of the re-timing circuits increases as well. In a 16-bit interface, skew needs to be managed between the individual bits in each byte, as well as between the data bytes. This adds another variable to the matrix of interface frequency, width, and clock speed when trying to balance implementation cost and complexity against the desired performance levels.

The serial RapidIO physical layer specifications call for data lines with 8B/10B embedded clocks that are recovered from special 'comma' characters transmitted within the data stream at regular intervals. This derived clock signal is used to lock the local clocks in phase and frequency to enable them to reliably recover the serial data from the RapidIO interface. In the 4x (4-lane) implementation, a considerable amount of skew must be tolerated among the bit lanes. This skew can be managed by receive FIFO structures. The choice of a 1x or 4x interface is critical in determining the implementation cost of an application.

14.2 IMPLEMENTING A RAPIDIO END POINT

14.2.1 End Point Overview

The RapidIO specification dictates that each interface on a RapidIO device must process both inbound and outbound transactions simultaneously. The inbound and outbound modules of the RapidIO interface can be designed separately or as one block. In either case, the inbound and outbound modules in the physical and transport layers must communicate.

For our sample end point, shown in Figure 14.2, the inbound and outbound are separate blocks, transferring packets in through the inbound module and routing them out to a user-defined interface (UDI) for processing within the end point.

The following list gives some sample features that can be implemented in a RapidIO end point:

- Full duplex operation
- Destination ID (DestID) mapping tables
- Large (16-bit) and small (8-bit) system device ID
- Multicast-event control symbol
- Receiver link level flow control/transmitter link level flow control
- Error management capability
- Port-write operation for error reporting
- Data plane extensions

    Sample features of a parallel end point:

- Full-duplex, double-data rate (DDR) RapidIO
- Operating frequencies of each interface can be configured separately (different clock rates)
- LVDS signaling

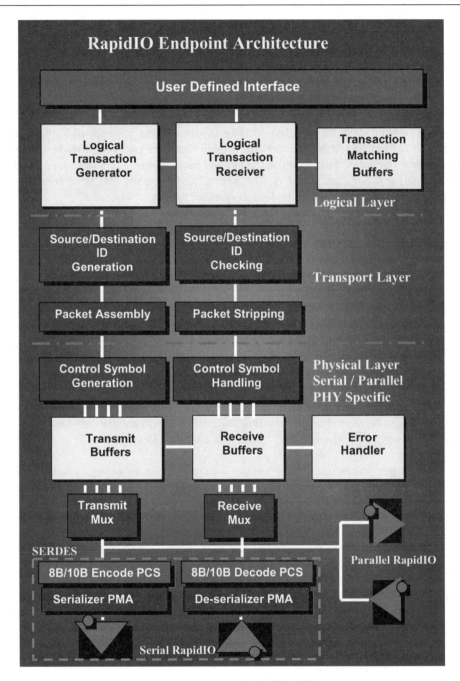

Figure 14.2 RapidIO end point block diagram

Sample features of a serial end point:

- Ports that can operate in 1x mode or 4x mode (different clock rates)
- SERDES with 8b/10b encoding/decoding circuit

14.2.2 Implementing RapidIO in an End Point

As with a switch design, the RapidIO Interface in an end point manages all RapidIO trans-actions to and from the end point device. The RapidIO interface can be separated into inbound and outbound modules. This optimizes individual handshaking and allows for concurrent bidirectional, or full-duplex, transfers. The inbound and outbound data flows combine both data and control information in a simple and organized format.

The end point needs to perform some type of transaction mapping between the RapidIO protocol and the devices connected to the other side of the end point, such as a PCI interface or a user-defined interface (UDI).

14.2.3 Transaction Routing: Terminating End Point

In a terminating end point, each end point has a base ID assigned to it to identify it from other devices in the system. The transport layer logic of an end point checks the DestID of all incoming packets against the BaseID register value to see if they are equivalent. All packets with different DestIDs should be ignored. A programmable control can be incorporated into the end point that disables the incoming DestID check and passes all incoming packets to the logical layer.

An error-handling block can be implemented for error detection and reporting. Two mechanisms can be used for error reporting: the error interrupt and the outbound port-write request. An outbound port-write would be sent to a specific DestID when an error condition has occurred.

Inbound port-writes can also be used to inform the host about error conditions in an external RapidIO device. The port-write packet payload is stored in the CSR registers and an interrupt would be generated to an external application for software to address.

The logical layer provides transaction mapping for the RapidIO packet into a user-defined interface (UDI). The UDI packets would typically not include any CRC or logic 0 padding fields specific to the RapidIO protocol. These fields would be automatically added for outbound packets or extracted from inbound packets by the logical layer of the end point.

The UDI packet format could also contain CRC error insertion fields to control error insertion options, for packets where the CRC field is generated by the end point. CRC checking is implemented in the physical layer. The physical layer could also allow for programmab-ility to suppress re-transmission of packets with CRC errors, but packets with errors should generally be submitted to the inbound UDI for further system error handling. A side band signal could be asserted to flag damaged packets.

14.2.4 Implementing Logical and Transport Layer Functions

The logical layer contains most of the functionality associated with the control of RapidIO transactions. The logical layer contains queue structures for holding transactions until they are

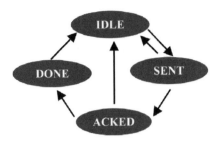

Figure 14.3 Example state machine

completed. The queues must have the ability to rerun a RapidIO transaction if required. The outbound queue services transactions going to the RapidIO network; the inbound queue services transactions coming from the RapidIO network. Every transaction is tracked with a state machine. An example state machine is shown in Figure 14.3.

Transactions always start off in the IDLE state. When they are scheduled for transmission and sent over RapidIO, they enter the SENT state. If the transaction needs rerunning because of resource limitations or error conditions, it is reset back to the IDLE state. If there is no need to rerun the transaction and the link acknowledges its receipt, the state machine tracking the transaction enters an ACKed state. If the transaction does not require a response, it is complete and can enter a DONE state immediately. When the transaction buffers are deallocated it transitions to the IDLE state. If the transaction requires a response and receives retry response, it enters the IDLE state and must run again. If the transaction receives a response of 'done' in the ACKed state, it is complete and enters the DONE state. When the transaction buffers are deallocated it transitions to the IDLE state.

Time-out counters are also used to track the lifetime of transactions over the RapidIO network. They may be used to detect errors in the system. For example, a time-out counter detects that an error has occurred if a read response transaction is not received within a predefined time frame. The time-out values are implementation dependent. When a time-out occurs, the system is notified of the error condition and the transaction is removed from the buffers.

The queues must obey the RapidIO transaction ordering and deadlock avoidance rules, as specified by the RapidIO physical layer specification. These are described in more detail in Chapters 4, 7 and 8.

The outbound queues also support alignment logic. Multiple aligned child RapidIO transactions can be generated from a single local unaligned parent transaction address. Unaligned transactions are split into RapidIO aligned byte lanes, using a simple lookup table. This means the outbound queues must gather responses for all child transactions before completing a parent transaction. For GSM transactions, the outbound queues may support multicast capabilities. The multicast capability allows queues to issue multiple instances of a single local transaction to multiple RapidIO targets, and gather their responses. This requires the outbound queues to have information about the RapidIO targets that are to receive the multicast transaction. The outbound queue generates the transaction data and most of the information contained in the transaction header.

The inbound and outbound queues can support mechanisms to translate a local address to a RapidIO system address. The inbound queues use a transaction identifier (TID) to communicate with the outbound queues and complete transactions. The inbound queues provide incoming transaction header information and data to the internal interface. Error detection and handling logic manages logical layer errors.

The transport layer appends source and destination ID information to the outbound packet header. The RapidIO fabric uses this ID information for routing a packet from source to destination over the network. Information indicating whether the packet supports small or large networks is also appended to the header, as specified by the TT fields. For incoming packets, the transport layer checks the network type and destination ID before processing the transaction. Error detection and handling logic manages transport layer errors. The transport layer implementation also supports packet assembly logic for outbound packets and packet stripping logic for incoming packets.

14.2.5 Implementing Serial Physical Layer Functions

The serial physical layer (PHY) is responsible for handling packets and control symbols. The RapidIO serial physical interface is defined to allow direct leverage of electrical drivers and receivers developed to support the XAUI 10-Gbit Ethernet interface. The main difference between the XAUI and the RapidIO physical interface specifications is the addition of lower speed modes for RapidIO. RapidIO supports 1.25 Gbaud and 2.50 Gbaud links, while XAUI only supports 3.125 Gbaud links. RapidIO also defines a low-voltage signaling mode intended to reduce power dissipation in short-reach applications.

The physical layer logic is responsible for a number of other functions, such as the generation and checking of error checking codes. For outbound packets a CRC-16 algorithm is used to generate and append a CRC checksum to packets. Packets are also padded with 0s if they are not aligned to a 32-bit boundary after appending the CRC checksum. A physical layer header, which includes packet sequence and service priority information, is appended to the packet. The packet sequence information is stored in case there is a packet retry or error recovery and the packet must be resent. On the inbound side, CRC and sequence numbers of the packets are checked. If the checks fail, an ACK-not-accepted is sent to indicate an error condition. If the checks pass and there are resource conflicts for handling the packet, an ACK-retry is sent to indicate a storage conflict. If the packet has no errors and no resource problems, an ACK-accepted is sent to indicate that reception of the packet over the link is complete.

Control symbol generators are responsible for generating packet start and termination information and are typically implemented as simple state machines. The control symbol generators are also responsible for 'piggybacking' incoming transaction ACK information, and embedding other outbound control symbols within outbound packet data streams to increase overall system performance. On the inbound side, control symbol handlers are responsible for extracting packet start, termination information, and embedded/piggybacked control information. The control symbols are protected by CRC-5 checksums.

The serial PHY contains lane alignment and synchronization state machines for the transmitter and receiver. The PHY also implements a state machine to handle idle sequence generation and other link-specific physical layer control.

Asynchronous interface circular queues provide synchronization of data between the physical layer clock domain and local clock domain.

The SERDES (serializer–de-serializer) contains 8B/10B encode functions on the outbound side to convert 8-bit data and control information to a 10-bit code (characters). Special control characters are used to control the serial data flow and to embed the clock information in the form of bit transitions on the serial data stream. The outbound serializer PMA (physical medium attachment) is used to serialize the 10-bit code (character) using a parallel-load serial-unload shift register on to one lane. The serial data stream is transmitted using differential drivers.

On the inbound side, serial data coming on the lanes are first received by differential receivers. They are deserialized using a serial-load and parallel-unload shift register and aligned properly using the 'align' special character. The 'comma' character is used to extract the clocks from the serial data stream. Phase-locked loops (PLL) are used to lock a local clock to the extracted clock and the local locked-in clock is used for recovering the data from the interface. The characters are 8B/10B decoded to generate data and control information that is processed by the upper layers.

14.2.6 Implementing Parallel Physical Layer Functions

The parallel PHY is responsible for handling packets and control symbols, just like the serial PHY.

For packets, the control implementation is similar to a serial PHY, except that the serial PHY allows a maximum of 31 outstanding transactions and the parallel PHY allows a maximum of 7 outstanding transactions over the link. This should not be confused with transactions outstanding over the network, as an implementation could theoretically have 256 outstanding transactions over the network while only having 5 transactions outstanding over any particular link. A physical link is just an interface between two adjacent devices in a network.

Control symbol generators and handlers are also present in the parallel PHY implementation. Error coverage is provided by a bit-wise complement scheme that aligns control symbols to 32-bit boundaries. Control symbol functionality is slightly different between the serial and parallel PHY to provide maximum benefits for each implementation.

Transmit structures that launch DDR (double-data rate) data on to the RapidIO interface must meet the skew requirements between clock and data lines as specified by the RapidIO AC specifications. The transmitter clock and data duty cycle must meet the specifications to guarantee error-free operation. Receive structures that receive DDR data from the RapidIO interface must be able to handle the skew requirements between the clock and data lines as specified by the RapidIO AC specifications. Special clock and data de-skew logic coupled with the training state machines guarantees data recovery at the specified speed.

14.2.7 Outbound Transaction Flow

This section presents the flow of a transaction through the layered structure of a RapidIO end point. Transactions from the internal interface to the RapidIO network flow through the outbound logical, transport and physical layers. Outbound transactions could be new requests or responses to incoming transactions.

At the logical layer, new requests may go through a process of address translation to convert a local address to a RapidIO system address. The new request may also pick up attributes to be used by the transaction for the target RapidIO address space window. Every new request transaction is assigned a unique transaction identifier (TID). Every response transaction is sent back to the requesting device with the same TID as the original request packet. In this way an incoming response transaction is matched with its outstanding request transaction.

Response time-out counters generate error interrupts when transactions are lost in the network. Error interrupts are also generated when a response does not match an outstanding request or when an error-response is received. If a response is required, the outbound queue holds a request transaction until it receives a response transaction, or until the response time-out counter expires, causing an error. Otherwise transactions in the outbound queue are completed when an ACK-accept is received from the adjacent device on the link. If the original transaction was unaligned, the outbound queue needs to gather responses for multiple packets before completion of the transaction. Transactions are scheduled by the outbound queue on a first-come first-served basis unless a retry condition exists. Transactions get re-scheduled to allow higher flow level transactions to bypass lower flow level transactions. Sophisticated deadlock-free schemes can be implemented that prevent starvation conditions, re-prioritizing requests over responses and re-prioritizing higher flow levels over lower flow levels.

At the transport layer, transactions are appended with source and destination ID information as well as the network size support information. The transactions are assembled and formatted as RapidIO packets ready for processing by the physical layer.

At the physical layer, control symbols are given priority over packet data so that control symbol data can be embedded in a packet stream. A CRC checksum is calculated for the packet and appended to it, as is the physical layer header with sequence and priority information. Packet start and termination control symbols are generated and merged with the packet data stream to identify packet boundaries as they are transmitted over the RapidIO link. For serial PHY, the packet and control symbol data bytes are fed to the SERDES. The SERDES encodes the data bytes by 8B/10B encoding and serializes the data on to a single lane, and clocks are embedded in the serial data stream. The SERDES also handles byte striping across lanes for multiple lane implementations.

14.2.8 Sample Outbound Module

A RapidIO end point can be designed to provide high-speed system interconnect to a user-defined interface (UDI). The outbound module includes a path from the RapidIO interface to the UDI. The left-hand column of functional blocks in Figure 14.2 shows a typical outbound module. The outbound module implements the following tasks:

- UDI to RapidIO transaction mapping
- RapidIO protocol requirements for packet transmission: control symbol generation, idle generation, hardware and software error recovery support
- Re-transmit queue and buffer management for retry support
- CRC generation
- Buffering of packets in the outbound queue
- Initialization and synchronization logic
- Serial RapidIO, 1x mode or 4x mode
- Serial RapidIO, reordering of packets with RapidIO rules in 1x mode, or priority rules in 4x mode
- Serial RapidIO, 8b/10b encoding
- Serial RapidIO, lane striping in 4x mode

As part of the outbound packet processing the following logic functional blocks will also be required to support proper RapidIO operation.

14.2.8.1 Re-transmit Queue

The outbound module contains packet buffers that hold outbound packets until they are acknowledged by the receiving end of the link. This module is responsible for sending the oldest, highest-priority packet after a packet retry is received.

14.2.8.2 Packet Generator

The packet generator assigns AckID values to the physical layer of outgoing packets.

14.2.8.3 Control Symbol Generator

The control symbol generator produces packet delimiter, status and acknowledge control symbols which are output to the link.

14.2.8.4 Logical Outbound Module

The output module of the end point can be designed to function in store and forward mode, where the transfer request is not sent to the physical layer until after the packet is completely written to the buffer by the UDI. As with the input side of the end point, the output module can allow for 'cut-through' operation where a packet is forwarded to the physical layer as soon as the packet header is received from the UDI. If an error is detected in the remaining portion of the packet, the end point must be capable of issuing a 'stomp' command to inform the physical layer of a corrupt packet and issue a retry command to the original packet source on the UDI.

Consideration should also be given to supporting a head of queue blocking avoidance algorithm, which reorders packets based on whether they can be accepted by their target port.

14.2.9 Inbound Transaction Flow

Transactions from the RapidIO port to the internal interface flow through the physical, transport and logical layers.

For a serial PHY, the incoming serial data is accumulated into characters (10-bit data). Clocks are also extracted from the serial data stream by custom circuits such as a PLL. The characters are decoded using 8B/10B and packet/control symbol data is generated. The packet data and control symbol data are handled independently. For packets, CRC and sequence numbers are checked. These errors are recoverable in hardware if the checks fail. ACK symbols are generated for the incoming packets and sent out on the RapidIO link. Control symbols are also checked for errors and are recoverable in hardware. If it has no errors, any incoming ACK control symbol is forwarded to the outbound physical layer for further processing.

The received packets are sent to the transport layer for striping. Checks are made to see if the destination ID field and device ID size field of the received packet matches the configuration of the target device. The header and data of the received transaction are then passed on to the logical layer processing logic.

The logical layer forwards information regarding response transactions to the outbound queue for completing outstanding transactions. New requests are processed one at a time, based on the absence of errors and resource conflicts for the given flow level. The transactions may go through a translation mechanism to convert RapidIO system addresses to local addresses, as well as picking up attributes for the local transaction. The inbound queue schedules incoming transactions to the local interface on a first-come first-served basis, unless the transactions cannot make forward progress on the local interface. If they cannot progress, transactions are re-scheduled to the local interface based on their flow levels.

14.2.10 Sample Inbound Module

The RapidIO data stream to the inbound module consists of data packets and control symbols that must be separated and routed to the correct destination. The incoming data packets are buffered and queued for processing by the higher-layer protocol logic. Inbound control symbols consist of either commands from the adjacent device or responses to prior outbound transactions from this device that require appropriate action.

14.2.10.1 Destination IDs

As with a switch design, the destination ID of incoming packets must be checked as soon as they arrive at the end point. The destination ID is used to direct the packet to the proper device, or destination, in the system.

A RapidIO end point can support an 8-bit or 16-bit destination ID. System designers must ensure that end points attached to the device support the appropriate sized destination IDs.

If the end point is used as a bridge, routing tables can be used to map incoming packets from RapidIO to another protocol (for example, PCI) and from another protocol to RapidIO.

14.2.10.2 End Point Inbound Module

A RapidIO end point can be designed to provide high-speed system interconnect to a user-defined interface (UDI). The inbound module must implement the following tasks:

- RapidIO protocol requirements: receiving packet and control symbols, detecting CRC and other error conditions, supporting hardware and software error recovery
- Initialization and synchronization support
- Alignment support in 4x mode (serial RapidIO)
- CRC checking
- Error reporting—outbound port-write and error interrupt generation
- CRC generation for maintenance responses and inbound port-write requests
- Inbound port-write support and interrupt generation
- Checking programmable DestID, selecting destination of the packet
- Deciding if the maintenance packet is accepted
- Buffering of packets
- CRC and logic 0 padding removal
- RapidIO to UDI mapping

In Serial RapidIO, packet data should be aggregated from 8-bit (1x mode) or 32-bit (4x mode) to a wider bus structure (64-bit bus for example) and written to a FIFO. By aggregating data to a wider bus structure it is possible to run the end point core clock at a lower rate than the link clock rate.

14.2.10.3 Transport Layer

This layer is responsible for routing packets through the UDI or to internal registers. Maintenance request packets, which have a DestID equal to the BaseID assigned to the endpoint device, are sent to the register space. Inbound maintenance response packets and all other packets are sent to the UDI.

A serial RapidIO device can be designed to support programmable control to disable DestID checking. In 4x mode the transport layer of lane 0 is the only one used.

14.2.10.4 Inbound Buffers

The inbound module should contain some number of packet buffers. Packet buffers hold incoming packets until the UDI is available. The number of buffers (and buffer depth) should be selected based on the UDIs ability to keep up with expected RapidIO traffic. The device can be designed to function in store and forward mode, where the transfer request is not sent to the logical layer until after the packet is completely written to the buffer by the UDI. The end point can also allow for 'cut-through' operation where a packet is forwarded to the UDI as soon as the packet header is received. If an error is detected in the remaining portion of the packet, the end point must issue a 'stomp' command to inform the UDI of a corrupt packet and issue a retry command to the original packet source.

14.2.10.5 Response, Port-write Out and Debug Packet Assembler

The inbound module is responsible for encapsulating register data and status into maintenance response port-write and debug packets. Response packets are generated by read or write requests to local registers. For read requests, response packets contain register data. Response packets are routed to the outbound flow of the port that received the read request. Port-write packets are generated by error conditions and are sent to a user programmable destination. The designer may also want to include some type of system debug capability by including a debug packet generator within the end point. Internal registers can be dedicated to supporting the content of debug packets. These debug packets are user defined and generated when a register bit is set.

14.2.10.6 Elastic Rx/Tx FIFO and Lane Alignment

For serial RapidIO endpoints, lane alignment needs to be given special consideration. The issue arises as the end point attempts to combine data from the four serial lanes of a 4x serial port. The RapidIO serial specification allows time for seven code groups of lane to lane skew. To deal with this skew, the end point could implement a series of elastic buffers (one per lane), which would also help in dealing with crossing clock boundaries (from the serial recovered clock to the end point core clock). Each of the buffers would need to be deep enough to accommodate time for seven code groups of lane skew. Data from the four lanes would be read out of the buffers using a common clock.

14.2.11 Acknowledgment Symbol History Queue

The outbound module of a device needs to track outstanding RapidIO packets that have not received an acknowledgment (ACK) symbol. A history queue can help in the generation of new AckIDs for packets heading out on RapidIO and keep the history of an AckID used for a particular transaction. This helps with packet re-transmission in case a retry is required.

A history queue implementation could have the following functionality:

- Generates a new AckID for a packet heading out on RapidIO, based on the AckID counter value
- Saves the history information of the AckID and the destination ID information
- Detects out-of-order received ACK symbols and takes appropriate action
- On an incoming ACK, generates the correct responses back to the transaction source to update the status bits
- Re-uses AckIDs as transactions are completed on the RapidIO interface
- Holds off transmission of packets on an ACK retry to re-start either with the same order or with a new order
- Times out transactions from the history queue to recover on a system error

14.3 SUPPORTING FUNCTIONS

14.3.1 RapidIO Controller Configuration

Most systems require both 'out-of-band' and 'In-band' configuration capabilities. Out-of-band configuration implies that the devices can be configured using other signaling or protocols beyond the mechanism provided within the RapidIO protocol. Conversely, in-band configuration relies solely on the mechanisms provided by the RapidIO protocol. In either case, it is desirable that all device configuration registers be accessible by either out-of-band or in-band signaling.

There are many standard approaches that can be used for out-of-band configuration of RapidIO devices. Two examples are:

- I2C Interface: the I2C Interface is commonly used on many device types and can be used for initializing internal end point registers after reset. The I2C interface can also be used to provide a method for performing random access master read and write operations to EEPROMs or to other I2C compatible slave devices. I2C is an acronym for intelligent I/O controller. It is an industry standard for a low-pin-count serial interface for device initialization and control.
- JTAG Interface: a RapidIO end point can also include a JTAG interface as an out-of-band configuration method. The JTAG interface would have the ability to access and configure the internal end point registers. Register access through JTAG can also be used during normal operation to perform extensive read and write accesses on the performance registers, without slowing down the normal traffic in the device. JTAG is an acronym for joint test access group. JTAG is an industry standard for a low-pin-count serial access to internal registers and other state bits of devices.

For in-band configuration, RapidIO maintenance request packets are used. Maintenance packets are the only packet types that access the registers internal to a device, for both read and write operations. An end point can be designed with an internal register bus that enables access to the registers while the device is operating. The register bus would be accessed through RapidIO maintenance packets.

14.3.2 Link Training

RapidIO training is a process that configures two adjacent parallel RapidIO ports for data transfer. The RapidIO training process occurs automatically during system initialization; user intervention is not required. This training process enables a RapidIO inbound module to align its input signals to properly sample data and frame signals transmitted from an adjacent device's outbound module. In a system where an 8-bit parallel RapidIO interface is connected to a 16-bit interface, training is also used to detect the usable width of the 16-bit interface.

The RapidIO training process requires that a pre-defined signal pattern, or training pattern, be applied to the RapidIO interface during initialization. The training pattern is sent after system power-up, system reset, error recovery, or (during normal operation) when the RapidIO inbound module loses the previously established alignment. For example, the alignment can be lost owing to excessive system noise or to a power supply fluctuation.

The training pattern is aligned to the 32-bit boundary, and it is easily distinguishable from control symbols and packet headers. The training pattern is a 64-bit pattern for 8-bit ports with the framing signal switching at the same time as the data bits. The format provides four beats of logic 1 that alternate with four beats of logic 0 for both the data bits and the framing signal of the interface widths. The frame signal does not have to transition high-to-low or low-to-high in phase with the data bits.

The following describes the training (or alignment) sequence:

- The first port begins the alignment sequence by transmitting the link-request/send-training control symbol followed by transmitting the training pattern 256 times to the attached port.
- At the same time, the first port tries to detect and align its input sampling window to the training pattern that is sent from the attached port. If the first port has completed transmitting the 256 iterations of the training pattern, but has not yet successfully adjusted its input sampling window, it again sends a link-request/send-training control symbol and another 256 iterations of the training pattern.
- When the attached port is operating, the first port finishes adjusting its input sampling window to the training pattern coming from the attached port. At this point, the first port sends one idle control symbol (instead of sending a link-request/send-training control symbol) between sending 256 iterations of the training pattern.
- The first port continues to send one idle control symbol after sending 256 iterations of its training pattern until it has received an idle control symbol from the attached port.
- The training pattern cannot be embedded in a packet or used to terminate a packet. All activity on the inbound module must stop gracefully before training patterns are issued.

Serial RapidIO end points have similar training protocols that must be supported.

14.3.3 Device Maintenance

A RapidIO end point supports the following Maintenance transaction types.

Maintenance read or write request: 4-byte access to configuration registers. The request is directed to the internal register bus. If more than one RapidIO ports are present on the end point device, the register bus may be common for all ports. If this is the case, an arbitration scheme is required to service requests from different ports. Since each port can independently accept read or write request to the register bus, the designer needs to determine how many outstanding maintenance transactions are allowed. Once the maximum number of outstanding transaction is reached, the RapidIO port retries all subsequent transactions.

Maintenance response: read response with 4-byte payload or write completion response. Responses are generated by the register block logic and inserted into the outbound queue (for CRC calculation) of the same port where the request was generated.

Maintenance outbound port-write request: the packet generated by the error handler logic. The port-write packet is forwarded to the port where an error condition had occurred. The end point device should also generate an interrupt to the UDI at the same time.

Maintenance inbound port-write request: the request generated by a remote RapidIO device to report an error condition. The end point device should load the inbound port-write payload into the status registers and generate an interrupt to the UDI.

Maintenance debug packet: a user-defined packet length that is generated on a by-port basis (under user register access) for debug purposes only.

A RapidIO switch claims maintenance packets when the hop count in the packet header is zero. The packet is claimed, de-packetized and the data is sent to the internal register bus. Upon receipt of a maintenance request packet, a maintenance response packet must be assembled in the inbound module and a CRC field generated. The source ID and the destination ID are reversed from the maintenance request packet; the source ID becomes the destination ID. When the packet is fully assembled, the inbound module makes a request to the switching fabric to transfer the response packet to the outbound module on the same RapidIO interface. When the outbound module receives the maintenance response packet, it sends the packet to the RapidIO source.

When a RapidIO switch inbound module receives a maintenance packet with a hop count not equal to zero, the hop count value is decremented and the packet is sent to the internal switching fabric for transmission to an outbound module. Because the hop count field is decremented, the inbound module must recalculate the CRC value prior to sending off the packet. This is the only instance of CRC recalculation within switches for RapidIO packets.

14.3.4 Performance Monitoring

Performance monitoring is a very useful RapidIO switch feature, as it allows observation of device and system status. Since RapidIO traffic can be initiated by unrelated sources and can experience congestion in any destination interfaces, it is desirable to implement performance monitoring on each port of a RapidIO switch.

System software can use the data gathered through performance monitoring to identify and correct situations that impact system performance. For instance, system software can be designed to routinely read the performance monitoring registers, analyze the traffic flow patterns, and re-route accordingly to avoid congestion.

The following are potential parameters that can be gathered as part of a device's performance monitoring:

- number of retries on CRC error
- number of CRC errors
- number of 32-bit words

- number of transactions
- number of reads/writes for each packet priority (0, 1, 2 and 3)
- queue depth for inbound and outbound buffer
- number of reordering commands

From these parameters the following system performance metrics can be calculated:

14.3.4.1 Traffic Efficiency

- packet rate (number of transactions/time)
- average packet size (number of 32-bit words/number of transactions)
- utilization ((packet rate × packet size)/max capacity)

The calculations of the packet rate, packet size, and utilization are typically done external to the switch. Registers within the switch could store the count for the number of transactions and the number of 32-bit words for each interface.

14.3.4.2 Bit Error Rate

- number of retries on CRC error
- number of CRC errors

In each switch interface, counters could be implemented to track the number of times a particular error occurs and store the data in internal switch registers. One of the registers could also contain a programmable threshold field, which when reached, would trigger an interrupt.

14.3.4.3 Throughput

The number of read and write transactions in each interface can be very important parameters when debugging RapidIO systems. This information can also be valuable when used by system software to dynamically re-route traffic around congested interfaces. The following parameters could be used to monitor the throughput on each RapidIO interface:

- number of reads for each priority level (0, 1, 2 and 3)
- number of writes for each priority level (0, 1, 2)

Each interface could have four counters for reads and four counters for writes (one for each priority level). The value of each counter would be reflected in the internal switch registers.

14.3.4.4 Bottleneck Detection

Monitoring the queue depth of the inbound and outbound modules can detect bottlenecked traffic in a RapidIO switch port. The number of packets waiting to be processed can be counted, and this count placed into internal registers. A queue depth watermark for the number of packets in the outbound and inbound buffers can be programmable. After the value in a programmable threshold is reached, an interrupt can be triggered to notify the system host of a potential bottleneck situation.

14.3.4.5 Congestion Detection

Packets need to be reordered when forward progress through the internal switching fabric is impeded. Packet reordering can be a sign of congestion in a RapidIO interface. A count of the number of times packets are reordered in each interface can be monitored and stored in internal registers. After the value in a programmable threshold is reached, an interrupt can be triggered indicating congestion on a particular interface.

14.3.5 Error Handling

Hardware-based error detection and recovery is a very important aspect of RapidIO systems. Compliance with the RapidIO specification requires that interfaces be designed with extensive error detection and recovery circuitry. Each packet deploys cyclic redundancy check (CRC) to detect bit error conditions. Because RapidIO does not allow dropped packets, every device maintains a copy of the sent packet until it has received a positive acknowledgement from the receiving device.

RapidIO errors that may occur in a system include packet errors, control symbol errors, and time-out scenarios. RapidIO has embedded provisions for error recovery in the case of recoverable errors, but non-recoverable errors require a higher level of resolution. In order to manage errors so they have a minimum effect on system performance, devices must implement a RapidIO error handler. If the error is recoverable, the handler needs to log data about the error and retry the transaction without external intervention. If the error is non-recoverable, the handler logs data about the error and communicates the information to another device, such as a host processor, that is capable of resolving the error condition.

14.3.5.1 Recoverable Errors

The RapidIO interfaces on a device are designed to detect two basic types of recoverable error: an error on a packet (inbound or outbound), and an error on a control symbol. Recoverable errors have provisions within the RapidIO protocol for being retried at the link without any system host intervention. In addition, resource conflicts such as an internal lack of queue space or other local problem can be retried and resolved with no system intervention. A switch can track recoverable errors in one of its registers. These registers would typically not report errors to the local processor until a threshold is exceeded.

Recoverable packet errors (inbound and outbound packets) and recoverable control symbol errors are described in Sections 14.3.5.2 and 14.3.5.3.

14.3.5.2 Inbound and Outbound Packet Errors

There are four basic types of RapidIO packet error:

Unexpected AckID: a new packet has an expected value for the AckID field at the receiver, so bit errors on this field are easily detected and the packet is not accepted.

Bad CRC: CRC protects the Prio, TT, and Ftype fields and two of the reserved bits, as well as the remainder of the transport layer and logical layer fields in a packet.

S-bit Parity: an error on the S-bit is detected with the redundant inverted S-bit parity. The S-bit is duplicated as in the control symbols to protect the packet from being interpreted as a control symbol.

Packet Exceeds Maximum Size: a packet that over-runs some defined boundary such as the maximum data payload (maximum data payload is 256 bytes, maximum total packet size is 276 bytes) or a transactional boundary.

14.3.5.3 Control Symbol Errors

There are three basic types of RapidIO control symbol error:

Corrupt Control Symbol:
True and Inverted 16-bit Halves Do Not Match: the entire aligned control symbol is protected by the bit-wise inversion of the control symbol used to align it to a 32-bit boundary. A corrupt control symbol is detected as a mismatch between the true and complement 16-bit halves of the aligned control symbol.
S-bit parity error: The S-bit, distinguishing a control symbol from a packet header, has an odd parity bit to protect a control symbol from being interpreted as a packet. An error on the S-bit is detected with the redundant inverted S parity bit.
Note: an indeterminate error is an S-bit parity error in which it is unclear whether the information being received is for a packet or a control symbol. These errors are handled as corrupt control symbols.
Uncorrupted Protocol Violating Control Symbols: a packet-accepted, packet-retry, or packet-not-accepted control symbol that is either:
Unsolicited
Has an unexpected AckID value
Time-out: counters that expire when the allotted time has elapsed without receiving the expected response from the system.
Note: a time-out on an acknowledge control symbol for a packet is treated like an acknowledge control symbol with an unexpected AckID value.

14.3.5.4 Non-recoverable Errors

Non-recoverable errors initiate re-synchronization with the adjacent RapidIO device. Once the devices negotiate the bus width, timing, and other necessary considerations, the system can transfer data across the local links and out across the fabric. Non-recoverable errors should be tracked in internal registers and the switch should be capable of generating interrupts for non-recoverable errors as required by the system. There are five basic types of RapidIO non-recoverable error:

Link Response Timeout: a link response is not received within the specified time-out interval.
Link Response Nonsensical AckID: AckID received with unexpected link response.
Error Recovery Threshold Error: error recovery threshold exceeded.
Retry Recovery Threshold Error: consecutive acknowledge retry threshold count exceeded.
Missing Idle After Training: idle not received after a requested training sequence completes.

14.3.5.5 Outbound Port-write

Port-write packets are generated by error conditions and need to be sent to a user programmable destination, generally a system host. Port Write request generation reports error and status information. The payload of the maintenance port-write packet contains the contents of several CSRs, the port on the device that encountered the error condition, and some optional implementation specific information.

The following port-write requirements must be supported in a switch design:

- Maintenance port-write requests are sent to a predetermined system host defined in the registers.
- The 16-byte data payload of the maintenance port-write packet contains the contents of several registers, the port on the device that encountered the error condition, and some optional implementation-specific information
- The port that received the packet detects physical layer errors. The error reporting logic of this port generates a port-write out request and submits it to the predetermined system host.
- The sending device sets the port-write pending status bit in the registers. Software indicates that it has seen the port-write operation by clearing the port-write pending status bit.
- The port-write generation feature can be disabled per device.
- The port-write out request is a packet that does not have guaranteed delivery and does not have an associated response. However, the serial RapidIO interface will continue to send port-write out requests until software clears the status registers. The programmable time out counter is defined in the registers.
- A write access in the registers stops the time-out counter of the physical layer's port-write out generation logic.

14.3.5.6 Inbound Port-write

End points should be designed to allow for inbound port-write requests. The number of outstanding requests a device can support is implementation specific. When only one outstanding request is supported and the device receives the maintenance port-write request packet with correct CRC and DestID (optional); it locks the payload in internal registers and generates an interrupt to the local processor. All subsequent inbound port-write requests are discarded until software unlocks the internal registers.

14.3.6 Interrupts

The RapidIO Specification dictates that interrupt programming models are beyond the scope of the RapidIO architecture. Interrupt programming models are typically associated with processing devices and not peripheral devices. It is likely that RapidIO-equipped devices will need to generate interrupts to signal exceptional events to processing devices in a system.

RapidIO devices may participate in the transmission of interrupts within a system. For example, in order to support PCI hardware interrupts between multiple bridges in a system, RapidIO mailbox messages can be used to transfer hardware interrupt data across the RapidIO fabric. A RapidIO bridge configured to receive PCI inbound hardware interrupts would generate special RapidIO outbound mailbox messages when it detects the assertion of a PCI interrupt signal on its PCI bus. A bridge configured to generate PCI outbound hardware interrupts would do so based on receiving these special RapidIO mailbox messages. In this way PCI hardware interrupts may be tunneled through a RapidIO fabric.

The remainder of this section provides information for consideration when implementing interrupts in a RapidIO device.

14.3.6.1 Sample Controller Interrupts

Some sample interrupts generated by a RapidIO controller could be:

- **Inbound buffer queue threshold interrupt**: Two registers would be used to track this interrupt. The first register would set the inbound buffer queue watermark. A second register would be used to monitor the state of the inbound buffer. New packets accumulate in the inbound buffers, destined for the switching fabric. When the number of buffers in use equals or exceeds the watermark, a bit in this register would be incremented by one. A threshold bit defines the number of times the count would be incremented before an error interrupt would be generated. This interrupt indicates there may be a problem with the throughput of data on this port.
- **Inbound CRC error threshold interrupt**: This register would contain a count of the CRC errors. It would define the number of CRC errors that can occur before an error interrupt would be generated.
- **Inbound reordering threshold interrupt**: When a packet cannot make forward progress in the inbound module, the switching fabric should reorder all the packets, based on packet priority. Each time the packets are reordered, the inbound reordering count could be incremented by one. The inbound reordering threshold would define the number of times the inbound reordering count would be incremented before an interrupt is generated. This interrupt would indicate there might be a problem with the throughput of data on this port.
- **Inbound destination ID not mapped interrupt**: This interrupt would be used to signal to the system host that an inbound packet was received with a destination ID that did not have a corresponding mapping in the routing table. The destination ID and switch port the packet was received on should be stored in internal registers.
- **Outbound buffer queue threshold interrupt**: Two registers could track this interrupt event. The first register would set the outbound buffer queue watermark for the number of packets in the outbound buffers. The second register would be used to monitor data congestion in the outbound buffer. New packets come from the switching fabric and accumulate in the outbound buffers. When the number of packets in the outbound buffers equals or exceeds the watermark set in the first register, the outbound buffer queue count would be incremented by one. If this count equals or exceeds the threshold, an interrupt would be generated.
- **Outbound retry threshold interrupt**: A value would be set in a register to dictate the maximum number of attempts to send a packet with a CRC error detected at destination. If the number of attempts to send the packet equals the value set, the packet would be discarded and an error reported to the system host.

14.3.6.2 Switching Fabric Interrupts

It may be desirable to generate an interrupt from within the switching fabric when a time-out is detected, because the requested destination is blocked. When this signal is asserted, it could be used to indicate that the requested transaction could not be completed and will be removed from the request queue. The following registers would be needed to control the interrupts:

- A register to control the transaction error signal. This enables or disables the transaction error interrupt, enables or disables the transaction error timer, and sets the number of clock cycles a request will wait for an acknowledge before a transaction error acknowledge.

- A register to contain a status bit for every port on the fabric. The status bits indicate on which port(s) a transaction error has occurred.

14.3.6.3 Terminating End Point

In a terminating end point an interrupt generation block would be designed into the device. Error interrupts would be generated simultaneously with outbound port-write generation. The error interrupts provide error reporting to a local processor, whereas the outbound port-write request packets provide error reporting to a remote host, if necessary. The error interrupts are generated under error detection conditions by the receiving port (as described previously). A reset interrupt would be generated when four RapidIO reset-device commands are received in a row without any other intervening packets or control symbols, except status control symbols. Status and enable registers should be included in the end point design to define the reset and interrupt functionality. Similarly, the end point device could be designed to generate interrupts based on receiving a multicast event control symbol or an inbound port-write packet at its inbound RapidIO port.

14.3.6.4 Bridging Device Interrupts

In our example of bridging RapidIO to PCI, an interrupt controller could be used to translate hardware interrupts to and from RapidIO mailbox messages. This translation ability allows multiple bridges to communicate interrupt data across a RapidIO fabric. In addition, it allows bridge devices to share interrupt information with other RapidIO devices in a RapidIO fabric.

Potential features of a RapidIO end point interrupt controller:

- Translates hardware interrupts to, and from, RapidIO mailbox messages.
- Can be designed to support multiple physical hardware interrupt lines.
  Each interrupt line could be independently programmed for direction (inbound or outbound).
  Each interrupt could generate, or respond to, a RapidIO mailbox message.
  Each interrupt could be routed to any other interrupt and/or RapidIO endpoint attached to a RapidIO fabric.
- Support outbound message signaled interrupt addresses (PCI terminology) and/or message translated interrupt addresses (RapidIO terminology).
- Support multiple pending hardware requests per outbound hardware interrupt.
- Handle hardware interrupts and message translated interrupts simultaneously from PCI to RapidIO.
- Handle PCI SERR/PME forwarding functions across a RapidIO fabric.

14.3.6.5 Mapping to Interrupt Output Pins

A RapidIO device or switch can be designed with any number of its own output interrupt pins. Interrupts that occur within a device can be routed to one of the device interrupt output pins. The following registers would be needed to control the interrupts:

- A register to indicate which port has generated an interrupt. It could show which part of the device has generated the interrupt when the routed interrupt output pin is signaled.
- A register to route the interrupt's sources (such as inbound and output modules or the internal switching fabric) to a particular interrupt output pin.

14.3.7 Reset

The RapidIO specification dictates that system reset is beyond the scope of the RapidIO architecture. However, the RapidIO specifications describe mechanisms for communicating reset information within maintenance packets. RapidIO controllers will need to respond appropriately to these packets. This section describes some considerations for resetting RapidIO devices.

- **Hardware reset**: this event would reset all of a RapidIO device's facilities.
- **Inbound RapidIO maintenance reset**: an external linked RapidIO device uses RapidIO link maintenance request packets containing reset commands to initiate the inbound RapidIO maintenance reset. According to the RapidIO specification, reset needs to be triggered when a switch receives at least four link maintenance request packets containing reset commands in a row, without any other intervening packets. After the fourth packet, the switch would generate an interrupt to signify the RapidIO software reset request. No reset action is taken by the RapidIO ports of the switch in response to the request. The system must provide a reset controller device on the board to determine how to process the reset request. The interrupt pin would be asserted when the software reset request is validated.
- **Outbound RapidIO maintenance reset**: a switch could reset an external linked RapidIO device using RapidIO link maintenance reset request packets it would generate from one of its RapidIO interfaces. The system would instruct the switch to send the reset packets to the external linked RapidIO device. Once the device receives the reset packets, it would determine how to process the reset request. According to the RapidIO specification, a device receiving a reset command in a link-request control symbol should not perform the reset function unless it has received at least four reset commands in a row without any other intervening packets or control symbols, except idle control symbols. Based on a system host setting appropriate values in the switch control registers, the switch would begin the link maintenance reset. The switch then issues the RapidIO maintenance reset sequence of at least four consecutive link maintenance control symbols containing reset commands to an external linked RapidIO device.

14.3.8 Clocking

14.3.8.1 Parallel Switch Clocking

A parallel RapidIO interface supports clock input and output options. The clock inputs are received by the switch's RapidIO high-speed inbound module from an adjacent RapidIO device. The clock outputs are transmitted by the switch's outbound module to an adjacent RapidIO device. The outbound module needs to drive the clock signals (differential signals TCLK_p and TCLK_n) to ensure it operates in an exact phase relationship with the adjacent device.

14.3.8.2 End Point Clocking

A parallel RapidIO Interface needs to support clock input and output options (serial RapidIO devices have 8b/10b embedded clocks in the serial links). The RapidIO inbound module receives the clock inputs from an adjacent RapidIO device. The inbound module receives the clock signals and would generate an internal clock that has an exact phase relationship with the adjacent device output clock. The outbound module transmits the clock outputs to an adjacent RapidIO device. The outbound module should drive the clock output such that it operates in an exact phase relationship with the adjacent device.

14.3.8.3 RapidIO Interface PLL Configuration

A parallel RapidIO interface can be designed with a clock generation phase-locked loop (PLL) configurable to clock the transmitter interface at the appropriate speed. The speed must be selected before the de-assertion of reset using a clock select signal. Once reset is de-asserted the speed would be latched internal to the switch. In our sample switch we would use an internal PLL to create clocking options for the core clock and the RapidIO interfaces.

14.3.8.4 Sample PLL Bypass

To provide greater system design flexibility, it may be desirable to bypass the PLL-based clock within the switch. The PLL bypass clock signals would need to be differential inputs of a high-speed LVDS-compatible clock input interface. The transmit clock, which is created from the bypass clock, is half the frequency of the bypass clock. This division by two would be performed within the switch to ensure the correct duty cycle of the transmit clock. As such, the frequency used for the PLL bypass clock must create a supported RapidIO transmit frequency.

14.3.8.5 Serial Switch Clocking

Using SERDES in a serial switch, it may be desirable for each SERDES channel to have its own receive and transmit clocks, so that individual ports could run at different frequencies. Data to and from each SERDES would need to be written to an elastic buffer that is clocked on the system clock domain (recovered from the 8b/10b link). The elastic buffers are used for crossing the clock boundary, lane de-skew and clock compensation. Control symbols must be decoded at the system clock speed in order to keep up with the link in 4x mode. Packet data would need to be aggregated to an internal parallel bus and passed, through a FIFO, into the internal fabric clock domain. The use of a parallel bus permits the internal fabric to run on a clock that is slower than the SERDES clock.

A serial switch can be designed with multiple ports that may operate at different link speeds. Our sample RapidIO Serial switch would support the following clock requirements:

- independent link speed for each port
- no dependency to system clocks of each port
- all gated clocks and a general clock must be aligned
- 50% duty cycle to all clocks

14.4 IMPLEMENTING A RAPIDIO SWITCH

This section describes how the RapidIO protocol can be implemented in a parallel or serial RapidIO switch. A switch's primary responsibility is routing packets across a RapidIO system from the source to the destination using information contained in the packet header. Switches do not modify the logical or transport layers of the packet; only the physical layer contained within the packet is modified at each link. Switches determine the network topology and are an integral part of overall system performance.

14.4.1 Switch Features

RapidIO is a packet-based protocol. Implementations of packet switching functions have been developed and designed for years. Packet switch architectures and design considerations are well understood. Generally, a switch first treats an incoming packet from a physical layer point of view; it deals with low-level aspects of link management, physical error conditions, buffer availability and other physical considerations. The switch then executes the routing function, as would any other packet-based protocol such as ATM or Ethernet.

As described in the RapidIO specification, each interface on a RapidIO switch must process both inbound and outbound transactions at the same time. The inbound and outbound modules of each interface can be designed as logically separate blocks or as a single block. Communication between the inbound and outbound ports in the physical and transport layers of RapidIO is easily supported with simple interfaces between the inbound and outbound modules. For our sample switch, the inbound and outbound ports are implemented as separate modules.

Figure 14.4 shows a four-port sample switch, with input/output modules, each with eight buffers and associated routing lookup tables (LUTs).

The following list provides some features that might be offered by a RapidIO switch:

- independent routing table for each interface (implementation dependent)
- non-blocking fabric (line-rate termination)
- performance and statistics gathering on each RapidIO interface
- large (16-bit) and small (8-bit) transport system support providing 64 k or 256 addressable endpoints
- multicast-event control symbol
- port mirroring
- receiver link level flow control
- transmitter link level flow control
- multicast
- error management capability
- port-write operations for error reporting
- priority-based buffer management
- hot-swap I/Os
- power management
- data plane extensions

Sample features of a parallel endpoint:

- full-duplex, double-data rate (DDR) RapidIO
- operating frequencies of each interface can be configured separately (different clock rates)
- LVDS signaling

Sample features of a serial end point:

- ports that can operate in 1x mode or 4x mode (different clock rates)
- SERDES with 8b/10b encoding/decoding circuit

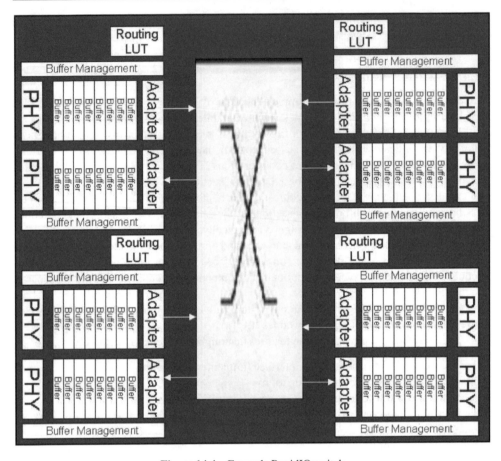

Figure 14.4 Example RapidIO switch

14.4.2 Switch Performance and Cost Implications

Some of the key factors that directly affect the performance of a packet switch from both a latency and throughput perspective are the internal fabric architecture, queuing structure, arbitration scheme, routing look-up and amount of buffering. While it might be possible to design a perfect switch, the realities of semiconductor device economics will quickly lead to switches that are not perfect, but offer reasonable trade-offs for latency, bandwidth, cost and reliability. On the cost side of the equation, both size and complexity of the design should be taken into account. Complexity also plays a role in the size and cost of a device, but perhaps more importantly, excess complexity can lead to added design and verification requirements that will translate into costly schedule delays.

14.4.3 Implementing RapidIO in a Switch

This section examines the considerations associated with implementing a RapidIO switch. A RapidIO switch provides connectivity for multiple RapidIO devices in a system. Switches can be connected to other switches or to end points.

A RapidIO switch, like switches for other protocols such as Ethernet or ATM, could use a non-blocking switch fabric architecture to connect multiple RapidIO interfaces together. The switching fabric would pass incoming RapidIO packets from the inbound module of one RapidIO interface to the outbound module of another RapidIO interface. The details of how a switch could be implemented are presented in the following sections.

14.4.3.1 Inbound and Outbound Modules

RapidIO switches use the physical layer to terminate a packet and the transport layer to determine the packet's destination; they do not typically interpret logical layer packet information. The physical layer fields, specifically the ackID field, is modified from hop to hop in a RapidIO network. The switch does not modify any of the information contained in the transport or logical layer fields of a typical packet. The only exception to this is when the RapidIO interface receives a maintenance transaction packet. Maintenance transactions may be targeted at switches. In which case, the switch must decode and operate on the contents of the maintenance transaction. Maintenance transactions also contain hop count fields that switches are required to update as they are forwarded through the network. This simplifies the design of a switch's RapidIO interface, as only support for maintenance transactions is provided at the logical layer.

One of the most important features of the inbound and outbound modules in our sample switch is transaction buffering. The inbound and outbound modules contain the transaction buffers. RapidIO packets are held in the transaction buffers before being transmitted or received. Considerable care must be taken when designing a buffering scheme, as optimal buffer management is critical in providing satisfactory RapidIO system performance. Both the inbound and the outbound module should contain enough buffering to avoid data starvation on the link. This means that there should be enough buffering to support possible outstanding unacknowledged transactions between two devices on a link.

14.4.3.2 Sample Inbound Module

In our sample switch, the inbound module is separate from the outbound module. It provides an interface between the RapidIO external interface signaling and the internal switching fabric. The inbound module of the sample switch performs the following functions:

- RapidIO interface protocol compliance: receiving packets and control symbols, detecting CRC and other error conditions, supporting hardware and software error recovery
- data capture and transmission
- data rate translation
- error checking and reporting, port-write and error interrupt generation
- control symbol decoding
- buffering of packets
- initialization and synchronization support

In a serial switch, the inbound module performs the following additional functions:

- lane de-striping (re-assembling data received by a 4x link)
- alignment support in 4x mode
- 8b/10b decoder; the 8b/10b decoder decodes 10-bit code group into 9-bit characters (8 bits of data and a control bit)

Each inbound module should contain buffers to hold packets while they wait to be sent to the internal fabric. The number and size of buffers included in a switch inbound module has a direct impact on how many transactions a switch can handle before becoming saturated. For example, if the inbound module of our sample switch contained a set of eight buffers and each buffer could hold one maximum sized RapidIO packet, then each inbound module could store up to eight maximum sized RapidIO packets in its buffers. The maximum packet size permitted by the current RapidIO specifications is 276 bytes. This includes all packet logical and transport layer header information, data payload and required CRC bytes. The data payload itself is a maximum of 256 bytes. The amount of buffering provided is implementation dependent. However, the buffer design must allow the switch port to receive and potentially reorder packets of increasingly higher priorities. This requirement is to allow high-priority packets (level 3 being the highest) to pass lower-priority packets, maintaining forward progress.

If a data packet goes into the inbound module and cannot make forward progress through the internal fabric, it is stored in the input buffers. When it is ready to be transferred through the switch, the inbound module sends a request to the internal fabric with a specific priority for the packet. The priority is carried in the PRIO field. The internal fabric examines the priority of the packet, and if adequate buffers are available in the output buffer for that particular priority, the transfer into the internal switching fabric is initiated.

If a packet can not make forward progress through the switching fabric because the target destination buffers are full and cannot accept the transaction, the input module needs to be able to reorder the transactions and present other packets to the switch fabric in order to avoid head of line blocking.

Indications of a stomp symbol are issued to the inbound module in order to cancel the current packet being stored, and data buffers associated with that packet are freed.

The inbound module can also be designed to operate in 'cut-through' mode. In this mode, when an incoming packet is received, a request is made to the internal switching fabric to start transferring the packet, without waiting until the entire packet has been received. This mode can minimize overall switch latency.

In cut-though mode, the transfer of a packet through the internal switching fabric begins before the packet's CRC is checked. If the CRC is found to have an invalid value after the transaction begins, the destination receiver is notified with a stomp symbol, the packet is discarded, and a request is sent to the initiator of the transaction to re-send the packet.

14.4.3.3 Destination IDs

The destination ID of incoming packets is checked as soon as the packets arrive at the switch. The destination ID is used to direct the packet to the proper device, or destination, in the system. In our sample switch, each interface would have a destination ID routing table that can be programmed with different values. The more destination IDs a routing table can hold the more flexible the packet routing, as each system end point can be mapped to specific outbound ports on the switch.

A RapidIO switch can support either 8-bit destination IDs, 16-bit destination IDs or both. System designers must ensure that end points attached to the device support the appropriate sized destination IDs for packets targeted to them.

If a routing table is used, the entries in the table point to one of the interfaces on the switch. The inbound module of our sample switch decodes the destination ID and transfers the packet into the internal switching fabric. The switching fabric transfers the packet to the outbound module. The outbound module sends the packet through the system to its destination.

14.4.3.4 Switch Packet Priority

Switches are not intended to decode transaction types. The only exception is for maintenance packets with a hop count equal to zero. As a result ordering decisions within a switch are independent of the transaction type.

The priority field establishes the relative importance of a packet. Priority is used in RapidIO for several purposes including transaction ordering and deadlock prevention. A higher-priority transaction or packet is serviced or transmitted before one of lower priority. A switch needs to support the four currently defined levels of priority (0 to 3), carried in the PRIO field of the packet. Packet priority is assigned by the end point processing element that initiates the packet (the source end point). Switch devices cannot modify packet priority. As discussed, a switch does not interpret the transaction field of a packet and cannot tell the difference between a request and a response packet.

Switches need to adhere to the following RapidIO delivery ordering rules. These rules ensure proper packet delivery.

1. A switch-processing element shall ensure that packets received at an inbound interface targeted to a common outbound interface shall not pass packets of same or higher priority.
2. A switch processing element shall not allow lower-priority non-maintenance packets to pass higher-priority non-maintenance packets with the same sourceID and destination ID as the packets flow through the switch.
3. A switch-processing element shall not change the order of packets that make up a transaction request flow (packets with the same sourceID, the same destination ID, the same priority and the same flowID).
4. A switch cannot alter the priority of a packet.
5. A switch provides a consistent path selection, based on the packet's destination ID field.

14.4.4 Port Mirroring Capability

Port (or interface) mirroring is a method of monitoring network traffic that duplicates each incoming packet on a selected interface of a switch. The original packet is sent to the destination interface mapped in the routing table, while the duplicate copy of the packet is sent to a pre-configured mirroring destination interface, where the packet can be studied.

Port mirroring is used as a diagnostic tool or debugging feature. It enables close tracking and analysis of switch performance. The nature of RapidIO signals (high speed, and low voltage) makes it very difficult to directly monitor interface signals. Port mirroring allows analysis of the packets after they have passed through the switch.

In our sample switch, each of the RapidIO interfaces could be configured for port mirroring, and each interface could be programmed separately to target a different destination interface. Each mirrored interface could have another port configured as the mirroring destination.

For example, in a device with four RapidIO interfaces, in order to mirror the traffic between ports 0 and 1, port 3 could be designated as the mirroring destination. Each interface should be able to enable or disable the option and program a mirroring destination interface. In order for proper handshaking to take place, the mirrored interface must be connected to a RapidIO endpoint.

14.4.5 Sample Outbound Module

In our sample switch, the outbound module is separate from the inbound module. It acts as an interface between the internal switching fabric and RapidIO signaling. The outbound module of the sample switch would perform the following functions:

- RapidIO protocol requirements for packet transmission: control symbol generation, idle generation, hardware and software error recovery support
- re-transmit queue and buffer management for retry support
- data capture and transmission
- data rate translation
- error checking
- control symbol encoding
- initialization and synchronization logic

In a serial switch, the outbound module performs the following additional functions:

- 8b/10b encoding
- lane striping in 4x mode

Each outbound module should contain buffers in order to hold packets while they wait to be sent out to the adjacent device. For example, if the outbound module of our sample switch contained a set of eight buffers and each buffer could hold one maximum sized RapidIO packet, then each outbound module could store up to eight maximum sized RapidIO packets in its buffers. The maximum packet size permitted by the RapidIO Interconnect Specification (Revision 1.2) is 276 bytes. This includes all packet logical and transport layer header information, data payload, and required CRC bytes. The data payload itself is a maximum of 256 bytes. The size and number of buffers in the output module is, again, implementation dependant. The designer needs to determine how many unacknowledged packets need to be stored in the switch during normal operation in order to determine the buffer size.

The outbound module in our sample switch also needs to perform all packet reordering. Reordering occurs when the destination device retries a packet. The outbound module keeps a copy of a packet until it is acknowledged, to ensure that the packet is re-sent when necessary. The module needs to reorder the outstanding packets in the queue by promoting the oldest packet with the highest priority.

14.4.6 Internal Switching Fabric

The switching fabric forms the core of a RapidIO switch device. In our sample switch, when the inbound module receives a packet, it performs destination ID mapping between the RapidIO devices connected to the interface, then transfers the packet to the switching fabric. When the switching fabric receives a transaction, the packet is sent to the destination interface identified by the destination ID.

The switching fabric in our sample switch would have the following features:

- separate data path for inbound and outbound traffic
- non-blocking architecture
- head-of-queue blocking prevention algorithm
- per-port selectable store and forward and cut-through algorithms to minimize data path latency
- fully aggregated bandwidth/data path supports: Serial 1x mode, Serial 4x mode, and Parallel
- packet buffers for inbound and outbound port
- support for asynchronous interfacing with RapidIO ports

14.4.7 Packet Routing: Routing Tables

The RapidIO specification dictates that the method by which packets are routed between the input and output of a switch is implementation dependent, as long as it is device-identifier-based packet routing, as outlined in the common transport layer of the RapidIO specification. One method recommended by the RapidIO specification is the use of routing or lookup tables.

There are many ways to implement routing tables in a device. The simplest method allows only a single path from every processing element to every other processing element, which is not ideal. A better solution is to use routing tables with multiple destination IDs that can be programmed with different values, allowing each interface in a device to route its packets separately.

With this method, a device connected to the switch can access only the routes mapped in the routing table of the interface to which it connects. This allows a system control device to protect itself from other system components, and to manage the addition of new components into the system. In a RapidIO switch with only a single routing table for all interfaces, all entities in the system have access to all other entities; there is no capability for limiting system access.

Multiple routing tables also allow for a highly-reliable system to physically isolate faulty components, to diagnose and test them in safety, and to bring replacement components into service. The system impact of device failure is reduced, since the failure will only affect a certain portion of the system.

End points also have the ability to create loopback functions through the various RapidIO interfaces in order to test the device operation. In a running system, with multiple mappings, it is possible to completely test all the interfaces in the device.

In our sample switch, the destination ID field is extracted from the incoming packet and compared with the values in the programmable routing table. The sample switch is designed with a separate routing table for each switch port, so each interface can route packets differently. The incoming packet is routed to its destination interface based on the comparison between the value in the destination ID field of the packet and the values programmed in the routing table. The packet can be routed to one of the following destinations:

- one of the other RapidIO switch ports
- multicast to multiple interfaces
- the RapidIO interface that received the transaction request; a RapidIO interface can target itself through a loopback transaction
- the register bus; RapidIO maintenance request packets are the only packet types that would access an internal register bus for read and write operations to internal switch registers

14.5 SUMMARY

The RapidIO architecture is a feature-rich technology enabling a wide variety of applications. While simple in structure it provides a powerful technology for the efficient communication of devices within a system. However, proper choices need to be made in selecting an optimal feature set for a given application. A good feature set coupled with a sound micro-architecture maximizes the benefits of using the RapidIO interconnect while minimizing the implementation cost and complexity. In terms of functionality, performance, and cost, RapidIO is superior to other architectures. However, this superiority can only be fully exploited by a well-designed implementation of the technology.

15

Implementation Benefits of the RapidIO Interconnect Technology in FPGAs

Nupur Shah

This chapter discusses how field programmable gate arrays (FPGAs) have moved to provide support for high-speed interconnects and how the RapidIO interconnect technology can be implemented in FPGAs either by customers or can be purchased as a pre-designed functional block from FPGA vendors. This chapter also shows how such implementations represent a fraction of the overall FPGA resources and capabilities and how FPGAs are becoming an increasingly important component of embedded systems.

Developers of embedded systems have, over the last few years, enjoyed a continuing increase in the speed and gate count of FPGA devices. FPGA vendors have pushed the capabilities of their products to the limits of what can be achieved with semiconductor technology. FPGA vendors are now typically the first vendors to market with new interconnect technologies. New high-speed interconnects based on parallel or serial differential signaled technology, such as the RapidIO interconnect, were identified as important emerging technologies for FPGAs to support, because of this the first RapidIO offerings available in the market were FPGA based.

The FPGA industry has evolved from being a quick and relatively easy solution for providing glue logic in low-volume systems to a mainstream solution for high-performance embedded processing systems that may be deployed in production systems with volumes of hundreds of thousands of units per year. FPGA vendors have realized that their technology should target the designers who want to prototype and produce systems that have high-speed capability and advanced system features. Designers have also discovered that FPGAs can be a way to support emerging technologies that are not readily available and that FPGAs can offer a level of customization that results in greater system flexibility and value.

A technology such as RapidIO offers a wide variety of advanced system features. These include hardware error recovery, software transparency and system scalability. Designers want to build systems that take advantage of these features. FPGAs facilitate the implementation of technologies such as RapidIO while still providing the flexibility of customization. The marriage of such an emerging technology with a programmable device is ideal.

RapidIO® The Embedded System Interconnect.   S. Fuller
© 2005 John Wiley & Sons, Ltd   ISBN: 0-470-09291-2

15.1 BUILDING THE ECOSYSTEM

Development of any new interconnect technology starts with a revolutionary idea that must take evolutionary steps to get adopted. Building an ecosystem to support a new interconnect is the first and foremost barrier that any new technology must cross to gain adoption.

For an interconnect ecosystem to exist three elements must be present: hardware components, validation tools, and software infrastructure. System hardware components such as hosts, switches, bridges and end points build the basic interconnect hardware. Validation components such as bus exercisers, bus analyzers, interoperability platforms and simulation models verify system connectivity and component functionality. A software infrastructure that supports system discovery, enumeration, management and error handling and recovery allows the designer to take advantage of the advanced system features of the interconnect technology. Figure 15.1 illustrates the interconnect ecosystem.

With the advanced I/O capability, high-speed logic matrix and re-programmability, an FPGA can be an enabler to this ecosystem. Figure 15.2 illustrates how an FPGA can be used to build hardware components, help certify validation tools and create a prototyping system that can be used as a basis on which to build a software infrastructure.

The realization that an FPGA can be an enabler to the ecosystem has led FPGA vendors to see themselves as a whole solution provider. This includes not only FPGA device families that offer a variety of options for I/O capability, density and speed, but also intellectual property (IP), software, support and training. The combination of these offerings facilitates the development of the ecosystem and exploits the time to market advantage of FPGAs.

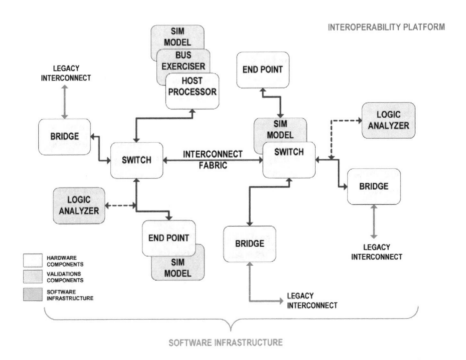

Figure 15.1   Interconnect ecosystem: hardware components, validation tools, software infrastructure

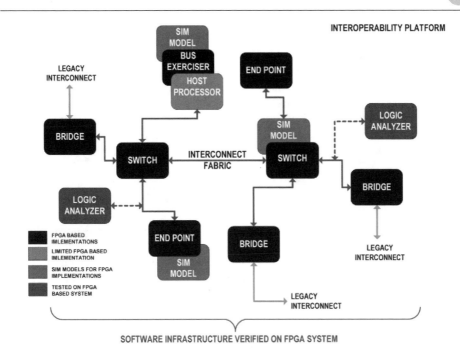

Figure 15.2 FPGA enablement of an interconnect ecosystem

15.2 ADVANCES IN FPGA TECHNOLOGY

15.2.1 I/O Connectivity Options

A large part of supporting any interconnect technology is providing the physical I/O capability. FPGA vendors realize that without physical I/O, enabling the ecosystem is next to impossible. We see two shifts in thinking when it comes to I/O support.

- FPGA vendors have become active in helping to define specifications such as RapidIO so that they can be realizable in programmable logic.
- A standards body, like the RapidIO Trade Association that governs the development of RapidIO, sees support by FPGA technology as a key requirement to an industry adopting any new I/O capability that would increase the interconnect bandwidth.

Given these shifts, FPGA vendors are now pushing the envelope in I/O design. For example, just as the industry is embracing 3.125 Gbps transfer rates, FPGA vendors have technology that can demonstrate transfer rates even higher than 3.125 Gbps. With their investigation and analysis of these higher bandwidth channels, these vendors are proactive in defining the I/O characteristics of a variety of standards including RapidIO.

While the push to higher transfer rates is driving the development of advanced I/O technology, FPGA vendors also see value in providing support for legacy I/O technology. As a result, a programmable device today can support an array of I/O technologies that allows the system designer to migrate to new technologies such as RapidIO while maintaining the investment in a legacy protocol.

15.2.2 Rich Logic Fabrics

Today's FPGA contains a variety of fabric features. The millions of available logic gates allow for the cost-effective implementation of technologies such as RapidIO with the integration of other system functionality all realizable on a single chip. In addition the FPGA provides a rich routing matrix, clocking network and embedded memory. System designers can build complete solutions that make use of technologies such as RapidIO before standard products would be available. Designers are also offered a variety of logic densities and performance capabilities that allow the selection of an FPGA to be tailored to the capability and price target of the design.

15.2.3 The IP Portfolio

As FPGA devices increase in density the time to market advantage for using an FPGA is actually diminishing. The reason for this is the significant effort required in any technology of designing, integrating, and verifying a design of several million gates. The engineering team required to develop a device of this density might be 20 to 30 engineers in size and the design schedule and verification schedule might approach a year in length.

FPGA vendors are investing resources into developing and verifying IP that will facilitate the development of these large system-on-a-chip (SoC) designs. In general if system designers are using a standard-based interconnect technology, they will be making a decision between several established and emerging system interconnects. To help reduce the effort of designing interconnect support for a large FPGA device, FPGA vendors have started providing pre-developed and verified IP to support these commonly used industry interconnects such as PCI and RapidIO.

Interconnect IP is usually provided as a soft core that can be targeted to fixed locations on a variety of devices. This allows the designer the benefit of reliability of a pre-verified function, guaranteed performance and the flexibility place it in one or more locations on any particular device.

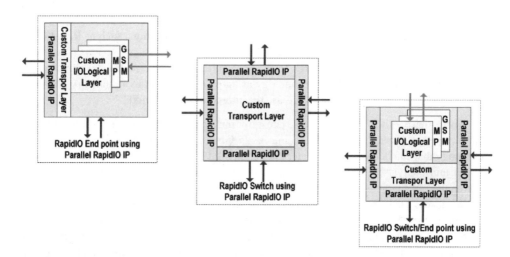

Figure 15.3   RapidIO physical layer IP implementation options

In the case of RapidIO, a layered protocol, vendors are providing RapidIO IP solutions that support one or more, including all of the layers of the RapidIO specification. This allows the designer the greatest amount of flexibility. A physical layer IP product can be used stand-alone to implement an end point, a switch or a switch/end point combination that can support logical I/O, message passing or even the GSM model. A physical layer IP product combined with a transport layer and logical I/O layer product can be combined to create a complete endpoint.

Creating IP blocks that align themselves with the layered nature of the RapidIO protocol offers the designer the option to use the RapidIO technology with greatest efficiency on available logic resources. Designers can choose to implement specific capabilities of the upper layer protocols based on the requirements of the system. For example, systems requiring message passing capability can build this functionality on top of the logical I/O–transport –physical layer structure or just the transport–physical layer structure, depending on the application's requirements.

15.3 MULTI PROTOCOL SUPPORT FOR THE EMBEDDED ENVIRONMENT

The enablement of any interconnect ecosystem in an embedded environment begins with the migration of legacy technology. For this reason the adoption of an open interconnect, such as RapidIO, into a system does not preclude the existence of other proprietary or other open standard interconnects.

FPGA vendors realize that systems will primarily be a mix of technology. Since programmable logic offers a variety of I/O support on the same device this allows heterogeneous systems to exist while enabling the adoption of new technologies. This is one reason that FPGAs today can support multiple differential, single-ended and serial-based standards on the same device.

One benefit to the RapidIO protocol is that it is defined to use industry standard I/O technology that is already supported on existing FPGAs. RapidIO defines an 8- or 16-bit LVDS interface as well as a XAUI-based serial implementation of one lane or four lanes running from 1.25 to 3.125 Gbps. This allows system designers to develop FPGA devices to support migration of their systems to RapidIO.

For example, with an FPGA, a system designer can build components that support a legacy protocol and RapidIO on the same chip without waiting for an application-specific standard product (ASSP) or building an application-specific integrated circuit (ASIC). In addition to supporting legacy protocols, system designers can also build bridge components that go from the RapidIO parallel environment to the RapidIO serial environment. Figure 15.4 illustrates how an FPGA can be used to assist in a system migration from legacy protocols to Parallel and Serial RapidIO interconnects.

Beyond I/O, RapidIO is an efficient interconnect standard that offers a variety of advanced system features that do not preclude its implementation in an FPGA. Overall, it has the benefit of a small silicon footprint that make it a cost-effective interconnect option.

Using the RapidIO physical layers as an example, there are three key features that make implementation in an FPGA sensible.

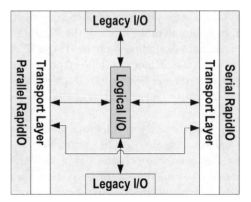

Sample FPGA Architecture
Supporting Parallel RapidIO, Serial RapidIO
& Legacy I/O Protocols

Figure 15.4   Support for legacy protocols and RapidIO

1. Simple handshake protocol
2. Low buffering overhead, 8 or 32 outstanding packets with a maximum packet size of 276 bytes
3. Efficient error coverage on packets and control symbols

15.4 SIMPLE HANDSHAKE

RapidIO offers a very simple handshake protocol that provides the reliability that is required by systems without the burden of complex logic to support it. Although the details of the standard are outside the scope of this chapter, there are some very basic rules behind the protocol.

- All transmitted packets are acknowledged with an acknowledgement control symbol.
- Packets that contain errors are negatively acknowledged. Once a packet is negatively acknowledged, the transmitter requests input status information from the receiver and negotiates the link back into operation.
- Packets that can not be accepted owing to buffering constraints are retried. Packets that are retried are automatically re-sent, based on the information transmitted in the retry control symbol.

These basic rules make up the framework of the handshake protocol and limit the complexity of logic design in the FPGA.

15.5 LOW BUFFERING OVERHEAD

In the RapidIO protocol, all outstanding packets are buffered until the receiver positively acknowledges them. Since the maximum packet size is 276 bytes and the protocol allows up to 8 outstanding packets in the parallel specification and 32 outstanding packets in the serial specification, buffering can be implemented with the on-chip FPGA memory.

FPGAs today offer large amount of on-chip memory. The required memory for a RapidIO end point design, including separate receive and transmit buffers, is typically only a small fraction of overall memory resources available on an FPGA. For example, mid-range FPGA device densities will offer on the order of 1–3 Mb of memory. For a RapidIO implementation that uses separate transmit and receive buffers and can hold eight packets for each direction would require about 36 kb of memory. Even with segmented buffering to allow more efficient addressing of the on-chip memory and less complexity in the data path resources, the overall percentage of required memory for a RapidIO endpoint is still quite small. For implementations that may have several ports, the option to go to higher-density devices with even more on-chip memory is also available.

15.6 EFFICIENT ERROR COVERAGE

With the built-in packet and control symbol error coverage, the protocol is able to offer hardware-based error recovery. Hardware-based error recovery allows systems to have higher reliability and lower system latency because packets can be checked for errors during each link transfer and immediately negatively acknowledged if necessary.

Data packets are covered by a 16-bit CRC that is embedded in the packet. Since the RapidIO protocol limits the maximum packet size to 276 bytes, a 16-bit CRC is sufficient to detect errors. Similar protocols utilize larger CRC polynomials to cover the data, even though the traffic may not utilize the maximum packet size allowed by those protocols. Larger packet size requirements on a protocol increase the size and complexity of the CRC logic to support protecting and checking the packet without necessarily being beneficial to the system requirements. With a 16-bit CRC, an FPGA implementation can implement the CRC function with less logic overhead and data path muxing.

In addition to data packets, RapidIO also defines an efficient mechanism to protect control symbols. In the parallel implementation, control symbols are transmitted with their complement to provide error coverage. In the serial implementation, control symbols are covered by a 5-bit CRC appended to the end of the control symbol.

All three error coverage mechanisms are simple and compact, allowing for reduced logic gate count to support them. These three features of the protocol offer the designer the ability to adopt a scalable and reliable transport in their system with a cost-effective solution based on an FPGA.

15.7 CONCLUSION

Regardless of whether designers choose to use an IP block offered by a FPGA vendor, to implement their own RapidIO design, or to use some combination of vendor-provided IP with custom-produced logic, the availability of FPGAs and pre-developed RapidIO technology enables system developers to quickly make use of the RapidIO interconnect within their systems.

16

Application of RapidIO to Mechanical Environments

David Wickliff

Elements of a mechanical environment include connectors, circuit board traces, pin assignments, board dimensions, etc. In this chapter we will examine some of the considerations of mapping RapidIO into mechanical environments. We will then survey some of the existing and emerging industry standard mechanical platforms that support the RapidIO interconnect standard.

16.1 HELPFUL FEATURES FOR MECHANICAL ENVIRONMENTS

Although the RapidIO interconnect standard does not specify a particular mechanical environment, the RapidIO technology has a number of features which ease the application of RapidIO into various mechanical environments. The foremost feature, from the perspective of easing the deployment of RapidIO in mechanical environments is the switched fabric architecture of RapidIO. This architectural approach avoids many of the mechanical environment issues presented by traditional multipoint bus architectures.

A multipoint bus has electrical signal paths (or channels) with multiple transmitters, receivers, connectors and associated circuit stubs. These create a multitude of impedance mismatches in the channel. These impedance mismatches lead to a multitude of signal reflections. Beyond the constant issue of signal reflections, when a card is inserted into or removed from a traditional bus system, the signaling channel will experience significant changes in its electrical characteristics. As signaling rates increase, these issues become increasingly hard or impossible to manage and force limits on the frequency at which a bus may operate or severely limit the number of devices that may occupy the bus or the number and type of connectors and printed circuit board materials that may be used in a system.

A RapidIO system with its switch fabric architecture offers, from a mechanical and signal integrity perspective, multiple point-to-point links. Each link is a relatively simple unidirectional point-to-point signaling channel comprised of a single set of transmitters and a single

RapidIO® The Embedded System Interconnect. S. Fuller
© 2005 John Wiley & Sons, Ltd ISBN: 0-470-09291-2

set of receivers. In most common environments this link will operate over fewer connectors and have fewer stubs affecting the signal integrity. Inserting or removing cards in a system should, to first order, affect only the point-to-point connections associated with that card. All the other signal channels, supporting other cards are not disturbed. Because of this the RapidIO point-to-point links will provide a much cleaner signaling environment than a traditional multipoint bus.

The parallel and serial physical layer electrical signaling specifications of RapidIO also ease the application to mechanical environments. The parallel electrical specification is based on LVDS, a well understood and broadly adopted signaling technology. The low output voltage swings and slow edge rates (dV/dt) reduce EMI issues. The differential signaling provides a high degree of noise immunity. The high per signal data rates of LVDS allows the same bandwidth with fewer pins when compared with other (especially single-ended) signaling technologies. This means fewer signal pins are needed on connectors and fewer traces are needed to route in the circuit boards. In general, the parallel RapidIO physical layer is best suited for relatively short printed circuit distances with up to one connector in the channel. It is an ideal interconnect for use on a single printed circuit board or between a base board and mezzanine board. With care, a backplane application of parallel RapidIO is possible.

The Serial RapidIO physical layer electrical specification makes use of a differential current steering technology similar to that used by IEEE 802.3 XUAI and Fiber Channel. The transmit clock is embedded with the data using the 8B/10B encoding scheme. Thus tight skew control between a separate clock signal and associated data signals is not required. This allows the operating frequency of the interface to scale to significantly higher speeds than is practical with the Parallel RapidIO interface. Because of the higher operating frequencies, the total number of signals required for a Serial RapidIO link at a given performance level is less than is required for the Parallel RapidIO link. This has the effect of further reducing pin count and trace routing complexity. Although not explicitly required by the RapidIO specifications, active compensation circuit techniques for frequency distortions can be used (e.g. pre- and post-equalization/emphasis) to increase signal robustness. These techniques are becoming increasing important at the higher frequencies and in backplane environments. Serial RapidIO transmitters are allowed to operate either at a reduced voltage swing to minimize overall required power or at a higher voltage swing to maximize signal strength and distance. These modes are known as the short-run and long-run transmitters, respectively. In general, the Serial RapidIO short-run transmitter is applicable to relatively short printed circuit distances with up to one connector. Serial short-run is ideal for use on a single board or across a mezzanine board connector. The serial long-run transmitter is better suited to longer distances, up to a meter, with two or more connectors. Thus, it is ideal for a backplane application.

16.2 CHANNEL CHARACTERISTICS

While these RapidIO switched electrical signaling features are helpful, the ultimate suitability of a signaling technology to a particular mechanical environment needs to be carefully evaluated. At these high signaling rates, the transmission line effects as well as the frequency-dependent loss of the channel must be considered. A combination of measurement, modeling and simulation should be used to validate the signal integrity of a given mechanical application.

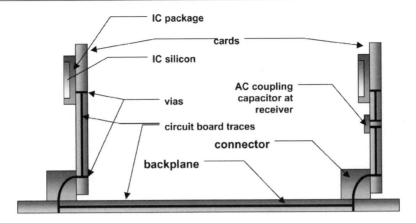

Figure 16.1   Backplane example

The important channel elements of a backplane application are illustrated in Figure 16.1. The effects of these elements on loss, reflection, crosstalk and skew should be evaluated. The printed circuit traces of the cards and backplane are the largest contributor to signal loss. Impedance discontinuities between IC package, vias, traces, and connectors and even within traces, connectors and IC packages can all generate signal reflections. Connectors, traces, vias and IC packages can all introduce crosstalk in the multiple signals of a system, thus further reducing a signal's integrity. Skew, both within a differential signal pair and between pairs, is another consideration. Finally, at these high frequencies even the stub-length effects of a via can be an issue.

A model of a complete channel can be assembled from models of the various elements present in the channel. A multiple channel model is needed to evaluate crosstalk and some of the skew issues. Individual element models are developed by detailed element simulations and/or characterizing real element test structures. In many cases, models of connectors, drivers and receivers are readily available from the respective manufacturers. Complete reference designs for backplane and connector systems from the manufacturers are becoming common as well.

16.3 INDUSTRY STANDARD MECHANICAL PLATFORMS SUPPORTING RAPIDIO

Platform standards help ensure a broad level of interoperability among a set of system components such as printed circuit boards, subracks and cabinets. Including a complete set of mechanical, electrical, protocol, and operating specifications in a platform standard achieves this interoperability. The specifications are either directly defined in the platform standard or are included by referencing other standards. A platform standard enables an industry to produce various system components, which in turn can be used by OEMs to readily build an application specific solution. VMEbus, CompactPCI and PMC are some well-known examples of platform standards.

The RapidIO interconnect is an interface standard that specifies protocol, electrical and management environments to ensure interoperability on the interface. From the beginning the RapidIO Trade Association made a conscious decision not to develop a complete platform specification, which would include mechanical environment specifications. Instead, the RapidIO Trade Association relies on and facilitates various other industry standard forums to incorporate the RapidIO protocol and electrical specifications. This approach is based on the realization that the embedded application market is very broad and diverse. No one mechanical platform could address the associated diverse mechanical packaging requirements.

In the following sections several industry standard mechanical platforms supporting RapidIO will be briefly examined. These are either officially completed standards or are proposed standards being developed at the time of this writing. Other RapidIO standard mechanical platforms are likely to emerge in the future. These platforms illustrate how RapidIO is being successfully applied to a variety of mechanical environments that use readily available and relatively cost-effective board materials and connector technologies. Table 16.1 summarizes the industry standard mechanical platforms discussed here.

These mechanical platforms can be useful for a number of other purposes, besides serving as a basis for assembling production products. One, they can be used as a way to rapidly prototype a system. This can be especially useful to enable early software development. They

Table 16.1 Some industry standard mechanical platforms that support RapidIO

Standard or proposed standard	Standard organization	Form factor	RapidIO physical layers supported per card	RapidIO fabric topologies supported
RapidIO HIP	RapidIO TA	Card ($\sim$300 cm^2 and greater) Baseboard (up to 4 cards)	Four Serial 1X/4X Two Parallel 8-bit One Parallel 16-bit	Any
CompactPCI Serial RapidIO (PICMG 2.18)	PICMG	Card ($\sim$350 cm^2) Backplane (up to 21 cards)	Four Serial 1X/4X @ 1.25 Gbps	Star, Dual Star, Mesh
AdvancedTCA Serial RapidIO (PICMG 3.5)	PICMG	Card ($\sim$900 cm^2) Backplane (up to 16 cards)	Fifteen Serial 1X/4X	Star, Dual Star, Full Mesh
Switched Mezzanine Card XMC (VITA 42.1 and 42.2)	VITA Standards Organization	Mezzanine ($\sim$110 cm^2)	Multiple Serial 1X/4X Two Parallel 8-bit One Parallel 8-bit	Any
Advanced Mezzaine Card (PICMG AMC)	PICMG	Mezzanine ($\sim$130 cm^2)	Multiple Serial 1X/4X Two Parallel 8-bit One Parallel 16-bit	Any
VME Switched Serial VXS (VITA 41.2)	VITA Standards Organization	Card ($\sim$350 cm^2) Backplane (up to 18 cards on fabric)	Two Serial 4X	Star, Dual Star
VME Advanced Module Format (VITA 46)	VITA Standards Organization	Card ($\sim$350 cm^2)	Multiple Serial 1X/4X	Start, Dual Star, Full Mesh

can be used as a reference design from which the design of an application-specific proprietary system platform can draw upon. In many cases, these industry standard platforms have had extensive channel simulations and real system validations. Finally, these standard platforms can be used as the infrastructure for interoperability testing between multiple devices from multiple vendors. As we will see, one standard RapidIO platform was explicitly developed for interoperability testing.

16.3.1 RapidIO Hardware Interoperability Platform (HIP)

The RapidIO hardware interoperability platform, or HIP, is an exception to the RapidIO Trade Association's strategy of not directly developing mechanical platform standards. It was developed by and is maintained by the trade association. The primary goal of the HIP specification is a simple and flexible mechanical platform that multiple RapidIO vendors can use for interoperability testing. The HIP specification references the PC industry ATX standard for its base motherboard and plug-in card mechanical specifications. To this specification an additional high-speed connector is added to carry RapidIO signals between a plug-in card and the baseboard. Overall, the HIP specification is very minimal and flexible. The intent is to ensure multi-vendor platform compatibility while encouraging innovation and adaptation to individual needs. Figure 16.2 is a photograph of the Tundra HIP motherboard. This photograph shows how the new square high-performance connector is placed below the traditional PCI connectors used in personal computers. The HIP motherboard definition can be used to provide switched connectivity and interoperability testing between different RapidIO equipped plug-in cards.

16.3.2 CompactPCI Serial RapidIO (PICMG 2.18)

The CompactPCI Serial RapidIO (PICMG 2.18) specification adds serial RapidIO as an alternative board-to-board interface for the popular CompactPCI platform standard. This provides a

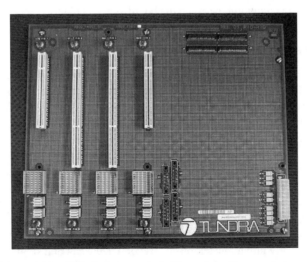

Figure 16.2   HIP platform example (Reproduced by permission of
Tundra Semiconductor Corporation, Ontario, Canada)

Figure 16.3   PICMG 2.18 processor board (Reproduced by permission of
Spectrum Signal Processing Inc.)

significant improvement in bandwidth compared with the traditional CompactPCI interfaces:
PCI, H.110 and Ethernet. These traditional interfaces are not required, but are also not
precluded. So, a heterogeneous CompactPCI system, supporting both new high bandwidth as
well as legacy, can be easily constructed. Figure 16.3 shows a PICMG 2.18 processor board.
This board, developed by Spectrum Signal Processing, transmits Serial RapidIO across the
existing connector and backplane structure where it can be connected to other RapidIO based
boards in the system.

The PICMG 2.18 standard effort chose to use the existing 2 mm CompactPCI connector.
This both ensures a level of mechanical compatibly and keeps connector costs reasonable.
After performing various signal integrity simulations it was determined that 1.25 Gbps was
the maximum reliable speed for a serial RapidIO differential pair on this platform. Each board
can have up to four bidirectional 4x Serial RapidIO links running at 1.25 Gbps for a total of
40 Gbps of aggregate bandwidth between the board and backplane.

16.3.3 AdvancedTCA Serial RapidIO (PICMG 3.5)

The advanced telecom computing architecture (ATCA or AdvancedTCA) is a recent family
of platform specifications targeted to requirements of carrier-grade communications equipment. PICMG 3.0 is the base standard, addressing mechanics, board dimensions, power distribution, connectors, and system management. This base does not specify an interface fabric for the
platform. Various subsidiary specifications provide choices for the fabric. Subsidiary specification PICMG 3.5 defines a serial RapidIO fabric over the ATCA backplane.

The high-speed differential connectors used in ATCA are capable of data rates in excess
of 5 Gbps per pair. Thus, 3.125 Gbps Serial RapidIO links are easily mapped on to the connectors. Each board supports up to fifteen 4x Serial RapidIO links. ATCA backplanes can have a
maximum of sixteen slots and can be routed as a full mesh, where every board slot has a link
to every other slot. Very-high-performance applications can use the full mesh for a potential
system aggregate bandwidth in the terabits per second. Dual star and single star topologies are
also supported as simple subsets of the full mesh topology.

Figure 16.4   Example ATCA chassis (Reproduced by permission of Mercury Computer Systems, Inc.)

16.3.4 Switched Mezzanine Card XMC (VITA 42)

The Switched Mezzanine Card (XMC) specification adds high-speed fabric interfaces to the widely popular PCI Mezzanine Card (PMC). This is achieved with a new high-speed connector. A traditional PMC module supports a 32/64-bit PCI bus on two or three connectors. An XMC module adds one or two new connectors. This allows both the traditional PCI bus and new high-speed fabrics to be supported simultaneously. For example, an XMC carrier board can contain both sets of connectors, allowing either legacy PMC modules or XMC modules to be used. XMC modules with only the new high-speed connectors are allowed as well. Figure 16.5

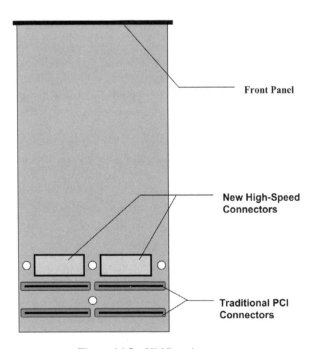

Figure 16.5   XMC card structure

shows the structure of the XMC card and the placement of the legacy PCI connectors and new high-performance connectors.

An early draft document called RapidIO Mezzanine Card (RMC) was developed in the RapidIO Trade Association and was used as the initial contribution in the development of the VITA XMC specifications. VITA 42.0 is the base specification, defining mechanical and connector elements. Subsidiary specifications define a mapping of a particular interface technology on to the base specification. VITA 42.1 defines the mapping for parallel RapidIO and VITA 42.2 defines the mapping for serial RapidIO.

16.3.5 Advanced Mezzanine Card Serial RapidIO (PICMG AMC)

The Advanced Mezzanine Card (AMC or AdvancedMC) specification is a mezzanine card format optimized for mounting on ATCA cards, but can be used on other card form factors as well. It contains a number of enhancements over earlier mezzanine card standards, including: hot-swapping and high-speed I/O connectors. Organized similar to the ATCA specifications, the AMC specifications include a base specification (PICMG AMC.0) with mechanical, thermal, power, connectors and management elements defined. Subsidiary specifications specify a mapping of a particular interface technology on to the base specification.

In the maximum configuration, the AMC connector supports 21 differential signal pairs into the card and 21 pairs out of the card (21 bidirectional ports). The connector's signal integrity supports a data rate in excess of 10 Gbps on each pair. Thus, multiple 3.125 Gbps 1x or 4x Serial RapidIO links are easily mapped on to the connectors. Two 8-bit parallel RapidIO interfaces or one 16-bit parallel RapidIO interface can also be supported on an AMC.

16.3.6 VME Switched Serial VXS for RapidIO (VITA 41.2)

The VME Switched Serial (VXS) specification adds a high-speed fabric interface to the venerable VME platform, while maintaining direct backwards compatibility with legacy VME64x boards. The base specification VITA 41.0 defines a new additional high-speed connector. A VXS backplane can support legacy VME bus boards, boards with both VME and a serial fabric, and boards with just the fabric interface. The subsidiary specification VITA 41.2 defines a serial RapidIO fabric on the new connector.

The high-speed differential connector used in VXS is capable of data rates in excess of 5 Gbps per pair. Thus, 3.125 Gbps Serial RapidIO links are easily mapped on to the connectors. Each payload board supports two 4x Serial RapidIO links. Two special fabric board slots are defined, each supporting up to eighteen 4x links, one link back to each payload board. The two fabric slots are also connected together with four more 4x links.

16.3.7 Advanced Module Format (VITA 46)

The Advanced Module Format platform retains many elements of VME while completely replacing the board-to-backplane connectors to support dense high-speed I/O. The board and chassis mechanics as well as the VME electrical and protocol interfaces are retained. Both VME bus signals and serial fabric signals are mapped on to the new connectors. While this approach does not support direct slot backward compatibility with legacy VME64x boards, a single chassis and backplane can be easily designed to support a heterogeneous mix of legacy slots and advanced slots. The high-speed differential connector used in VITA 46 is capable of data rates in excess of 5 Gbps per pair. Thus, 3.125 Gbps Serial RapidIO links are easily mapped on to the connectors. A minimal fabric configuration supports four 4x Serial RapidIO links on a board. In a fabric-only configuration more than twenty 4x Serial RapidIO links can be supported on a board.

16.4 SUMMARY

As system performance requirements of embedded equipment increases, there are several concurrent efforts in the industry to either migrate or replace existing bus-based mechanical standards with new standards that support the high-speed differential signaling used by the RapidIO interconnect technology and other similar technologies. Existing design techniques and board and connector technologies will impose limits with respect to the frequency and hence bandwidth available on the differential links. These limits are driving several well-funded efforts focused on the creation of new PCB materials, layout techniques and new connector technologies to support signaling speeds up to 10 Gbps for a single differential signal pair.

Appendix A

RapidIO Logical and Transport Layer Registers

Table A.1 shows the register map for the approved RapidIO specifications. There are two classes of registers defined in the RapidIO architecture. There are capability registers (CARs) and command and status registers (CSRs). CARs are read-only registers, which provide information to the system describing the capabilities of the RapidIO device. The CSRs are read/write registers that are used to configure a RapidIO device and to provide current status on the operation or performance of the RapidIO device.

Table A.1 Integrated register map

Configuration space byte offset	Register name (word 0)	Register name (word 1)
0x0	Device identity CAR	Device information CAR
0x8	Assembly identity CAR	Assembly information CAR
0x10	Processing element features CAR	Switch port information CAR
0x18	Source operations CAR	Destination operations CAR
0x20–38	Reserved	
0x3c	Data streaming information CAR	Reserved
0x40	Mailbox CSR	Write-port CSR Doorbell CSR
0x48	Data streaming logical layer control CSR	Processing element logical layer control CSR
0x50	Reserved	
0x58	Local configuration space high base address CSR	Local configuration space base address CSR
0x60	Base device ID CSR	Reserved
0x68	Host base device ID lock CSR	Component tag CSR
0x70–F8		
0x100–FFF8	Extended features space	
0x10000–FFFFF8	Implementation-defined space	

RapidIO® The Embedded System Interconnect.   S. Fuller
© 2005 John Wiley & Sons, Ltd   ISBN: 0-470-09291-2

Table A.2 Configuration space reserved access behavior

Byte offset	Space name	Item	Initiator behavior	Target behavior
0x0–3C	Capability register space (CAR space, this space is read-only)	Reserved bit	Read – ignore returned value[1]	Read – return logic 0
		Implementation-defined bit	Write – read – ignore returned value unless implementation-defined function understood	Write – ignored Read – return implementation-defined value
		Reserved register	Write – read – ignore returned value	Write – ignored Read – return logic 0s
0x40–FC	Command and Status Register Space (CSR Space)	Reserved bit	Write – read – ignore returned value	Write – ignored Read – return logic 0
			Write – preserve current value[2]	Write – ignored
		Implementation-defined bit	Read – ignore returned value unless implementation-defined function understood	Read – return implementation-defined value
			Write – preserve current value if implementation-defined function not understood	Write – implementation-defined
		Reserved register	Read – ignore returned value	Read – return logic 0s
0x100–FFFC	Extended features space	Reserved bit	Write – read – ignore returned value	Write – ignored Read – return logic 0
			Write – preserve current value	Write – ignored
		Implementation-defined bit	Read – ignore returned value unless implementation-defined function understood	Read – return implementation-defined value
			Write – preserve current value if implementation-defined function not understood	Write – implementation-defined
		Reserved register	Read – ignore returned value	Read – return logic 0s
			Write –	Write – ignored
0x10000–FFFFFC	Implementation-defined space	Reserved bit and register	All behavior implementation-defined	

[1] Do not depend on reserved bits being a particular value; use appropriate masks to extract defined bits from the read value
[2] All register writes shall be in the form: read the register to obtain the values of all reserved bits, merge in the desired values for defined bits to be modified, and write the register, thus preserving the value of all reserved bits

Any register offsets not defined are considered reserved unless otherwise stated. Other registers required for a processing element are defined in other applicable read and write accesses to reserved register offsets terminate normally and do not cause an error condition in the target device. Writes to CAR (read-only) registers will also terminate normally and not cause an error condition in the target device.

This map combines the register definitions for the currently approved RapidIO specifications.

A.1 RESERVED REGISTER AND BIT BEHAVIOR

Table A.2 describes the required behavior for accesses to reserved register bits and reserved registers for the RapidIO register space.

A.2 CAPABILITY REGISTERS (CARs)

Every processing element shall contain a set of registers that allows an external processing element to determine its capabilities through maintenance read operations. All registers are 32 bits wide and are organized and accessed in 32-bit (4-byte) quantities, although some processing elements may optionally allow larger accesses. CARs are read-only. CARs are big-endian with bit 0 and word 0 respectively the most significant bit and word.

A.2.1 Device Identity CAR (Offset 0x0 Word 0)

The DeviceVendorIdentity field identifies the vendor that manufactured the device containing the processing element; see Table A.3. A value for the DeviceVendorIdentity field is uniquely assigned to a device vendor by the registration authority of the RapidIO Trade Association.

The DeviceIdentity field is intended to uniquely identify the type of device from the vendor specified by the DeviceVendorIdentity field. The values for the DeviceIdentity field are assigned and managed by the respective vendor.

Table A.3 Bit settings for device identity CAR

Bit	Field name	Description
0–15	DeviceIdentity	Device identifier
16–31	DeviceVendorIdentity	Device vendor identifier

A.2.2 Device Information CAR (Offset 0x0 Word 1)

The DeviceRev field is intended to identify the revision level of the device; see Table A.4. The value for the DeviceRev field is assigned and managed by the vendor specified by the DeviceVendorIdentity field.

Table A.4 Bit settings for device information CAR

Bit	Field name	Description
0–31	DeviceRev	Device revision level

A.2.3 Assembly Identity CAR (Offset 0x8 Word 0)

The AssyVendorIdentity field identifies the vendor that manufactured the assembly or subsystem containing the device; see Table A.5. A value for the AssyVendorIdentity field is uniquely assigned to an assembly vendor by the registration authority of the RapidIO Trade Association.

The AssyIdentity field is intended to uniquely identify the type of assembly from the vendor specified by the AssyVendorIdentity field. The values for the AssyIdentity field are assigned and managed by the respective vendor.

Table A.5 Bit settings for assembly identity CAR

Bit	Field name	Description
0–15	AssyIdentity	Assembly identifier
16–31	AssyVendorIdentity	Assembly vendor identifie

A.2.4 Assembly Information CAR (Offset 0x8 Word 1)

This register contains additional information about the assembly; see Table A.6.

Table A.6 Bit settings for assembly information CAR

Bit	Field name	Description
0–15	AssyRev	Assembly revision level
16–31	ExtendedFeaturesPtr	Pointer to the first entry in the extended features list

A.2.5 Processing Element Features CAR (Offset 0x10 Word 0)

This register identifies the major functionality provided by the processing element; see Table A.7.

Table A.7 Bit settings for processing element features CAR

Bit	Field name	Description
0	Bridge	PE can bridge to another interface. Examples are PCI, proprietary processor buses, DRAM, etc.
1	Memory	PE has physically addressable local address space and can be accessed as an end point through non-maintenance (i.e. non-coherent read and write) operations. This local address space may be limited to local configuration registers, or could be on-chip SRAM, etc.
2	Processor	PE physically contains a local processor or similar device that executes code. A device that bridges to an interface that connects to a processor does not count (see bit 0 above)
3	Switch	PE can bridge to another external RapidIO interface – an internal port to a local end point does not count as a switch port. For example, a device with two RapidIO ports and a local end point is a two port switch, not a three-port switch, regardless of the internal architecture.
4–7		Reserved
8	Mailbox 0	PE supports inbound mailbox 0
9	Mailbox 1	PE supports inbound mailbox 1
10	Mailbox 2	PE supports inbound mailbox 2
11	Mailbox 3	PE supports inbound mailbox 3
12	Doorbell	PE support inbound doorbells
13–23		Reserved
24	Flow control support	Support for flow control extensions 0b0　Does not support flow control extensions 0b1　Supports flow control extensions
25–26		Reserved
27	Common transport large system support	0b0　PE does not support common transport large systems 0b1　PE supports common transport large systems
28	Extended features	PE has extended features list; the extended features pointer is valid
29–31	Extended addressing support	Indicates the number address bits supported by the PE both as a source and target of an operation. All PEs shall at minimum support 34 bit addresses 0b111　PE supports 66-, 50- and 34-bit addresses 0b101　PE supports 66- and 34-bit addresses 0b011　PE supports 50- and 34-bit addresses 0b001　PE supports 34-bit addresses All other encodings reserved

A.2.6 Switch Port Information CAR (Offset 0x10 Word 1)

This register defines the capabilities of a RapidIO switch; see Table A.8. This register is valid only if bit 3 is set in the processing element features CAR.

Table A.8   Bit settings for switch port information CAR

Bit	Field name	Description
0–15		Reserved
16–23	PortTotal	The total number of RapidIO ports on the processing element
		0b00000000 Reserved
		0b00000001 1 port
		0b00000010 2 ports
		0b00000011 3 ports
		0b00000100 4 ports
		. . .
		0b11111111 255 ports
24–31	PortNumber	This is the port number from which the maintenance read operation accessed this register. Ports are numbered starting with 0x00

A.2.7 Source Operations CAR (Offset 0x18 Word 0)

This register defines the set of RapidIO IO logical operations that can be issued by this processing element; see Table A.9. It is assumed that a processing element can generate I/O logical maintenance read and write requests if it is required to access CARs and CSRs in other processing elements. The source operations CAR is applicable for end point devices only. RapidIO switches are required to be able to route any packet.

Table A.9   Bit settings for source operations CAR

Bit	Field name	Description
0	Read	PE can support a read operation
1	Instruction read	PE can support an instruction read operation
2	Read-for-ownership	PE can support a read-for-ownership operation
3	Data cache invalidate	PE can support a data cache invalidate operation
4	Castout	PE can support a castout operation
5	Data cache flush	PE can support a data cache flush operation
6	I/O read	PE can support an I/O read operation
7	Instruction cache invalidate	PE can support an instruction cache invalidate operation
8	TLB invalidate-entry	PE can support a TLB invalidate-entry operation
9	TLB invalidate-entry sync	PE can support a TLB invalidate-entry sync operation
10–12		Reserved
13	Data streaming	PE can support a data streaming operation
14–15	Implementation defined	Defined by the device implementation
16	Non-coherent read	PE can support a non-coherent read operation
17	Non-coherent write	PE can support a non-coherent write operation
18	Streaming-write	PE can support a streaming-write operation
19	Write-with-response	PE can support a write-with-response operation
20	Data message	PE can support a data message operation
21	Doorbell	PE can support a doorbell operation

Table A.9 (*continued*)

Bit	Field name	Description
22		Reserved
23	Atomic (test-and-swap)	PE can support an atomic test-and-swap operation
24	Atomic (increment)	PE can support an atomic increment operation
25	Atomic (decrement)	PE can support an atomic decrement operation
26	Atomic (set)	PE can support an atomic set operation
27	Atomic (clear)	PE can support an atomic clear operation
28		Reserved
29	Port-write	PE can support a port-write operation
30–31	Implementation-defined	Defined by the device implementation

A.2.8 Destination Operations CAR (Offset 0x18 Word 1)

This register defines the set of RapidIO operations that can be supported by this processing element; see Table A.10. It is required that all processing elements can respond to maintenance read and write requests in order to access these registers. The destination operations CAR is applicable for end point devices only.

Table A.10   Bit settings for destination operations CAR

Bit	Field name	Description
0	Read	PE can support a read operation
1	Instruction read	PE can support an instruction read operation
2	Read-for-ownership	PE can support a read-for-ownership operation
3	Data cache invalidate	PE can support a data cache invalidate operation
4	Castout	PE can support a castout operation
5	Data cache flush	PE can support a data cache flush operation
6	I/O read	PE can support an I/O read operation
7	Instruction cache invalidate	PE can support an instruction cache invalidate operation
8	TLB invalidate-entry	PE can support a TLB invalidate-entry operation
9	TLB invalidate-entry sync	PE can support a TLB invalidate-entry sync operation
10–12		Reserved
13	Data streaming	PE can support a data streaming operation
14–15	Implementation-defined	Defined by the device implementation
16	Read	PE can support a read operation
17	Write	PE can support a write operation
18	Streaming-write	PE can support a streaming-write operation
19	Write-with-response	PE can support a write-with-response operation
20	Data message	PE can support a data message operation
21	Doorbell	PE can support a doorbell operation
22		Reserved
23	Atomic (test-and-swap)	PE can support an atomic test-and-swap operation
24	Atomic (increment)	PE can support an atomic increment operation
25	Atomic (decrement)	PE can support an atomic decrement operation
26	Atomic (set)	PE can support an atomic set operation
27	Atomic (clear)	PE can support an atomic clear operation
28		Reserved
29	Port-write	PE can support a port-write operation
30–31	Implementation-defined	Defined by the device implementation

A.2.9 Data Streaming Information CAR (Configuration Space Offset 0x3C)

This register defines the data streaming capabilities of a processing element; see Table A.11. It is required for destination end point devices.

Table A.11   Bit settings for data streaming information CAR

Bit	Field name	Description
0–15	MaxPDU	Maximum PDU. The maximum PDU size in bytes supported by the destination end point 0x0000 64 kbyte 0x0001 1 byte 0x0002 2 byte 0x0003 3 byte . . . 0xFFFF 64kbytes – 1
16–31	SegSupport	Segmentation support. The number of segmentation contexts supported by the destination end point 0x0000 64 k segmentation contexts 0x0001 1 segmentation context 0x0002 2 segmentation contexts 0x0003 3 segmentation contexts 0x0004 4 segmentation contexts . . . 0xFFFF 64 k – 1 segmentation contexts

A.3 COMMAND AND STATUS REGISTERS (CSRs)

All RapidIO end points or switches contain a set of command and status registers (CSRs) that allow an external device to control and determine the status of a target device's internal hardware. All registers are 32 bits wide and are organized and accessed in the same way as the CARs.

A.3.1 Mailbox CSR (Offset 0x40 Word 0)

The mailbox command and status register is accessed if an external processing element wishes to determine the status of this processing elements's mailbox hardware, if any is present; see Table A.12. It is not necessary to examine this register before sending a message since the RapidIO protocol shall accept, retry, or send an error response message, depending upon the status of the addressed mailbox. This register is read-only.

A.3.2 Write-port or Doorbell CSR (Offset 0x40 Word 1)

The write-port CSR is accessed if an external processing element wishes to determine the status of this processing element's write-port hardware; see Table A.13. It is not necessary to examine this register before sending a port-write transaction since the protocol will behave appropriately, depending upon the status of the hardware. This register is read-only.

Table A.12   Bit settings for mailbox CSR

Bit	Field name	Description
0	Mailbox 0 available	Mailbox 0 is initialized and ready to accept messages. If not available, all incoming message transactions return error responses
1	Mailbox 0 full	Mailbox 0 is full. All incoming message transactions return retry responses
2	Mailbox 0 empty	Mailbox 0 has no outstanding messages
3	Mailbox 0 busy	Mailbox 0 is busy receiving a message operation. New message operations return retry responses
4	Mailbox 0 failed	Mailbox 0 had an internal fault or error condition and is waiting for assistance. All incoming message transactions return error responses
5	Mailbox 0 error	Mailbox 0 encountered a message operation or transaction of an unacceptable size. All incoming message transactions return error responses
6–7		Reserved
8	Mailbox 1 available	Mailbox 1 is initialized and ready to accept messages. If not available, all incoming message transactions return error responses
9	Mailbox 1 full	Mailbox 1 is full. All incoming message transactions return retry responses
10	Mailbox 1 empty	Mailbox 1 has no outstanding messages
11	Mailbox 1 busy	Mailbox 1 is busy receiving a message operation. New message operations return retry responses
12	Mailbox 1 failed	Mailbox 1 had an internal fault or error condition and is waiting for assistance. All incoming message transactions return error responses
13	Mailbox 1 error	Mailbox 1 encountered a message operation or transaction of an unacceptable size. All incoming message transactions return error responses
14–15		Reserved
16	Mailbox 2 available	Mailbox 2 is initialized and ready to accept messages. If not available, all incoming message transactions return error responses
17	Mailbox 2 full	Mailbox 2 is full. All incoming message transactions return retry responses
18	Mailbox 2 empty	Mailbox 2 has no outstanding messages
19	Mailbox 2 busy	Mailbox 2 is busy receiving a message operation. New message operations return retry responses
20	Mailbox 2 failed	Mailbox 2 had an internal fault or error condition and is waiting for assistance. All incoming message transactions return error responses
21	Mailbox 2 error	Mailbox 2 encountered a message operation or transaction of an unacceptable size. All incoming message transactions return error responses
22–23		Reserved
24	Mailbox 3 available	Mailbox 3 is initialized and ready to accept messages. If not available, all incoming message transactions return error responses
25	Mailbox 3 full	Mailbox 3 is full. All incoming message transactions return retry responses
26	Mailbox 3 empty	Mailbox 3 has no outstanding messages
27	Mailbox 3 busy	Mailbox 3 is busy receiving a message operation. New message operations return retry responses
28	Mailbox 3 failed	Mailbox 3 had an internal fault or error condition and is waiting for assistance. All incoming message transactions return error responses
29	Mailbox 3 error	Mailbox 3 encountered a message operation or transaction of an unacceptable size. All incoming message transactions return error responses
30–31		Reserved

Table A.13   Bit settings for write-port or doorbell CSR

Bit	Field name	Description
0	Doorbell available	Doorbell hardware is initialized and ready to accept doorbell messages. If not available, all incoming doorbell transactions return error responses
1	Doorbell full	Doorbell hardware is full. All incoming doorbell transactions return retry responses
2	Doorbell empty	Doorbell hardware has no outstanding doorbell messages
3	Doorbell busy	Doorbell hardware is busy queueing a doorbell message. Incoming doorbell transactions may or may not return a retry response depending upon the implementation of the doorbell hardware in the PE
4	Doorbell failed	Doorbell hardware has had an internal fault or error condition and is waiting for assistance. All incoming doorbell transactions return error responses
5	Doorbell error	Doorbell hardware has encountered an Doorbell transaction that is found to be illegal for some reason. All incoming doorbell transactions return error responses
6–23		Reserved
24	Write port available	Write port hardware is initialized and ready to accept a port-write transaction. If not available, all incoming port-write transactions will be discarded
25	Write port full	Write port hardware is full. All incoming port-write transactions will be discarded
26	Write port empty	Write port hardware has no outstanding port-write transactions
27	Write port busy	Write port hardware is busy queueing a port-write transaction. Incoming port-write transactions may or may not be discarded, depending upon the implementation of the write port hardware in the PE
28	Write port failed	Write port hardware has had an internal fault or error condition and is waiting for assistance. All incoming port-write transactions will be discarded
29	Write port error	Write port hardware has encountered a port-write transaction that is found to be illegal for some reason. All incoming port-write transactions will be discarded
30–31		Reserved

The doorbell CSR is accessed if an external processing element wishes to determine the status of this processing element's doorbell hardware if the target processing element supports these operations. It is not necessary to examine this register before sending a doorbell message since the protocol shall behave appropriately, depending upon the status of the hardware.

A.3.3 Data Streaming Logical Layer Control CSR (Offset 0x48 Word 0)

The data streaming logical layer control CSR is used for general command and status information for the logical interface; see Table A.14.

Table A.14 Bit settings for data streaming logical layer control CSR

Bit	Field name	Description
0–25		Reserved
26–31	MTU	Maximum transmission unit. Controls the data payload size for segments of an encapsulated PDU. Only single-segment PDUs and end segments are permitted to have a data payload that is less this value. The MTU can be specified in increments of 4 bytes. Support for the entire range is required
		0b000000 32 byte block size
		0b000001 36 byte block size
		0b000010 40 byte block size
		. . .
		0b111000 256 byte block size
		0b111001 Reserved
		. . .
		0b111111 Reserved
		All other encodings reserved

Table A.15 Bit settings for processing element logical layer control CSR

Bit	Field name	Description
0–28		Reserved
29–31	Extended addressing control	Controls the number of address bits generated by the PE as a source and processed by the PE as the target of an operation
		0b100 PE supports 66-bit addresses
		0b010 PE supports 50-bit addresses
		0b001 PE supports 34-bit addresses (default)
		All other encodings reserved

A.3.4 Processing Element Logical Layer Control CSR (Offset 0x48 Word 1)

The processing element logical layer control CSR is used for general command and status information for the logical interface; see Table A.15.

A.3.5 Local Configuration Space Base Address 0 CSR (Offset 0x58 Word 0)

The local configuration space base address 0 register specifies the most significant bits of the local physical address double-word offset for the processing element's configuration register space; see Table A.16.

A.3.6 Local Configuration Space Base Address 1 CSR (Offset 0x58 Word 1)

The local configuration space base address 1 register specifies the least significant bits of the local physical address double-word offset for the processing element's configuration register space, allowing the configuration register space to be physically mapped in the processing

Table A.16 Bit settings for local configuration space base address 0 CSR

Bit	Field name	Description
0		Reserved
1–16	LCSBA	Reserved for a 34-bit local physical address Reserved for a 50-bit local physical address Bits 0–15 of a 66-bit local physical address
17–31	LCSBA	Reserved for a 34-bit local physical address Bits 0–14 of a 50-bit local physical address Bits 16–30 of a 66-bit local physical address

Table A.17 Bit settings for local configuration space base address 1 CSR

Bit	Field name	Description
0	LCSBA	Reserved for a 34-bit local physical address Bit 15 of a 50-bit local physical address Bit 31 of a 66-bit local physical address
1–31	LCSBA	Bits 0–30 of a 34-bit local physical address Bits 16–46 of a 50-bit local physical address Bits 32–62 of a 66-bit local physical address

element; see Table A.17. This register allows configuration and maintenance of a processing element through regular read and write operations rather than maintenance operations. The double-word offset is right-justified in the register.

A.3.7 Base Device ID CSR (Offset 0x60 Word 0)

The base device ID CSR contains the base device ID values for the processing element; see Table A.18. A device may have multiple device ID values, but these are not defined in a standard CSR.

Table A.18 Bit settings for base device ID CSR

Bits	Name	Reset value	Description
0–7			Reserved
8–15	Base_deviceID	See footnote[1]	This is the base ID of the device in a small common transport system (end point devices only)
16–31	Large_base_deviceID	See footnote[2]	This is the base ID of the device in a large common transport system (only valid for end point device and if bit 27 of the processing element features CAR is set)

[1] The Base_deviceID reset value is implementation dependent
[2] The Large_base_deviceID reset value is implementation dependent

A.3.8 Host Base Device ID Lock CSR (Offset 0x68 Word 0)

The host base device ID lock CSR contains the base device ID value for the processing element in the system that is responsible for initializing this processing element; see Table A.19. The Host_base_deviceID field is a write-once/resetable field which provides a lock function. Once the Host_base_deviceID field is written, all subsequent writes to the field are ignored, except in the case that the value written matches the value contained in the field. In this case, the register is re-initialized to 0xFFFF. After writing the Host_base_deviceID field a processing element must then read the host base device ID lock CSR to verify that it owns the lock before attempting to initialize this processing element.

Table A.19 Bit settings for host base device ID lock CSR

Bits	Name	Reset value	Description
0–15			Reserved
16–31	Host_base_deviceID	0xFFFF	This is the base device ID for the PE that is initializing this PE

A.3.9 Component Tag CSR (Offset 0x68 Word 1)

The component tag CSR contains a component tag value for the processing element and can be assigned by software when the device is initialized; see Table A.20. It is especially useful for labeling and identifying devices that are not end points and do not have device ID registers.

Table A.20 Bit settings for component ID CSR

Bits	Name	Reset value	Description
0–31	component_tag	All 0s	This is a component tag for the PE

A.4 EXTENDED FEATURES DATA STRUCTURE

The RapidIO capability and command and status registers implement an extended capability data structure. If the extended features bit (bit 28) in the processing element features register is set, the extended features pointer is valid and points to the first entry in the extended features data structure. This pointer is an offset into the standard 16 Mbyte capability register (CAR) and command and status register (CSR) space and is accessed with a maintenance read operation in the same way as when accessing CARs and CSRs.

The extended features data structure is a singly linked list of double-word structures. Each of these contains a pointer to the next structure (EF_PTR) and an extended feature type identifier (EF_ID). The end of the list is determined when the next extended feature pointer has a value of logic 0. All pointers and extended features blocks shall index completely into the extended features space of the CSR space, and all shall be aligned to a double-word boundary so the three least significant bits shall equal logic 0. Pointer values not in extended features space or improperly aligned are illegal and shall be treated as the end of the data structure. Figure A.1 shows an example of an extentended features data structure.

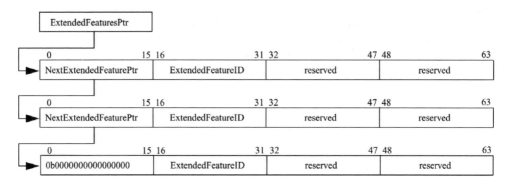

Figure A.1 Example extended features data structure

Appendix B

Serial Physical Layer Registers

This section specifies the 1x/4x LP-Serial Command and Status Register (CSR) set. All registers in the set are 32 bits long and aligned to a 32-bit boundary. These registers allow an external processing element to determine the capabilities, configuration, and status of a processing element using this 1x/4x LP-Serial physical layer. The registers can be accessed using the maintenance operations defined in Chapter 3.

These registers are located in the 1x/4x LP-Serial physical features block, which is an extended features block in the extended features space. The block may exist in any position in the extended features data structure and may exist in any portion of the extended features address space implemented by a device. (The extended features space is located at byte offsets 0x0100 through 0xFFFC of the device configuration space.)

Register offsets into the block that are not defined are reserved unless otherwise stated. Read and write accesses to reserved register offsets shall terminate normally and shall not cause an error condition in the target device.

This appendix specifies only the registers and register bits that comprise the 1x/4x LP-Serial Command and Status Register set. Refer to the other appendices for the specification of the complete set of registers and register bits required for a given device.

Table B.1 describes the required behavior for accesses to reserved register bits and reserved registers for the RapidIO extended features register space.

This chapter is divided into three sections, each addressing a different type of RapidIO device. The three types are generic end points, generic end points with software error recovery support and switch devices.

RapidIO® The Embedded System Interconnect. S. Fuller
© 2005 John Wiley & Sons, Ltd ISBN: 0-470-09291-2

Table B.1   Extended feature space reserved access behavior

Byte offset	Space name	Item	Initiator behavior	Target behavior
0x100– FFFC	Extended Features Space	Reserved bit	Read – ignore returned value[1]	Read – return logic 0
			Write – preserve current value[2]	Write – ignored
		Implementation-defined bit	Read – ignore returned value unless implementation-defined function understood	Read – return implementation-defined value
			Write – preserve current value if implementation-defined function not understood	Write – implementation-defined
		Reserved register	Read – ignore returned value	Read – return logic 0s
			Write –	Write – ignored

[1]   Do not depend on reserved bits being a particular value; use appropriate masks to extract defined bits from the read value
[2]   All register writes shall be in the form: read the register to obtain the values of all reserved bits, merge in the desired values for defined bits to be modified, and write the register, thus preserving the value of all reserved bits

B.1 GENERIC END POINT DEVICES

This section describes the 1x/4x LP-Serial registers for a generic end point device. This extended features register block is assigned extended features block ID 0x0001.

B.1.1 Register Maps

Table B.2 shows the register map for generic RapidIO 1x/4x LP-Serial end point devices. The block offset is the offset relative to the 16-bit extended features pointer (EF_PTR) that points to the beginning of the block.

The address of a byte in the block is calculated by adding the block byte offset to EP_PTR that points to the beginning of the block. This is denoted as [EF_PTR + xx] where xx is the block byte offset in hexadecimal.

This register map is currently defined only for devices with up to 16 RapidIO ports, but can be extended or shortened if more or less port definitions are required for a device. For example, a device with four RapidIO ports is only required to use register map space corresponding to offsets [EF_PTR + 0x00] through [EF_PTR + 0xB8]. Register map offset [EF_PTR + 0xC0] can be used for another extended features block.

B.1.2 Command and Status Registers (CSRs)

Refer to Table B.1 for the required behavior for accesses to reserved registers and register bits.

Table B.2 Serial physical layer register map: generic end point device

Block byte offset	Register name (word 0)	Register name (word 1)
0x0	1x/4x LP-Serial port maintenance block header	
0x8–18	Reserved	
0x20	Port link time-out control CSR	Port response time-out control CSR
0x28	Reserved	
0x30	Reserved	
0x38	Reserved	Port general control CSR
0x40	Reserved	
0x48	Reserved	
0x50	Reserved	
0x58	Port 0 error and status CSR	Port 0 control CSR
0x60	Reserved	
0x68	Reserved	
0x70	Reserved	
0x78	Port 1 error and status CSR	Port 1 control CSR
0x80–218	Assigned to port 2–14 CSRs	
0x220	Reserved	
0x228	Reserved	
0x230	Reserved	
0x238	Port 15 error and status CSR	Port 15 control CSR

B.1.2.1 Port Maintenance Block Header 0 (Block Offset 0x0 Word 0)

The port maintenance block header 0 register contains the EF_PTR to the next EF_BLK and the EF_ID that identifies this as the generic end point port maintenance block header; see Table B.3.

Table B.3 Bit settings for port maintenance block header 0

Bit	Name	Reset value	Description
0–15	EF_PTR		Hard-wired pointer to the next block in the data structure, if one exists
16–31	EF_ID	0x0001	Hard-wired extended features ID

B.1.2.2 Port Maintenance Block Header 1 (Block Offset 0x0 Word 1)

The port maintenance block header 1 register is reserved; see Table B.4.

Table B.4   Bit settings for port maintenance block header 1

Bit	Name	Reset value	Description
0–31			Reserved

B.1.2.3 Port Link Time-out Control CSR (Block Offset 0x20 Word 0)

The port link time-out control register contains the time-out timer value for all ports on a device; see Table B.5. This time-out is for link events, such as sending a packet to receiving the corresponding acknowledge, and sending a link-request to receiving the corresponding link-response. The reset value is the maximum time-out interval, and represents between 3 and 6 seconds.

Table B.5   Bit settings for port link time-out control CSR

Bit	Name	Reset value	Description
0–23	Time-out value	All 1s	Time-out interval value
24–31			Reserved

B.1.2.4 Port Response Time-out Control CSR (Block Offset 0x20 Word 1)

The port response time-out control register contains the time-out timer count for all ports on a device; see Table B.6. This time-out is for sending a request packet to receiving the corresponding response packet. The reset value is the maximum time-out interval, and represents between 3 and 6 seconds.

Table B.6   Bit settings for port response time-out control CSR

Bit	Name	Reset value	Description
0–23	Time-out value	All 1s	Time-out interval value
24–31			Reserved

B.1.2.5 Port General Control CSR (Block Offset 0x38 Word 1)

The bits accessible through the port general control CSR are bits that apply to all ports on a device; see Table B.7. There is a single copy of each such bit per device. These bits are also accessible through the port general control CSR of any other physical layers implemented on a device.

Table B.7 Bit settings for port general control CSRs

Bit	Name	Reset value	Description
0	Host	See footnote[1]	A Host device is a device that is responsible for system exploration, initialization, and maintenance. Agent or slave devices are typically initialized by host devices
			0b0 agent or slave device
			0b1 host device
1	Master enable	See footnote[2]	The master enable bit controls whether or not a device is allowed to issue requests into the system. If the master enable is not set, the device may only respond to requests.
			0b0 processing element cannot issue requests
			0b1 processing element can issue requests
2	Discovered	See footnote[3]	This device has been located by the processing element responsible for system configuration
			0b0 the device has not been previously discovered
			0b1 the device has been discovered by another processing element
3–31			Reserved

[1] The host reset value is implementation dependent
[2] The master enable reset value is implementation dependent
[3] The Discovered reset value is implementation dependent

B.1.2.6 Port n Error and Status CSRs (Offsets 0x58, 78, . . . , 238 Word 0)

These registers are accessed when a local processor or an external device wishes to examine the port error and status information; see Table B.8.

B.1.2.7 Port n Control CSR (Block Offsets 0x58, 78, . . . , 238 Word 1)

The port n control registers contain control register bits for individual ports on a processing element; see Table B.9.

B.2 GENERIC END POINT DEVICES: SOFTWARE-ASSISTED ERROR RECOVERY OPTION

This section describes the 1x/4x LP-Serial registers for a general end point device that supports software assisted error recovery. This is most useful for devices that for whatever reason do not want to implement error recovery in hardware and to allow software to generate link-request control symbols and see the results of the responses. This extended features register block is assigned extended features block ID=0x0002.

Table B.8 Bit settings for port *n* error and status CSRs

Bit	Name	Reset value	Description
0–10			Reserved
11	Output Retry-encountered	0b0	Output port has encountered a retry condition. This bit is set when bit 13 is set. Once set, remains set until written with a logic 1 to clear
12	Output Retried	0b0	Output port has received a packet-retry control symbol and can not make forward progress. This bit is set when bit 13 is set and is cleared when a packet-accepted or a packet-not-accepted control symbol is received (read-only)
13	Output Retry-stopped	0b0	Output port has received a packet-retry control symbol and is in the 'output retry-stopped' state (read-only)
14	Output Error-encountered	0b0	Output port has encountered (and possibly recovered from) a transmission error. This bit is set when bit 15 is set. Once set, remains set until written with a logic 1 to clear
15	Output Error-stopped	0b0	Output is in the 'output error-stopped' state (read-only)
16–20			Reserved
21	Input Retry-stopped	0b0	Input port is in the 'input retry-stopped' state (read-only)
22	Input Error-encountered	0b0	Input port has encountered (and possibly recovered from) a transmission error. This bit is set when bit 23 is set. Once set, remains set until written with a logic 1 to clear
23	Input Error-stopped	0b0	Input port is in the 'input error-stopped' state (read-only)
24–26			Reserved
27	Port-write pending	0b0	Port has encountered a condition which required it to initiate a maintenance port-write operation. This bit is only valid if the device is capable of issuing a maintenance port-write transaction. Once set remains set until written with a logic 1 to clear
28			Reserved
29	Port error	0b0	Input or output port has encountered an error from which hardware was unable to recover. Once set, remains set until written with a logic 1 to clear
30	Port OK	0b0	The input and output ports are initialized and the port is exchanging error-free control symbols with the attached device (read-only)
31	Port uninitialized	0b1	Input and output ports are not initialized. This bit and bit 30 are mutually exclusive (read-only)

Table B.9　Bit settings for port n control CSRs

Bit	Name	Reset value	Description
0–1	Port width	See footnote[1]	Hardware width of the port (read-only): 0b00　single-lane port 0b01　four-lane port 0b10–0b11　reserved
2–4	Initialized port width	See footnote[2]	Width of the ports after initialized (read only): 0b000　single-lane port, lane 0 0b001　single-lane port, lane 2 0b010　four-lane port 0b011–0b111　reserved
5–7	Port width override	0b000	Soft port configuration to override the hardware size: 0b000　no override 0b001　reserved 0b010　force single lane, lane 0 0b011　force single lane, lane 2 0b100–0b111　reserved
8	Port disable	0b0	Port disable: 0b0　port receivers/drivers are enabled 0b1　port receivers/drivers are disabled and are unable to receive/transmit to any packets or control symbols
9	Output port enable	See footnote[3]	Output port transmit enable: 0b0　port is stopped and not enabled to issue any packets except to route or respond to I/O logical maintenance packets, depending upon the functionality of the processing element. Control symbols are not affected and are sent normally 0b1　port is enabled to issue any packets
10	Input port enable	See footnote[4]	Input port receive enable: 0b0　port is stopped and only enabled to route or respond I/O logical maintenance packets, depending upon the functionality of the processing element. Other packets generate packet-not-accepted control symbols to force an error condition to be signaled by the sending device. Control symbols are not affected and are received and handled normally 0b1　port is enabled to respond to any packet
11	Error checking disable	0b0	This bit disables all RapidIO transmission error checking 0b0　error checking and recovery is enabled 0b1　error checking and recovery is disabled Device behavior when error checking and recovery is disabled and an error condition occurs is undefined
12	Multicast-event participant	See footnote[5]	Send incoming multicast-event control symbols to this port (multiple port devices only)
13–30			Reserved
31	Port Type		This indicates the port type, parallel or serial (read only) 0b0　parallel port 0b1　serial port

---

[1]　The port width reset value is implementation dependent
[2]　The initialized port width reset value is implementation dependent
[3]　The output port enable reset value is implementation dependent
[4]　The input port enable reset value is implementation dependent
[5]　The multicast-event participant reset value is implementation dependent

Table B.10 LP-serial register map: generic end point devices (SW-assisted)

Block byte offset	Register name (word 0)	Register name (word 1)
0x0	1x/4x LP-Serial port maintenance block Header	
0x8–18	Reserved	
0x20	Port link time-out control CSR	Port response time-out control CSR
0x28	Reserved	
0x30	Reserved	
0x38	Reserved	Port general control CSR
0x40	Port 0 link maintenance request CSR	Port 0 link maintenance response CSR
0x48	Port 0 local ackID status CSR	Reserved
0x50	Reserved	
0x58	Port 0 error and status CSR	Port 0 control CSR
0x60	Port 1 link maintenance request CSR	Port 1 link maintenance response CSR
0x68	Port 1 local ackID status CSR	Reserved
0x70	Reserved	
0x78	Port 1 error and status CSR	Port 1 control CSR
0x80–218	Assigned to port 2–14 CSRs	
0x220	Port 15 link maintenance request CSR	Port 15 link maintenance response CSR
0x228	Port 15 local ackID status CSR	Reserved
0x230	Reserved	
0x238	Port 15 error and status CSR	Port 15 control CSR

B.2.1 Register Map

Table B.10 shows the register map for generic RapidIO 1x/4x LP-Serial end point devices with software-assisted error recovery. The block offset is the offset based on the extended features pointer (EF_PTR) to this block. This register map is currently only defined for devices with up to 16 RapidIO ports, but can be extended or shortened if more or less port definitions are required for a device. For example, a device with four RapidIO ports is only required to use register map space corresponding to offsets [EF_PTR + 0x00] through [EF_PTR + 0xB8]. Register map offset [EF_PTR + 0xC0] can be used for another extended features block.

B.2.2 Command and Status Registers (CSRs)

Refer to Table B.1 for the required behavior for accesses to reserved registers and register bits.

B.2.2.1 Port Maintenance Block Header 0 (Block Offset 0x0 Word 0)

The port maintenance block header 0 register contains the EF_PTR to the next EF_BLK and the EF_ID that identifies this as the generic end point port maintenance block header; see Table B.11.

Table B.11 Bit settings for port maintenance block header 0

Bit	Name	Reset value	Description
0–15	EF_PTR		Hard-wired pointer to the next block in the data structure, if one exists
16–31	EF_ID	0x0002	Hard-wired extended features ID

B.2.2.2 Port Maintenance Block Header 1 (Block Offset 0x0 Word 1)

The port maintenance block header 1 register is reserved; see Table B.12.

Table B.12 Bit settings for port maintenance block header 1

Bit	Name	Reset value	Description
0–31			Reserved

B.2.2.3 Port Link Time-out Control CSR (Block Offset 0x20 Word 0)

The port link time-out control register contains the time-out timer value for all ports on a device; see Table B.13. This time-out is for link events such as sending a packet to receiving the corresponding acknowledge and sending a link-request to receiving the corresponding link-response. The reset value is the maximum time-out interval, and represents between 3 and 6 seconds.

Table B.13 Bit settings for port link time-out control CSR

Bit	Name	Reset value	Description
0–23	Time-out value	All 1s	Time-out interval value
24–31			Reserved

B.2.2.4 Port Response Time-out Control CSR (Block Offset 0x20 Word 1)

The port response time-out control register contains the time-out timer count for all ports on a device; see Table B.14. This time-out is for sending a request packet to receiving the corresponding response packet. The reset value is the maximum time-out interval, and represents between 3 and 6 seconds.

Table B.14 Bit settings for port response time-out control CSR

Bit	Name	Reset value	Description
0–23	Time-out value	All 1s	Time-out interval value
24–31			Reserved

B.2.2.5 Port General Control CSR (Block Offset 0x38 Word 1)

The bits accessible through the Port General Control CSR are bits that apply to all ports on a device; see Table B.15. There is a single copy of each such bit per device. These bits are also accessible through the port general control CSR of any other physical layers implemented on a device.

Table B.15   Bit settings for port general control CSRs

Bit	Name	Reset value	Description
0	Host	See footnote[1]	A host device is a device that is responsible for system exploration, initialization, and maintenance. Agent or slave devices are initialized by host devices 0b0 agent or slave device 0b1 host device
1	Master enable	See footnote[2]	The master enable bit controls whether or not a device is allowed to issue requests into the system. If the master enable is not set, the device may only respond to requests 0b0 processing element cannot issue requests 0b1 processing element can issue requests
2	Discovered	See footnote[3]	This device has been located by the processing element responsible for system configuration 0b0 the device has not been previously discovered 0b1 the device has been discovered by another processing element
3–31			Reserved

[1] The host reset value is implementation dependent
[2] The master enable reset value is implementation dependent
[3] The discovered reset value is implementation dependent

B.2.2.6 Port n Link Maintenance Request CSRs (Offsets 0x40, 60, . . . , 220 Word 0)

The port link maintenance request registers are accessible both by a local processor and an external device; see Table B.16. A write to one of these registers generates a link-request control symbol on the corresponding RapidIO port interface.

Table B.16   Bit settings for port *n* link maintenance request CSRs

Bit	Name	Reset value	Description
0–28			Reserved
29–31	Command	0b000	Command to be sent in the link-request control symbol. If read, this field returns the last written value

B.2.2.7 Port n Link Maintenance Response CSRs (0x40, 60, . . . , 220 Word 1)

The port link maintenance response registers are accessible both by a local processor and an external device; see Table B.17. A read to this register returns the status received in a link-response control symbol. This register is read-only.

Table B.17 Bit settings for port *n* link maintenance response CSRs

Bit	Name	Reset value	Description
0	response_valid	0b0	If the link-request causes a link-response, this bit indicates that the link-response has been received and the status fields are valid If the link-request does not cause a link-response, this bit indicates that the link-request has been transmitted This bit automatically clears on read
1–21			Reserved
22–26	ackID_status	0b00000	ackID status field from the link-response control symbol
27–31	link_status	0b00000	link status field from the link-response control symbol

B.2.2.8 Port n Local ackID CSRs (Block Offsets 0x48, 68, . . . , 228 Word 0)

The port link local ackID status registers are accessible both by a local processor and an external device; see Table B.18. A read to this register returns the local ackID status for both the out and input ports of the device.

Table B.18 Bit settings for port *n* local ackID status CSRs

Bit	Name	Reset value	Description
0–2			Reserved
3–7	Inbound_ackID	0b00000	Input port next expected ackID value
8–15			Reserved
19–23	Outstanding_ackID	0x00000	Output port unacknowledged ackID status. Next expected acknowledge control symbol ackID field that indicates the ackID value expected in the next received acknowledge control symbol
24–26			Reserved
27–31	Outbound_ackID	0b00000	Output port next transmitted ackID value. Software writing this value can force retransmission of outstanding unacknowledged packets in order to manually implement error recovery

B.2.2.9 Port n Error and Status CSRs (Block Offset 0x58, 78, . . . , 238 Word 0)

These registers are accessed when a local processor or an external device wishes to examine the port error and status information; see Table B.19.

B.2.2.10 Port n Control CSR (Block Offsets 0x58, 78, . . . , 238 Word 1)

The port *n* control registers contain control register bits for individual ports on a processing element; see Table B.20.

Table B.19 Bit settings for port *n* error and status CSRs

Bit	Name	Reset value	Description
0–4			Reserved
5	Output Packet-dropped	0b0	Output port has discarded a packet. Once set remains set until written with a logic 1 to clear
6	Output Failed-encountered	0b0	Output port has encountered a failed condition, meaning that the port's failed error threshold has been reached in the port *n* error rate threshold register. Once set remains set until written with a logic 1 to clear
7	Output Degraded-encountered	0b0	Output port has encountered a degraded condition, meaning that the port's degraded error threshold has been reached in the Port *n* Error Rate Threshold register. Once set remains set until written with a logic 1 to clear
8–10			Reserved
11	Output Retry-encountered	0b0	Output port has encountered a retry condition. This bit is set when bit 13 is set. Once set, remains set until written with a logic 1 to clear
12	Output Retried	0b0	Output port has received a packet-retry control symbol and can not make forward progress. This bit is set when bit 13 is set and is cleared when a packet-accepted or a packet-not-accepted control symbol is received (read-only)
13	Output Retry-stopped	0b0	Output port has received a packet-retry control symbol and is in the 'output retry-stopped' state (read-only)
14	Output Error-encountered	0b0	Output port has encountered (and possibly recovered from) a transmission error. This bit is set when bit 15 is set. Once set, remains set until written with a logic 1 to clear
15	Output Error-stopped	0b0	Output is in the 'output error-stopped' state (read-only)
16–20			Reserved
21	Input Retry-stopped	0b0	Input port is in the 'input retry-stopped' state (read-only)
22	Input Error-encountered	0b0	Input port has encountered (and possibly recovered from) a transmission error. This bit is set when bit 23 is set. Once set, remains set until written with a logic 1 to clear
23	Input Error-stopped	0b0	Input port is in the 'input error-stopped' state (read-only)
24–26			Reserved
27	Port-write Pending	0b0	Port has encountered a condition which required it to initiate a maintenance port-write operation. This bit is only valid if the device is capable of issuing a maintenance port-write transaction. Once set remains set until written with a logic 1 to clear
28			Reserved
29	Port error	0b0	Input or output port has encountered an error from which hardware was unable to recover. Once set, remains set until written with a logic 1 to clear
30	Port OK	0b0	The input and output ports are initialized and the port is exchanging error-free control symbols with the attached device (read-only)
31	Port uninitialized	0b1	Input and output ports are not initialized. This bit and bit 30 are mutually exclusive (read-only)

Table B.20 Bit settings for port *n* control and status register

Bit	Name	Reset value	Description
0–1	Port width	See footnote[1]	Hardware width of the port (read-only): 0b00 single-lane port 0b01 four-lane port 0b10–0b11 reserved
2–4	Initialized Port Width	See footnote[2]	Width of the ports after initialized (read only): 0b000 single-lane port, lane 0 0b001 single-lane port, lane 2 0b010 four-lane port 0b011–0b111 reserved
5–7	Port width override	0b000	Soft port configuration to override the hardware size: 0b000 no override 0b001 reserved 0b010 force single lane, lane 0 0b011 force single lane, lane 2 0b100–0b111 reserved
8	Port disable	0b0	Port disable: 0b0 port receivers/drivers are enabled 0b1 port receivers/drivers are disabled and are unable to receive/transmit to any packets or control symbols
9	Output port enable	See footnote[3]	Output port transmit enable: 0b0 port is stopped and not enabled to issue any packets except to route or respond to I/O logical maintenance packets, depending upon the functionality of the processing element. Control symbols are not affected and are sent normally 0b1 port is enabled to issue any packets
10	Input port enable	See footnote[4]	Input port receive enable: 0b0 port is stopped and only enabled to route or respond I/O logical maintenance packets, depending upon the functionality of the processing element. Other packets generate packet-not-accepted control symbols to force an error condition to be signaled by the sending device Control symbols are not affected and are received and handled normally 0b1 port is enabled to respond to any packet
11	Error checking disable	0b0	This bit disables all RapidIO transmission error checking 0b0 error checking and recovery is enabled 0b1 error checking and recovery is disabled Device behavior when error checking and recovery is disabled and an error condition occurs is undefined
12	Multicast-event participant	See footnote[5]	Send incoming multicast-event control symbols to this port (multiple port devices only)
13–30			Reserved
31	Port type		This indicates the port type, parallel or serial (read only) 0b0 Parallel port 0b1 serial port

1 The port width reset value is implementation dependent
2 The initialized port width reset value is implementation dependent
3 The output port enable reset value is implementation dependent
4 The input port enable reset value is implementation dependent
5 The multicast-event participant reset value is implementation dependent

Table B.21   LP-serial register map: switch devices

Block byte offset	Register name (Word 0)	Register name (Word 1)
0x0	1x/4x LP-Serial port maintenance block header	
0x8–18	Reserved	
0x20	Port link time-out control CSR	Reserved
0x28	Reserved	
0x30	Reserved	
0x38	Reserved	Port general control CSR
0x40	Reserved	
0x48	Reserved	
0x50	Reserved	
0x58	Port 0 error and status CSR	Port 0 control CSR
0x60	Reserved	
0x68	Reserved	
0x70	Reserved	
0x78	Port 1 error and status CSR	Port 1 control CSR
0x80–218	Assigned to port 2–14 CSRs	
0x220	Reserved	
0x228	Reserved	
0x230	Reserved	
0x238	Port 15 error and status CSR	Port 15 control CSR

B.2.3 Generic Switch Devices

This section describes the 1x/4x LP-Serial registers for devices that do not contain end point functionality (i.e. switches). This extended features register block uses extended features block ID=0x0003.

Table B.21 shows the register map for generic RapidIO 1x/4x LP-Serial end point-free devices. The block offset is the offset based on the extended features pointer (EF_PTR) to this block. This register map is currently only defined for devices with up to 16 RapidIO ports, but can be extended or shortened if more or less port definitions are required for a device. For example, a device with four RapidIO ports is only required to use register map space corresponding to offsets [EF_PTR + 0x00] through [EF_PTR + 0xB8]. Register map offset [EF_PTR + 0xC0] can be used for another extended features block.

B.2.4 Command and Status Registers (CSRs)

Refer to Table B.1 for the required behavior for accesses to reserved registers and register bits.

B.2.4.1 Port Maintenance Block Header 0 (Block Offset 0x0 Word 0)

The port maintenance block header 0 register contains the EF_PTR to the next EF_BLK and the EF_ID that identifies this as the generic end point port maintenance block header; see Table B.22.

B.2.4.2 Port Maintenance Block Header 1 (Block Offset 0x0 Word 1)

The port maintenance block header 1 register is reserved; see Table B.23.

Table B.22 Bit settings for port maintenance block header 0

Bit	Name	Reset value	Description
0–15	EF_PTR		Hard-wired pointer to the next block in the data structure, if one exists
16–31	EF_ID	0x0003	Hard-wired extended features ID

Table B.23 Bit settings for port maintenance block header 1

Bit	Name	Reset value	Description
0–31			Reserved

B.2.4.3 Port Link Time-out Control CSR (Block Offset 0x20 Word 0)

The port link time-out control register contains the time-out timer value for all ports on a device; see Table B.24. This time-out is for link events such as sending a packet to receiving the corresponding acknowledge and sending a link-request to receiving the corresponding link-response. The reset value is the maximum time-out interval, and represents between 3 and 6 seconds.

Table B.24 Bit settings for port link time-out control CSR

Bit	Name	Reset value	Description
0–23	Time-out value	All 1s	Time-out interval value
24–31			Reserved

B.2.4.4 Port General Control CSR (Block Offset 0x38 Word 1)

The bits accessible through the Port General Control CSR are bits that apply to all ports on a device; see Table B.25. There is a single copy of each such bit per device. These bits are also accessible through the port general control CSR of any other physical layers implemented on a device.

Table B.25 Bit settings for port general control CSRs

Bit	Name	Reset value	Description
0–1			Reserved
2	Discovered	See footnote[1]	This device has been located by the processing element responsible for system configuration 0b0 the device has not been previously discovered 0b1 the device has been discovered by another processing element
0–31			Reserved

[1] The discovered reset value is implementation dependent

Table B.26   Bit settings for port *n* error and status CSRs

Bit	Name	Reset value	Description
0–10			Reserved
11	Output Retry-encountered	0b0	Output port has encountered a retry condition. This bit is set when bit 13 is set. Once set, remains set until written with a logic 1 to clear
12	Output Retried	0b0	Output port has received a packet-retry control symbol and can not make forward progress. This bit is set when bit 13 is set and is cleared when a packet-accepted or a packet-not-accepted control symbol is received (read-only)
13	Output Retry-stopped	0b0	Output port has received a packet-retry control symbol and is in the 'output retry-stopped' state (read-only)
14	Output Error-encountered	0b0	Output port has encountered (and possibly recovered from) a transmission error. This bit is set when bit 15 is set. Once set, remains set until written with a logic 1 to clear
15	Output Error-stopped	0b0	Output is in the 'output error-stopped' state (read-only)
16–20			Reserved
21	Input Retry-stopped	0b0	Input port is in the 'input retry-stopped' state (read-only)
22	Input Error-encountered	0b0	Input port has encountered (and possibly recovered from) a transmission error. This bit is set when bit 23 is set. Once set, remains set until written with a logic 1 to clear
23	Input Error-stopped	0b0	Input port is in the 'input error-stopped' state (read-only)
24–26			Reserved
27	Port-write Pending	0b0	Port has encountered a condition which required it to initiate a maintenance port-write operation. This bit is only valid if the device is capable of issuing a maintenance port-write transaction. Once set remains set until written with a logic 1 to clear
28			Reserved
29	Port error	0b0	Input or output port has encountered an error from which hardware was unable to recover. Once set, remains set until written with a logic 1 to clear
30	Port OK	0b0	The input and output ports are initialized and the port is exchanging error-free control symbols with the attached device (read-only)
31	Port uninitialized	0b1	Input and output ports are not initialized. This bit and bit 30 are mutually exclusive (read-only)

B.2.4.5 Port n Error and Status CSRs (Block Offsets 0x58, 78, ... , 238 Word 0)

These registers are accessed when a local processor or an external device wishes to examine the port error and status information; see Table B.26.

B.2.4.6 Port n Control CSR (Block Offsets 0x58, 78, ... , 238 Word 1)

The port *n* control registers contain control register bits for individual ports on a processing element; see Table B.27.

Table B.27 Bit settings for port *n* control CSRs

Bit	Name	Reset value	Description
0–1	Port width	See footnote[1]	Hardware width of the port (read-only): 0b00 single-lane port 0b01 four-lane port 0b10–0b11 reserved
2–4	Initialized port width	See footnote[2]	Width of the ports after initialized (read only): 0b000 single-lane port, lane 0 0b001 single-lane port, lane 2 0b010 four-lane port 0b011–0b111 reserved
5–7	Port width override	0b000	Soft port configuration to override the hardware size: 0b000 no override 0b001 reserved 0b010 force single lane, lane 0 0b011 force single lane, lane 2 0b100–0b111 reserved
8	Port disable	0b0	Port disable: 0b0 port receivers/drivers are enabled 0b1 port receivers/drivers are disabled and are unable to receive/transmit to any packets or control symbols
9	Output port enable	See footnote[3]	Output port transmit enable: 0b0 port is stopped and not enabled to issue any packets except to route or respond to I/O logical maintenance packets, depending upon the functionality of the processing element. Control symbols are not affected and are sent normally 0b1 port is enabled to issue any packets
10	Input port enable	See footnote[4]	Input port receive enable: 0b0 port is stopped and only enabled to route or respond I/O logical maintenance packets, depending upon the functionality of the processing element. Other packets generate packet-not-accepted control symbols to force an error condition to be signaled by the sending device Control symbols are not affected and are received and handled normally 0b1 port is enabled to respond to any packet
11	Error checking disable	0b0	This bit disables all RapidIO transmission error checking 0b0 error checking and recovery is enabled 0b1 error checking and recovery is disabled Device behavior when error checking and recovery is disabled and an error condition occurs is undefined
12	Multicast-event participant	See footnote[5]	Send incoming multicast-event control symbols to this port (multiple port devices only)
13–30			Reserved
31	Port type		This indicates the port type, parallel or serial (read only) 0b0 parallel port 0b1 serial port

[1] The port width reset value is implementation dependent
[2] The initialized port width reset value is implementation dependent
[3] The output port enable reset value is implementation dependent
[4] The input port enable reset value is implementation dependent
[5] The multicast-event participant reset value is implementation dependent

Appendix C

Parallel Physical Layer Registers

This appendix describes the command and status register (CSR) set that allows an external processing element to determine the capabilities, configuration, and status of a processing element using the parallel physical layer specification. This chapter describes only registers or register bits defined by this specification. Refer to the other RapidIO logical, transport, and physical specifications of interest to determine a complete list of registers and bit definitions. All registers are 32 bits and aligned to a 32-bit boundary.

These registers utilize the extended features blocks and can be accessed by maintenance operations. Any register offsets not defined are considered reserved for this specification unless otherwise stated. Read and write accesses to reserved register offsets will terminate normally and not cause an error condition in the target device. The extended features pointer (EF_PTR) contains the offset of the first extended features block in the extended features data structure for a device. The 8/16 LP-LVDS physical features block can exist in any position in the extended features data structure.

Table C.1 describes the required behavior for accesses to reserved register bits and reserved registers for the RapidIO Extended Features register space.

This appendix is divided into three sections, each addressing a different type of RapidIO device.

C.1 GENERIC END POINT DEVICES

This section describes the 8/16 LP-LVDS registers for a general end point device. This extended features register block is assigned extended features block ID=0x0001.

RapidIO® The Embedded System Interconnect.   S. Fuller
© 2005 John Wiley & Sons, Ltd   ISBN: 0-470-09291-2

Table C.1   Extended feature space reserved access behavior

Byte offset	Space name	Item	Initiator behavior	Target behavior
0x100–FFFC	Extended features space	Reserved bit	Read – ignore returned value[1]	Read – return logic 0
			Write – preserve current value[2]	Write – ignored
		Implementation-defined bit	Read – ignore returned value unless implementation-defined function understood	Read – return implementation-defined value
			Write – preserve current value if implementation-defined function not understood	Write – implementation-defined
		Reserved register	Read – ignore returned value	Read – return logic 0s
			Write –	Write – ignored

[1] Do not depend on reserved bits being a particular value; use appropriate masks to extract defined bits from the read value
[2] All register writes shall be in the form: read the register to obtain the values of all reserved bits, merge in the desired values for defined bits to be modified, and write the register, thus preserving the value of all reserved bits

C.1.1 Register Map

Table C.2 shows the register map for generic RapidIO 8/16 LP-LVDS end point devices. The block offset is the offset based on the extended features pointer (EF_PTR) to this block. This register map is currently defined only for devices with up to 16 RapidIO ports, but can be extended or shortened if more or less port definitions are required for a device. For example, a device with four RapidIO ports is only required to use register map space corresponding to offsets [EF_PTR + 0x00] through [EF_PTR + 0x98]. Register map offset [EF_PTR + 0xA0] can be used for another extended features block.

C.1.2 Command and Status Registers (CSRs)

Refer to Table C.1 for the required behavior for accesses to reserved registers and register bits.

C.1.2.1 Port Maintenance Block Header 0 (Block Offset 0x0 Word 0)

The port maintenance block header 0 register contains the EF_PTR to the next EF_BLK and the EF_ID that identifies this as the generic end point port maintenance block header; see Table C.3.

C.1.2.2 Port Maintenance Block Header 1 (Block Offset 0x0 Word 1)

The port maintenance block header 1 register is reserved; see Table C.4.

Table C.2   Physical 8/16 LP-LVDS register map

Block byte offset	Register name (Word 0)	Register name (Word 1)
0x0	8/16 LP-LVDS port maintenance block header	
0x8–18	Reserved	
0x20	Port link time-out control CSR	Port response time-out control CSR
0x28	Reserved	
0x30	Reserved	
0x38	Reserved	Port general control CSR
0x40	Reserved	
0x48	Reserved	
0x50	Reserved	
0x58	Port 0 error and status CSR	Port 0 control CSR
0x60	Reserved	
0x68	Reserved	
0x70	Reserved	
0x78	Port 1 error and status CSR	Port 1 control CSR
0x80–218	Assigned to port 2–14 CSRs	
0x220	Reserved	
0x228	Reserved	
0x230	Reserved	
0x238	Port 15 error and status CSR	Port 15 control CSR

Table C.3   Bit settings for port maintenance block header 0

Bit	Name	Reset value	Description
0–15	EF_PTR		Hard-wired pointer to the next block in the data structure, if one exists
16–31	EF_ID	0x0001	Hard-wired extended features ID

Table C.4   Bit settings for port maintenance block header 1

Bit	Name	Reset value	Description
0–31			Reserved

C.1.2.3 Port Link Time-out Control CSR (Block Offset 0x20 Word 0)

The port link time-out control register contains the time-out timer value for all ports on a device; see Table C.5. This time-out is for link events such as sending a packet to receiving the corresponding acknowledge, and sending a link-request to receiving the corresponding link-response. The reset value is the maximum time-out interval, and represents between 3 and 5 seconds.

Table C.5   Bit settings for port link time-out control CSR

Bit	Name	Reset value	Description
0–23 24–31	Time-out_value	All 1s	Time-out interval value Reserved

C.1.2.4 Port Response Time-out Control CSR (Block Offset 0x20 Word 1)

The port response time-out control register contains the time-out timer count for all ports on a device; see Table C.6. This time-out is for sending a request packet to receiving the corresponding response packet. The reset value is the maximum time-out interval, and represents between 3 and 5 seconds.

Table C.6   Bit settings for port response time-out control CSR

Bit	Name	Reset value	Description
0–23 24–31	Time-out_value	All 1s	Time-out interval value Reserved

C.1.2.5 Port General Control CSR (Block Offset 0x38 Word 1)

The bits accessible through the Port General Control CSR are bits that apply to all ports on a device; see Table C.7. There is a single copy of each such bit per device. These bits are also accessible through the Port General Control CSR of any other physical layers implemented on a device.

Table C.7   Bit settings for port general control CSRs

Bit	Name	Reset value	Description
0	Host	See footnote[1]	A host device is a device that is responsible for system exploration, initialization, and maintenance. Agent or slave devices are typically initialized by host devices 0b0 agent or slave device 0b1 host device
1	Master enable	See footnote[2]	The master enable bit controls whether or not a device is allowed to issue requests into the system. If the master enable is not set, the device may only respond to requests 0b0 processing element cannot issue requests 0b1 processing element can issue requests
2	Discovered	See footnote[3]	This device has been located by the processing element responsible for system configuration 0b0 the device has not been previously discovered 0b1 the device has been discovered by another processing element
3–31			Reserved

[1] The host reset value is implementation dependent
[2] The master enable reset value is implementation dependent
[3] The discovered reset value is implementation dependent

Table C.8 Bit settings for port *n* error and status CSRs

Bit	Name	Reset value	Description
0–10			Reserved
11	Output Retry-encountered	0b0	Output port has encountered a retry condition. This bit is set when bit 13 is set. Once set remains set until written with a logic 1 to clear
12	Output Retried	0b0	Output port has received a packet-retry control symbol and can not make forward progress. This bit is set when bit 13 is set and is cleared when a packet-accepted or a packet-not-accepted control symbol is received (read-only)
13	Output Retry-stopped	0b0	Output port has received a packet-retry control symbol and is in the 'output retry-stopped' state (read-only)
14	Output Error-encountered	0b0	Output port has encountered (and possibly recovered from) a transmission error. This bit is set when bit 15 is set. Once set remains set until written with a logic 1 to clear
15	Output Error-stopped	0b0	Output port is in the 'output error-stopped' state (read-only)
16–20			Reserved
21	Input Retry-stopped	0b0	Input port is in the 'input retry-stopped' state (read-only)
22	Input Error-encountered	0b0	Input port has encountered (and possibly recovered from) a transmission error. This bit is set when bit 23 is set. Once set remains set until written with a logic 1 to clear
23	Input Error-stopped	0b0	Input port is in the 'input error-stopped' state (read-only)
24–26			Reserved
27	Port-write pending	0b0	Port has encountered a condition which required it to initiate a maintenance port-write operation.This bit is only valid if the device is capable of issuing a maintenance port-write transaction. Once set remains set until written with a logic 1 to clear
28	Port present	0b0	The port is receiving the free-running clock on the input port
29	Port error	0b0	Input or output port has encountered an error from which hardware was unable to recover. Once set remains set until written with a logic 1 to clear
30	Port OK	0b0	Input and output ports are initialized and can communicate with the adjacent device. This bit and bit 31 are mutually exclusive (read-only)
31	Port uninitialized	0b1	Input and output ports are not initialized and is in training mode. This bit and bit 30 are mutually exclusive (read-only)

C.1.2.6 Port n Error and Status CSRs (Block Offsets 0x58, 78, . . . , 238 Word 0)

These registers are accessed when a local processor or an external device wishes to examine the port error and status information; see Table C.8.

C.1.2.7 Port n Control CSR (Block Offsets 0x58, 78, . . . , 238 Word 1)

The port *n* control registers contain control register bits for individual ports on a processing element; see Table C.9.

Table C.9 Bit settings for port n control CSRs

Bit	Name	Reset value	Description
0	Output port width	See footnote[1]	Operating width of the port (read-only): 0b0 8-bit port 0b1 16-bit port
1	Output port enable	See footnote[2]	Output port transmit enable: 0b0 port is stopped and not enabled to issue any packets except to route or respond to I/O logical maintenance packets, depending upon the functionality of the processing element. Control symbols are not affected and are sent normally 0b1 port is enabled to issue any packets
2	Output port driver disable	0b0	Output port driver disable: 0b0 output port drivers are turned on and will drive the pins normally 0b1 output port drivers are turned off and will not drive the pins This is useful for power management
3			Reserved
4	Input port width	See footnote[3]	Operating width of the port (read-only): 0b0 8-bit port 0b1 16-bit port
5	Input port enable	See footnote[4]	Input port receive enable: 0b0 port is stopped and only enabled to route or respond I/O logical maintenance packets, depending upon the functionality of the processing element. Other packets generate packet-not-accepted control symbols to force an error condition to be signaled by the sending device. Control symbols are not affected and are received and handled normally 0b1 port is enabled to respond to any packet
6	Input port receiver disable	0b0	Input port receiver enable: 0b0 input port receivers are enabled 0b1 input port receivers are disabled and are unable to receive to any packets or control symbols
7			Reserved
8	Error checking disable	0b0	This bit disables all RapidIO transmission error checking 0b0 error checking and recovery is enabled 0b1 error checking and recovery is disabled Device behavior when error checking and recovery is disabled and an error condition occurs is undefined
9	Multicast-event participant	See footnote[5]	Send incoming multicast-event control symbols to this port (multiple port devices only)
10–12			Reserved
13	Flow control participant	0b0	Enable flow control transactions 0b0 do not route or issue flow control transactions to this port 0b1 route or issue flow control transactions to this port
14–27			Reserved
28	Stop on port failed-encountered enable	0b0	This bit is used with the drop packet enable bit to force certain behavior when the error rate failed threshold has been met or exceeded.

Table C.9   (continued)

Bit	Name	Reset value	Description
29	Drop packet enable	0b0	This bit is used with the stop on port failed-encountered enable bit to force certail behavior when the error rate failed Threshold has been met or exceeded
30	Port lockout	0b0	When this bit is cleared, the packets that may be received and issued are controlled by the state of the Output Port Enable and Input Port Enable bits in the Port *n* Control CSR When this bit is set, this port is stopped and is not enabled to issue or receive any packets; the input port can still follow the training procedure and can still send and respond to link-requests; all received packets return packet-not-accepted control symbols to force an error condition to be signaled by the sending device
31	Port type		This indicates the port type, parallel or serial (read only) 0b0 parallel port 0b1 serial port

[1] The output port width reset value is implementation dependent
[2] The output port enable reset value is implementation dependent
[3] The input port width reset value is implementation dependent
[4] The input port enable reset value is implementation dependent
[5] The multicast-event participant reset value is implementation dependent

C.2 GENERIC END POINT DEVICES: SOFTWARE-ASSISTED ERROR RECOVERY OPTION

This section describes the 8/16 LP-LVDS registers for a general end point device that supports software assisted error recovery. This is most useful for devices that for whatever reason do not want to implement error recovery in hardware and to allow software to generate link-request control symbols and see the results of the responses. This extended features register block is assigned extended features block ID=0x0002.

C.2.1 Register Map

Table C.10 shows the register map for generic RapidIO 8/16 LP-LVDS end point devices with software-assisted error recovery. The block offset is the offset based on the extended features pointer (EF_PTR) to this block. This register map is currently defined only for devices with up to 16 RapidIO ports, but can be extended or shortened if more or less port definitions are required for a device. For example, a device with four RapidIO ports is only required to use register map space corresponding to offsets [EF_PTR + 0x00] through [EF_PTR + 0x98]. Register map offset [EF_PTR + 0xA0] can be used for another extended features block.

Table C.10   Physical 8/16 LP-LVDS register map

Block byte offset	Register name (Word 0)	Register name (Word 1)
0x0	8/16 LP-LVDS port Maintenance block header	
0x8–18	Reserved	
0x20	Port link time-Out control CSR	Port response time-out control CSR
0x28	Reserved	
0x30	Reserved	
0x38	Reserved	Port General Control CSR
0x40	Port 0 link maintenance request CSR	Port 0 link maintenance response CSR
0x48	Port 0 local ackID status CSR	Reserved
0x50	Reserved	
0x58	Port 0 error and status CSR	Port 0 control CSR
0x60	Port 1 link maintenance request CSR	Port 1 link maintenance response CSR
0x68	Port 1 local ackID status CSR	Reserved
0x70	Reserved	
0x78	Port 1 error and status CSR	Port 1 control CSR
0x80–218	Assigned to port 2–14 CSRs	
0x220	Port 15 link maintenance request CSR	Port 15 link maintenance response CSR
0x228	Port 15 local ackID status CSR	Reserved
0x230	Reserved	
0x238	Port 15 error and status CSR	Port 15 control CSR

C.2.2 Command and Status Registers (CSRs)

Refer to Table C.1 for the required behavior for accesses to reserved registers and register bits.

C.2.2.1 Port Maintenance Block Header 0 (Block Offset 0x0 Word 0)

The port maintenance block header 0 register contains the EF_PTR to the next EF_BLK and the EF_ID that identifies this as the generic end point port maintenance block header; see Table C.11.

Table C.11   Bit settings for port maintenance block header 0

Bit	Name	Reset value	Description
0–15	EF_PTR		Hard-wired pointer to the next block in the data structure, if one exists
16–31	EF_ID	0x0002	Hard-wired extended features ID

C.2.2.2 Port Maintenance Block Header 1 (Block Offset 0x0 Word 1)

The port maintenance block header 1 register is reserved; see Table C.12.

Table C.12 Bit settings for port maintenance block header 1

Bit	Name	Reset value	Description
0–31			Reserved

C.2.2.3 Port Link Time-out Control CSR (Block Offset 0x20 Word 0)

The port link time-out control register contains the time-out timer value for all ports on a device; see Table C.13. This time-out is for link events such as sending a packet to receiving the corresponding acknowledge and sending a link-request to receiving the corresponding link-response. The reset value is the maximum time-out interval, and represents between 3 and 5 seconds.

Table C.13 Bit settings for port link time-out control CSR

Bit	Name	Reset value	Description
0–23	Time-out_value	All 1s	Time-out interval value
24–31			Reserved

C.2.2.4 Port Response Time-out Control CSR (Block Offset 0x20 Word 1)

The port response time-out control register contains the time-out timer count for all ports on a device; see Table C.14. This time-out is for sending a request packet to receiving the corresponding response packet. The reset value is the maximum time-out interval, and represents between 3 and 5 seconds.

Table C.14 Bit settings for port response time-out control CSR

Bit	Name	Reset value	Description
0–23	Time-out_value	All 1s	Time-out interval value
24–31			Reserved

C.2.2.5 Port General Control CSR (Block Offset 0x38 Word 1)

The bits accessible through the Port General Control CSR are bits that apply to all ports on a device; see Table C.15. There is a single copy of each such bit per device. These bits are also accessible through the port general control CSR of any other physical layers implemented on a device.

Table C.15 Bit settings for port general control CSRs

Bit	Name	Reset value	Description
0	Host	See footnote[1]	A host device is a device that is responsible for system exploration, initialization, and maintenance. Agent or slave devices are initialized by host devices 0b0 agent or slave device 0b1 host device
1	Master enable	See footnote[2]	The master enable bit controls whether or not a device is allowed to issue requests into the system. If the master enable is not set, the device may only respond to requests 0b0 processing element cannot issue requests 0b1 processing element can issue requests
2	Discovered	See footnote[3]	This device has been located by the processing element responsible for system configuration 0b0 the device has not been previously discovered 0b1 the device has been discovered by another processing element
3–31			Reserved

[1] The Host reset value is implementation dependent
[2] The Master Enable reset value is implementation dependent
[3] The Discovered reset value is implementation dependent

C.2.2.6 Port n Link Maintenance Request CSRs (Block Offsets 0x40, 60, . . . , 220 Word 0)

The port link maintenance request registers are accessible both by a local processor and an external device; see Table C.16. A write to one of these registers generates a link-request control symbol on the corresponding RapidIO port interface.

Table C.16 Bit settings for port n link maintenance request CSRs

Bit	Name	Reset value	Description
0–28			Reserved
29–31	Command	0b000	Command to be sent in the link-request control symbol. If read, this field returns the last written value

C.2.2.7 Port n *Link Maintenance Response CSRs (Block Offsets 0x40, 60, . . . , 220 Word 1)*

The port link maintenance response registers are accessible both by a local processor and an external device; see Table C.17. A read to this register returns the status received in a link-response control symbol. This register is read-only.

Table C.17   Bit settings for port *n* link maintenance response CSRs

Bit	Name	Reset value	Description
0	response_valid	0b0	If the link-request causes a link-response, this bit indicates that the link-response has been received and the status fields are valid
			If the link-request does not cause a link-response, this bit indicates that the link-request has been transmitted
			This bit automatically clears on read
1–24			Reserved
25–27	ackID_status	0b000	ackID status field from the link-response control symbol
28–31	link_status	0b0000	Link status field from the link-response control symbol

C.2.2.8 Port n *Local ackID Status CSRs (Block Offsets 0x48, 68, . . . , 228 Word 0)*

The port link local ackID status registers are accessible both by a local processor and an external device; see Table C.18. A read to this register returns the local ackID status for both the out and input ports of the device.

Table C.18   Bit settings for port *n* local ackID status CSRs

Bit	Name	Reset value	Description
0–4			Reserved
5–7	Inbound_ackID	0b000	Input port next expected ackID value
8–15			Reserved
16–23	Outstanding_ackID	0x00	Output port unacknowledged ackID status. A set bit indicates that the corresponding ackID value has been used to send a packet to an attached device, but a corresponding acknowledge control symbol has not been received. 0b1xxx_xxxx indicates ackID 0, 0bx1xx_xxxx indicates ackID 1, 0bxx1x_xxxx indicates ackID 2, etc. This field is read-only
24–28			Reserved
29–31	Outbound_ackID	0b000	Output port next transmitted ackID value. Software writing this value can force re-transmission of outstanding unacknowledged packets in order to manually implement error recovery

C.2.2.9 Port n *Error and Status CSRs (Block Offsets 0x58, 78, . . . , 238 Word 0)*

These registers are accessed when a local processor or an external device wishes to examine the port error and status information; see Table C.19.

Table C.19   Bit settings for port *n* error and status CSRs

Bit	Name	Reset value	Description
0–10			Reserved
11	Output Retry-encountered	0b0	Output port has encountered a retry condition. This bit is set when bit 13 is set. Once set remains set until written with a logic 1 to clear
12	Output Retried	0b0	Output port has received a packet-retry control symbol and can not make forward progress. This bit is set when bit 13 is set and is cleared when a packet-accepted or a packet-not-accepted control symbol is received (read-only)
13	Output Retry-stopped	0b0	Output port has received a packet-retry control symbol and is in the 'output retry-stopped' state (read-only).
14	Output Error-encountered	0b0	Output port has encountered (and possibly recovered from) a transmission error. This bit is set when bit 15 is set. Once set remains set until written with a logic 1 to clear
15	Output Error-stopped	0b0	Output port is in the 'output error-stopped' state (read-only)
16–20			Reserved
21	Input Retry-stopped	0b0	Input port is in the 'input retry-stopped' state (read-only)
22	Input Error-encountered	0b0	Input port has encountered (and possibly recovered from) a transmission error. This bit is set when bit 23 is set. Once set remains set until written with a logic 1 to clear
23	Input Error-stopped	0b0	Input port is in the 'input error-stopped' state (read-only)
24–26			Reserved
27	Port-write pending	0b0	Port has encountered a condition which required it to initiate a maintenance port-write operation. This bit is only valid if the device is capable of issuing a maintenance port-write transaction. Once set remains set until written with a logic 1 to clear
28	Port present	0b0	The port is receiving the free-running clock on the input port
29	Port error	0b0	Input or output port has encountered an error from which hardware was unable to recover. Once set remains set until written with a logic 1 to clear
30	Port OK	0b0	Input and output ports are initialized and can communicate with the adjacent device. This bit and bit 31 are mutually exclusive (read-only)
31	Port uninitialized	0b1	Input and output ports are not initialized and is in training mode. This bit and bit 30 are mutually exclusive (read-only)

C.2.2.10 Port n Control CSR (Block Offsets 0x58, 78, . . . , 238 Word 1)

The port *n* control registers contain control register bits for individual ports on a processing element; see Table C.20.

Table C.20   Bit settings for port *n* control CSRs

Bit	Name	Reset value	Description
0	Output port width	See footnote[1]	Operating width of the port (read-only): 0b0 8-bit port 0b1 16-bit port
1	Output port enable	See footnote[2]	Output port transmit enable: 0b0 port is stopped and not enabled to issue any packets except to route or respond to I/O logical maintenance packets, depending upon the functionality of the processing element. Control symbols are not affected and are sent normally 0b1 port is enabled to issue any packets
2	Output port driver disable	0b0	Output port driver disable: 0b0 output port drivers are turned on and will drive the pins normally 0b1 output port drivers are turned off and will not drive the pins This is useful for power management
3			Reserved
4	Input port width	See footnote[3]	Operating width of the port (read-only): 0b0 8-bit port 0b1 16-bit port
5	Input port enable	See footnote[4]	Input port receive enable: 0b0 port is stopped and only enabled to route or respond I/O logical maintenance packets, depending upon the functionality of the processing element. Other packets generate packet-not-accepted control symbols to force an error condition to be signaled by the sending device. Control symbols are not affected and are received and handled normally 0b1 port is enabled to respond to any packet
6	Input port receiver disable	0b0	Input port receiver enable: 0b0 input port receivers are enabled 0b1 input port receivers are disabled and are unable to receive to any packets or control symbols
7			Reserved
8	Error checking disable	0b0	This bit disables all RapidIO transmission error checking 0b0 error checking and recovery is enabled 0b1 error checking and recovery is disabled Device behavior when error checking and recovery is disabled and an error condition occurs is undefined
9	Multicast-event participant	See footnote[5]	Send incoming multicast-event control symbols to this port (multiple port devices only)
10–30			Reserved
31	Port type		This indicates the port type, parallel or serial (read only) 0b0 parallel port 0b1 Serial port

[1] The output port width reset value is implementation dependent
[2] The output port enable reset value is implementation dependent
[3] The input port width reset value is implementation dependent
[4] The input port enable reset value is implementation dependent
[5] The multicast-event participant reset value is implementation dependent

C.3 SWITCH DEVICES

This section describes the 8/16 LP-LVDS registers for generic devices that do not contain end point functionality. Typically these devices are switches. This extended features register block uses extended features block ID=0x0003.

C.3.1 Register Map

Table C.21 shows the register map for generic RapidIO 8/16 LP-LVDS end point-free devices. The block offset is the offset based on the extended features pointer (EF_PTR) to this block. This register map is currently defined only for devices with up to 16 RapidIO ports, but can be extended or shortened if more or less port definitions are required for a device. For example, a device with four RapidIO ports is only required to use register map space corresponding to offsets [EF_PTR + 0x00] through [EF_PTR + 0x98]. Register map offset [EF_PTR + 0xA0] can be used for another extended features block.

Table C.21   Physical 8/16 LP-LVDS Register Map

Block byte offset	Register name (Word 0)	Register name (Word 1)
0x0	8/16 LP-LVDS port maintenance block header	
0x8–18	Reserved	
0x20	Port link time-out control CSR	Reserved
0x28	Reserved	
0x30	Reserved	
0x38	Reserved	Port general control CSR
0x40	Reserved	
0x48	Reserved	
0x50	Reserved	
0x58	Port 0 error and status CSR	Port 0 control CSR
0x60	Reserved	
0x68	Reserved	
0x70	Reserved	
0x78	Port 1 error and status CSR	Port 1 control CSR
0x80–218	Assigned to port 2–14 CSRs	
0x220	Reserved	
0x228	Reserved	
0x230	Reserved	
0x238	Port 15 error and status CSR	Port 15 control CSR

C.3.2 Command and Status Registers (CSRs)

Refer to Table C.1 for the required behavior for accesses to reserved registers and register bits.

C.3.2.1 Port Maintenance Block Header 0 (Block Offset 0x0 Word 0)

The port maintenance block header 0 register contains the EF_PTR to the next EF_BLK and the EF_ID that identifies this as the generic end point port maintenance block header; see Table C.22.

Table C.22 Bit settings for port maintenance block header 0

Bit	Name	Reset value	Description
0–15	EF_PTR		Hard-wired pointer to the next block in the data structure, if one exists
16–31	EF_ID	0x0003	Hard-wired Extended Features ID

C.3.2.2 Port Maintenance Block Header 1 (Block Offset 0x0 Word 1)

The port maintenance block header 1 register is reserved; see Table C.23.

Table C.23 Bit settings for port maintenance block header 1

Bit	Name	Reset value	Description
0–31			Reserved

C.3.2.3 Port Link Time-out Control CSR (Block Offset 0x20 Word 0)

The port link time-out control register contains the time-out timer value for all ports on a device; see Table C.24. This time-out is for link events such as sending a packet to receiving the corresponding acknowledge and sending a link-request to receiving the corresponding link-response. The reset value is the maximum time-out interval, and represents between 3 and 5 seconds.

Table C.24 Bit settings for port link time-out control CSR

Bit	Name	Reset value	Description
0–23	Time-out_value	All 1s	Time-out interval value
24–31			Reserved

C.3.2.4 Port General Control CSR (Block Offset 0x38 Word 1)

The bits accessible through the port general control CSR are bits that apply to all ports on a device; see Table C.25. There is a single copy of each such bit per device. These bits are also accessible through the port general control CSR of any other physical layers implemented on a device.

Table C.25 Bit settings for port general control CSRs

Bit	Name	Reset value	Description
0–1			Reserved
2	Discovered	See footnote[1]	This device has been located by the processing element responsible for system configuration 0b0 the device has not been previously discovered 0b1 the device has been discovered by another processing element
0–31			Reserved

[1] The discovered reset value is implementation dependent

Table C.26   Bit settings for port *n* error and status CSRs

Bit	Name	Reset value	Description
0–4			Reserved
5	Output Packet-dropped	0b0	Output port has discarded a packet. Once set remains set until written with a logic 1 to clear
6	Output Failed-encountered	0b0	Output port has encountered a failed condition, meaning that the port's failed error threshold has been reached in the port *n* error rate threshold register. Once set remains set until written with a logic 1 to clear
7	Output Degraded-encountered	0b0	Output port has encountered a degraded condition, meaning that the port's degraded error threshold has been reached in the port *n* error rate threshold register. Once set remains set until written with a logic 1 to clear
8–10			Reserved
11	Output Retry-encountered	0b0	Output port has encountered a retry condition. This bit is set when bit 13 is set. Once set remains set until written with a logic 1 to clear
12	Output Retried	0b0	Output port has received a packet-retry control symbol and can not make forward progress. This bit is set when bit 13 is set and is cleared when a packet-accepted or a packet-not-accepted control symbol is received (read-only)
13	Output Retry-stopped	0b0	Output port has received a packet-retry control symbol and is in the 'output retry-stopped' state (read-only)
14	Output Error-encountered	0b0	Output port has encountered (and possibly recovered from) a transmission error. This bit is set when bit 15 is set. Once set remains set until written with a logic 1 to clear
15	Output Error-stopped	0b0	Output port is in the 'output error-stopped' state (read-only)
16–20			Reserved
21	Input Retry-stopped	0b0	Input port is in the 'input retry-stopped' state (read-only)
22	Input Error-encountered	0b0	Input port has encountered (and possibly recovered from) a transmission error. This bit is set when bit 23 is set. Once set remains set until written with a logic 1 to clear
23	Input Error-stopped	0b0	Input port is in the 'input error-stopped' state (read-only)
24–26			Reserved
27	Port-write pending	0b0	Port has encountered a condition which required it to initiate a maintenance port-write operation. This bit is only valid if the device is capable of issuing a maintenance port-write transaction. Once set remains set until written with a logic 1 to clear
28	Port present	0b0	The port is receiving the free-running clock on the input port
29	Port error	0b0	Input or output port has encountered an error from which hardware was unable to recover. Once set remains set until written with a logic 1 to clear
30	Port OK	0b0	Input and output ports are initialized and can communicate with the adjacent device. This bit and bit 31 are mutually exclusive (read-only)
31	Port uninitialized	0b1	Input and output ports are not initialized and is in training mode. This bit and bit 30 are mutually exclusive (read-only)

C.3.2.5 Port n Error and Status CSRs (Block Offsets 0x58, 78, . . . , 238 Word 0)

These registers are accessed when a local processor or an external device wishes to examine the port error and status information; see Table C.26.

C.3.2.6 Port n Control CSR (Block Offsets 0x58, 78, . . . , 238 Word 1)

The port *n* control registers contain control register bits for individual ports on a processing element; see Table C.27.

Table C.27   Bit settings for port *n* control CSRs

Bit	Name	Reset value	Description
0	Output port width	See footnote[1]	Operating width of the port (read-only): 0b0 8-bit port 0b1 16-bit port
1	Output port enable	See footnote[2]	Output port transmit enable: 0b0 port is stopped and not enabled to issue any packets except to route or respond to I/O logical maintenance packets, depending upon the functionality of the processing element. Control symbols are not affected and are sent normally 0b1 port is enabled to issue any packets
2	Output port driver disable	0b0	Output port driver disable: 0b0 output port drivers are turned on and will drive the pins normally 0b1 output port drivers are turned off and will not drive the pins This is useful for power management
3			Reserved
4	Input port width	See footnote[3]	Operating width of the port (read-only): 0b0 8-bit port 0b1 16-bit port
5	Input port enable	See footnote[4]	Input port receive enable: 0b0 port is stopped and only enabled to route or respond I/O logical maintenance packets, depending upon the functionality of the processing element. Other packets generate packet-not-accepted control symbols to force an error condition to be signaled by the sending device. Control symbols are not affected and are received and handled normally 0b1 port is enabled to respond to any packet
6	Input port receiver disable	0b0	Input port receiver enable: 0b0 input port receivers are enabled 0b1 input port receivers are disabled and are unable to receive to any packets or control symbols
7–8			Reserved
9	Multicast-event participant	See footnote[5]	Send incoming multicast-event control symbols to this output port (multiple port devices only)

Table C.27   (*continued*)

Bit	Name	Reset value	Description
10	Flow control participant	0b0	Enable flow control transactions 0b0 do not route or issue flow control transactions to this port 0b1 route or issue flow control transactions to this port
11–27			Reserved
28	Stop on port failed-encountered enable	0b0	This bit is used with the drop-packet enable bit to force certain behavior when the error rate failed threshold has been met or exceeded
29	Drop-packet enable	0b0	This bit is used with the stop on port failed-encountered enable bit to force certail behavior when the error rate failed threshold has been met or exceeded
30	Port lockout	0b0	When this bit is cleared, the packets that may be received and issued are controlled by the state of the output port enable and input port enable bits in the port *n* Control CSR When this bit is set, this port is stopped and is not enabled to issue or receive any packets; the input port can still follow the training procedure and can still send and respond to link-requests; all received packets return packet-not-accepted control symbols to force an error condition to be signaled by the sending device
31	Port type		This indicates the port type, parallel or serial (read only) 0b0 parallel port 0b1 serial port

[1] The output port width reset value is implementation dependent
[2] The output port enable reset value is implementation dependent
[3] The input port width reset value is implementation dependent
[4] The input port enable reset value is implementation dependent
[5] The multicast-event participant reset value is implementation dependent

Appendix D

Error Management Extensions Registers

This appendix describes the error management extended features block. It also describes additions to the existing standard physical layer registers.

D.1 ADDITIONS TO EXISTING REGISTERS

The following bits are added to the parallel and serial logical layer specification port n control CSRs (Table D.1).

The following bits are added to the parallel and serial specification Port n Error and Status CSRs (Table D.2).

D.2 NEW ERROR MANAGEMENT REGISTER

This section describes the extended features block (EF_ID=0h0007) that allows an external processing element to manage the error status and reporting for a processing element. This appendix describes only registers or register bits defined by this extended features block. All registers are 32 bits and aligned to a 32-bit boundary.

Table D.1   Bit settings for port n control CSRs

Bit	Name	Reset value	Description
28	Stop on port failed-encountered enable	0b0	This bit is used with the drop-packet enable bit to force certain behavior when the error rate failed Threshold has been met or exceeded.
29	Drop-packet enable	0b0	This bit is used with the stop on port failed-encountered enable bit to force certail behavior when the error rate failed threshold has been met or exceeded.
30	Port lockout	0b0	When this bit is cleared, the packets that may be received and issued are controlled by the state of the output port enable and input port enable bits in the port n control CSR When this bit is set, this port is stopped and is not enabled to issue or receive any packets; the input port can still follow the training procedure and can still send and respond to link-requests; all received packets return packet-not-accepted control symbols to force an error condition to be signaled by the sending device

RapidIO® The Embedded System Interconnect.   S. Fuller
© 2005 John Wiley & Sons, Ltd   ISBN: 0-470-09291-2

Table D.2   Bit settings for port *n* error and status CSRs

Bit	Name	Reset value	Description
5	Output packet-dropped	0b0	Output port has discarded a packet. Once set remains set until written with a logic 1 to clear
6	Output failed-encountered	0b0	Output port has encountered a failed condition, meaning that the port's failed error threshold has been reached in the port *n* error rate threshold register. Once set remains set until written with a logic 1 to clear
7	Output degraded-encountered	0b0	Output port has encountered a degraded condition, meaning that the port's degraded error threshold has been reached in the port *n* error rate threshold register. Once set remains set until written with a logic 1 to clear

Table D.3   Extended feature space reserved access behavior

Byte offset	Space name	Item	Initiator behavior	Target behavior
0x100–FFFC	Extended features space	Reserved bit	Read – ignore returned value[1] Write – preserve current value[2]	Read – return logic 0 Write – ignored
		Implementation-defined bit	Read – ignore returned value unless implementation-defined function understood	Read – return implementation-defined value
			Write – preserve current value if implementation-defined function not understood	Write – implementation-defined
		Reserved register	Read – ignore returned value Write –	Read – return logic 0s Write – ignored

[1] Do not depend on reserved bits being a particular value; use appropriate masks to extract defined bits from the read value
[2] All register writes shall be in the form: read the register to obtain the values of all reserved bits, merge in the desired values for defined bits to be modified, and write the register, thus preserving the value of all reserved bits

Table D.3 describes the required behavior for accesses to reserved register bits and reserved registers for the RapidIO extended features register space.

D.2.1 Register Map

Table D.4 shows the register map for the error management registers. This register map is currently only defined for devices with up to 16 RapidIO ports, but can be extended or shortened if more or less port definitions are required for a device. For example, a device with four RapidIO ports is only required to use register map space corresponding to offsets [EF_PTR+0x00] through [EF_PTR+0x138]. Register map offset [EF_PTR+0x140] can be used for another extended features block.

Table D.4   Error reporting register map

Block byte offset	Register name (Word 0)	Register name (Word 1)
0x0	Error reporting block header	Reserved
0x8	Logical/transport layer error detect CSR	Logical/transport layer error enable CSR
0x10	Logical/transport layer high address capture CSR	Logical/transport layer address capture CSR
0x18	Logical/transport layer device ID CSR	Logical/transport layer control capture CSR
0x20	Reserved	
0x28	Port-write target deviceID CSR	Packet time-to-live CSR
0x30	Reserved	
0x38	Reserved	
0x40	Port 0 error detect CSR	Port 0 error rate enable CSR
0x48	Port 0 attributes error capture CSR	Port 0 packet/control symbol error capture CSR 0
0x50	Port 0 packet error capture CSR 1	Port 0 packet error capture CSR 2
0x58	Port 0 packet error capture CSR 3	Reserved
0x60	Reserved	
0x68	Port 0 error rate CSR	Port 0 error rate threshold CSR
0x70	Reserved	
0x78	Reserved	
0x80	Port 1 error detect CSR	Port 1 error rate enable CSR
0x88	Port 1 error capture attributes CSR	Port 1 packet/control symbol error capture CSR 0
0x90	Port 1 packet error capture CSR 1	Port 1 packet error capture CSR 2
0x98	Port 1 packet error capture CSR 3	Reserved
0xA0	Reserved	
0xA8	Port 1 error rate CSR	Port 1 error rate threshold CSR
0xB0	Reserved	
0xB8	Reserved	
0xC0–3F8	Assigned to port 2–14 CSRs	
0x400	Port 15 error detect CSR	Port 15 error rate enable CSR
0x408	Port 15 error capture attributes CSR	Port 15 packet/control symbol error capture CSR 0
0x410	Port 15 packet error capture CSR 1	Port 15 packet error capture CSR 2
0x418	Port 15 packet error capture CSR 3	Reserved
0x420	Reserved	
0x428	Port 15 error rate CSR	Port 15 error rate threshold CSR
0x430	Reserved	
0x438	Reserved	

D.2.2 Command and Status Registers (CSRs)

D.2.2.1 *Error Reporting Block Header (Block Offset 0x0 Word 0)*

The error reporting block header register contains the EF_PTR to the next EF_BLK and the EF_ID that identifies this as the error reporting block header; see Table D.5.

Table D.5   Bit settings for error reporting block header

Bit	Name	Reset value	Description
0–15	EF_PTR		Hard-wired pointer to the next block in the data structure, if one exists
16–31	EF_ID	0x0007	Hard-wired extended features ID

D.2.2.2   Logical/Transport Layer Error Detect CSR (Block Offset 0x08 Word 0)

This register indicates the error that was detected by the logical or transport logic layer; see Table D.6. Multiple bits may get set in the register if simultaneous errors are detected during the same clock cycle that the errors are logged.

Table D.6   Bit settings for logical/transport layer error detect CSR

Bit	Name	Reset value	Description
0	IO error response	0b0	Received a response of 'error' for an IO logical layer request (end point device only)
1	Message error response	0b0	Received a response of 'error' for an MSG logical layer request (end point device only)
2	GSM error response	0b0	Received a response of 'error' for a GSM logical layer request (end point device only)
3	Message format error	0b0	Received message packet data payload with an invalid size or segment (MSG logical, end point device only)
4	Illegal transaction decode	0b0	Received illegal fields in the request/response packet for a supported transaction (IO/MSG/GSM logical, switch or endpoint device)
5	Illegal transaction target error	0b0	Received a packet that contained a destination ID that is not defined for this end point. End points with multiple ports and a built-in switch function may not report this as an error (transport, end point device only)
6	Message request time-out	0b0	A required message request has not been received within the specified time-out interval (MSG logical, end point device only)
7	Packet response time-out	0b0	A required response has not been received within the specified time-out interval (IO/MSG/GSM logical, end point device only)
8	Unsolicited response	0b0	An unsolicited/unexpected response packet was received (IO/MSG/GSM logical; only maintenance response for switches, switch or end point device)
9	Unsupported transaction	0b0	A transaction is received that is not supported in the destination operations CAR (IO/MSG/GSM logical; only maintenance port-write for switches, switch or end point device)
10–23			Reserved
24–31	Implementation-specific error	0x00	An implementation-specific error has occurred (switch or end point device)

D.2.2.3 Logical/Transport Layer Error Enable CSR (Block Offset 0x08 Word 1)

This register contains the bits that control if an error condition locks the logical/transport layer error detect and capture registers and is reported to the system host; see Table D.7.

Table D.7 Bit settings for logical/transport layer error enable CSR

Bit	Name	Reset value	Description
0	IO error response enable	0b0	Enable reporting of an IO error response. Save and lock original request transaction capture information in all logical/transport layer capture CSRs (end point device only)
1	Message error response enable	0b0	Enable reporting of a message error response. Save and lock transaction capture information in all logical/transport layer capture CSRs (end point device only)
2	GSM error response enable	0b0	Enable reporting of a GSM error response. Save and lock original request address in logical/transport layer address capture CSRs. Save and lock transaction capture information in all logical/transport layer device ID and control CSRs (end point device only)
3	Message format error enable	0b0	Enable reporting of a message format error. Save and lock transaction capture information in logical/transport layer device ID and control capture CSRs (end point device only)
4	Illegal transaction decode enable	0b0	Enable reporting of an illegal transaction decode error save and lock transaction capture information in logical/transport layer device ID and control capture CSRs (switch or end point device)
5	Illegal transaction target error enable	0b0	Enable reporting of an illegal transaction target error. Save and lock transaction capture information in logical/transport layer device ID and control capture CSRs (end point device only)
6	Message request time-out enable	0b0	Enable reporting of a message request time-out error. Save and lock transaction capture information in logical/transport layer device ID and control capture CSRs for the last message request segment packet received (end point device only)
7	Packet response time-out error enable	0b0	Enable reporting of a packet response time-out error. Save and lock original request address in logical/transport layer address capture CSRs. Save and lock original request destination ID in logical/transport layer device ID capture CSR (end point device only)
8	Unsolicited response error enable	0b0	Enable reporting of an unsolicited response error. Save and lock transaction capture information in logical/transport layer device ID and control capture CSRs (switch or end point device)
9	Unsupported transaction error enable	0b0	Enable reporting of an unsupported transaction error. Save and lock transaction capture information in logical/transport layer device ID and control capture CSRs (switch or end point device)
10–23			Reserved
24–31	Implementation-specific error enable	0x00	Enable reporting of an implementation-specific error has occurred. Save and lock capture information in appropriate logical/transport layer capture CSRs

D.2.2.4 Logical/Transport Layer High Address Capture CSR (Block Offset 0x10 Word 0)

This register contains error information; see Table D.8. It is locked when a logical/transport error is detected and the corresponding enable bit is set. This register is required only for end point devices that support 66- or 50-bit addresses.

Table D.8 Bit settings for logical/transport layer high address capture CSR

Bit	Name	Reset value	Description
0–31	address[0–31]	All 0s	Most significant 32 bits of the address associated with the error (for requests, for responses if available)

D.2.2.5 Logical/Transport Layer Address Capture CSR (Block Offset 0x10 Word 1)

This register contains error information; see Table D.9. It is locked when a logical/transport error is detected and the corresponding enable bit is set.

Table D.9 Bit settings for logical/transport layer address capture CSR

Bit	Name	Reset value	Description
0–28	address[32–60]	All 0s	Least significant 29 bits of the address associated with the error (for requests, for responses if available)
29			Reserved
30–31	xamsbs	0b00	Extended address bits of the address associated with the error (for requests, for responses if available)

D.2.2.6 Logical/Transport Layer Device ID Capture CSR (Block Offset 0x18 Word 0)

This register contains error information; see Table D.10. It is locked when an error is detected and the corresponding enable bit is set.

Table D.10 Bit settings for logical/transport layer device ID capture CSR

Bit	Name	Reset value	Description
0–7	MSB destinationID	0x00	Most significant byte of the destinationID associated with the error (large transport systems only)
8–15	destinationID	0x00	The destinationID associated with the error
16–23	MSB sourceID	0x00	Most significant byte of the sourceID associated with the error (large transport systems only)
24–31	sourceID	0x00	The sourceID associated with the error

D.2.2.7 Logical/Transport Layer Control Capture CSR (Block Offset 0x18 Word 1)

This register contains error information; see Table D.11. It is locked when a logical/transport error is detected and the corresponding enable bit is set.

Table D.11 Bit settings for logical/transport layer header capture CSR 1

Bit	Name	Reset value	Description
0–3	ftype	0x0	Format type associated with the error
4–7	ttype	0x0	Transaction type associated with the error
8–15	msg info	0x00	letter, mbox, and msgseg for the last message request received for the mailbox that had an error (message errors only)
16–31	Implementation-specific	0x0000	Implementation-specific information associated with the error

D.2.2.8 Port-write Target deviceID CSR (Block Offset 0x28 Word 0)

This register contains the target deviceID to be used when a device generates a maintenance port-write operation to report errors to a system host; see Table D.12.

Table D.12 Bit settings for port-write target deviceID CSR

Bit	Name	Reset value	Description
0–7	deviceID_msb	0x00	This is the most significant byte of the port-write target deviceID (large transport systems only)
8–15	deviceID	0x00	This is the port-write target deviceID
16–31			Reserved

D.2.2.9 Packet Time-to-live CSR (Block Offset 0x28 Word 1)

The Packet time-to-live register specifies the length of time that a packet is allowed to exist within a switch device; see Table D.13. The maximum value of the time-to-live variable (0xFFFF) shall correspond to 100 ms $\pm34\%$. The resolution (minimum step size) of the time-to-live variable shall be (maximum value of time-to-live)/($2^{16}-1$). The reset value is all logic 0s, which disables the time-to-live function so that a packet never times out. This register is not required for devices without switch functionality.

Table D.13 Bit settings for packet time-to-live CSR

Bit	Name	Reset value	Description
0–15	Time-to-live value	0x0000	Maximum time that a packet is allowed to exist within a switch device
16–31			Reserved

D.2.2.10 Port n Error Detect CSR (Block Offset 0x40, 80, . . . , 400, Word 0)

The port n error detect register indicates transmission errors that are detected by the hardware; see Table D.14.

Table D.14   Bit settings for port n error detect CSR

Bit	Name	Reset value	Description
0	Implementation-specific error	0b0	An implementation-specific error has been detected
1–7			Reserved
8	Received S-bit error	0b0	Received a packet/control symbol with an S-bit parity error (parallel)
9	Received corrupt control symbol	0b0	Received a control symbol with a bad CRC value (serial) Received a control symbol with a true/complement mismatch (parallel)
10	Received acknowledge control symbol with unexpected ackID	0b0	Received an acknowledge control symbol with an unexpected ackID (packet-accepted or packet_retry)
11	Received packet-not-accepted control symbol	0b0	Received packet-not-accepted acknowledge control symbol
12	Received packet with unexpected ackID	0b0	Received packet with unexpected ackID value, out-of-sequence ackID
13	Received packet with bad CRC	0b0	Received packet with a bad CRC value
14	Received packet exceeds 276 bytes	0b0	Received packet which exceeds the maximum allowed size
15–25			Reserved
26	Non-outstanding ackID	0b0	Link_response received with an ackID that is not outstanding
27	Protocol error	0b0	An unexpected packet or control symbol was received
28	Frame toggle edge error	0b0	Frame signal toggled on falling edge of receive clock (parallel)
29	Delineation error	0b0	Frame signal toggled on non-32-bit boundary (parallel) Received unaligned /SC/ or /PD/ or undefined code-group (serial)
30	Unsolicited acknowledge control symbol	0b0	An unexpected acknowledge control symbol was received
31	Link time-out	0b0	An acknowledge or link-response control symbol is not received within the specified time-out interval

D.2.2.11 Port n Error Rate Enable CSR (Block Offset 0x40, 80, . . . , 400, Word 1)

This register contains the bits that control when an error condition is allowed to increment the error rate counter in the port n error rate threshold register and lock the port n error capture registers; see Table D.15.

Table D.15 Bit settings for port *n* error rate enable CSR

Bit	Name	Reset value	Description
0	Implementation-specific error enable	0b0	Enable error rate counting of implementation-specific errors
1–7			Reserved
8	Received *S*-bit error enable	0b0	Enable error rate counting of a packet/control symbol with an *S*-bit parity error (parallel)
9	Received control symbol with bad CRC enable	0b0	Enable error rate counting of a corrupt control symbol
10	Received out-of-sequence acknowledge control symbol enable	0b0	Enable error rate counting of an acknowledge control symbol with an unexpected ackID
11	Received packet-not-accepted control symbol enable	0b0	Enable error rate counting of received packet-not-accepted control symbols
12	Received packet with unexpected ackID enable	0b0	Enable error rate counting of packet with unexpected ackID value – out-of-sequence ackID
13	Received packet with bad CRC enable	0b0	Enable error rate counting of packet with a bad CRC value
14	Received packet exceeds 276 bytes enable	0b0	Enable error rate counting of packet which exceeds the maximum allowed size
15–25			Reserved
26	Non-outstanding ackID enable	0b0	Enable error rate counting of link-responses received with an ackID that is not outstanding
27	Protocol error enable	0b0	Enable error rate counting of protocol errors
28	Frame toggle edge error enable	0b0	Enable error rate counting of frame toggle edge errors
29	Delineation error	0b0	Enable error rate counting of delineation errors
30	Unsolicited acknowledge control symbol	0b0	Enable error rate counting of unsolicited acknowledge control symbol errors
31	Link time-out	0b0	Enable error rate counting of link time-out errors

D.2.2.12 Port n Error Capture Attributes CSR (Block Offset 0x48, 88, . . . , 408, Word 0)

The error capture attribute register indicates the type of information contained in the port *n* error capture registers; see Table D.16. In the case of multiple detected errors during the same clock cycle one of the errors must be reflected in the error type field. The error that is reflected is implementation dependent.

D.2.2.13 Port n Packet/Control Symbol Error Capture CSR 0 (Block Offset 0x48, 88, . . . , 408, Word 1)

Captured control symbol information includes the true and complement of the control symbol; see Table D.17. This is exactly what arrives on the RapidIO interface with bits 0–7 of the capture register containing the least significant byte of the 32-bit quantity. This register contains the first 4 bytes of captured packet symbol information.

Table D.16 Bit settings for port *n* attributes error capture CSR

Bit	Name	Reset value	Description
0–1	Info type	0b00	Type of information logged 00 packet 01 control symbol (only error capture register 0 is valid) 10 implementation specific (capture register contents are implementation specific) 11 undefined (*S*-bit error), capture as if a packet (parallel physical layer only)
2			Reserved
3–7	Error type	0x00	The encoded value of the bit in the port *n* error detect CSR that describes the error captured in the port *n* error capture CSRs
8–27	Implementation-dependent	All 0s	Implementation-dependent error information
28–30			Reserved
31	Capture valid info	0b0	This bit is set by hardware to indicate that the packet/control symbol capture registers contain valid information. For control symbols, only capture register 0 will contain meaningful information

Table D.17 Bit settings for port *n* packet/control symbol error capture CSR 0

Bit	Name	Reset value	Description
0–31	Capture 0	All 0s	True and complement of control symbol (parallel) or control character and control symbol (serial) or bytes 0–3 of packet header

D.2.2.14 Port n Packet Error Capture CSR 1 (Block Offset 0x50, 90, . . . , 410, Word 0)

Error capture register 1 contains bytes 4–7 of the packet header; see Table D.18.

Table D.18 Bit settings for port *n* packet error capture CSR 1

Bit	Name	Reset value	Description
0–31	Capture 1	All 0s	Bytes 4–7 of the packet header

D.2.2.15 Port n Packet Error Capture CSR 2 (Block Offset 0x50, 90, . . . , 410, Word 1)

Error capture register 2 contains bytes 8–11 of the packet header; see Table D.19.

Table D.19 Bit settings for port *n* packet error capture CSR 2

Bit	Name	Reset value	Description
0–31	Capture 2	All 0s	Bytes 8–11of the packet header

D.2.2.16 Port n Packet Error Capture CSR 3 (Block Offset 0x58, 98, . . . , 418, Word 0)

Error capture register 3 contains bytes 12–15 of the packet header; see Table D.20.

Table D.20 Bit settings for port *n* packet error capture CSR 3

Bit	Name	Reset value	Description
0–31	Capture 3	All 0s	Bytes 12–15 of the packet header

D.2.2.17 Port n Error Rate CSR (Block Offset 0x68, A8, . . . , 428, Word 0)

The port *n* error rate register is a 32-bit register used with the port *n* error rate threshold register to monitor and control the reporting of transmission errors, shown in Table D.21.

Table D.21 Bit settings for port *n* error rate CSR

Bit	Name	Reset value	Description
0–7	Error rate bias	0x80	These bits provide the error rate bias value 0x00 do not decrement the error rate counter 0x01 decrement every 1ms (±34%) 0x02 decrement every 10ms (±34%) 0x04 decrement every 100ms (±34%) 0x08 decrement every 1s (±34%) 0x10 decrement every 10s (±34%) 0x20 decrement every 100s (±34%) 0x40 decrement every 1000s (±34%) 0x80 decrement every 10000s (±34%) other values are reserved
8–13			Reserved
14–15	Error rate recovery	0b00	These bits limit the incrementing of the error rate counter above the failed threshold trigger 0b00 count only 2 errors above 0b01 count only 4 errors above 0b10 count only 16 error above 0b11 do not limit incrementing the error rate count
16–23	Peak error rate	0x00	This field contains the peak value attained by the error rate counter
24–31	Error rate counter	0x00	These bits maintain a count of the number of transmission errors that have been detected by the port, decremented by the error rate bias mechanism, to create an indication of the link error rate

D.2.2.18 Port n Error Rate Threshold CSR (Block Offset 0x68, A8, . . . , 428, Word 1)

The port *n* error rate threshold register is a 32-bit register used to control the reporting of the link status to the system host; see Table D.22.

Table D.22   Bit settings for port *n* error rate threshold CSR

Bit	Name	Reset value	Description
0–7	Error rate failed threshold trigger	0xFF	These bits provide the threshold value for reporting an error condition due to a possibly broken link 0x00 disable the error rate failed threshold trigger 0x01 set the error reporting threshold to 1 0x02 set the error reporting threshold to 2 . . . 0xFF set the error reporting threshold to 255
8–15	Error rate degraded threshold trigger	0xFF	These bits provide the threshold value for reporting an error condition due to a degrading link 0x00 disable the error rate degraded threshold trigger 0x01 set the error reporting threshold to 1 0x02 set the error reporting threshold to 2 . . . 0xFF set the error reporting threshold to 255
16–31			Reserved

Index